Reading the World

Science and Culture in the Nineteenth Century

Bernard Lightman, Editor

Reading

the

British Practices
of Natural History,
1760-1820

World

Edwin D. Rose

University of Pittsburgh Press

Published by the University of Pittsburgh Press, Pittsburgh, Pa., 15260
Copyright © 2025, University of Pittsburgh Press
All rights reserved
Manufactured in the United States of America
Printed on acid-free paper
10 9 8 7 6 5 4 3 2 1

Cataloging-in-Publication data is available from the Library of Congress

ISBN 13: 978-0-8229-4851-3
ISBN 10: 0-8229-4851-6

Cover art: Sydney Parkinson's illustration of the *Alstromeria salsilla* (*Bomarea edulis.*
Herbert). Courtesy of the Trustees of the Natural History Museum, London.
Cover design: Melissa Dias-Mandoly

To my parents

Contents

Acknowledgments

The intellectual and practical construction of books is a collaborative process. Although practices, processes, and motivations have changed somewhat since the eighteenth century, this current book has engaged numerous people to whom I am greatly indebted. First and foremost, I thank Jim Secord. I first discussed the origins of this book project with Jim in the Department of History and Philosophy of Science at the University of Cambridge in early 2016. Jim's encouragement, enthusiasm, knowledge, generosity, and detailed reading of this work shaped both this book and my historical practice. Nick Jardine proved to be an exceptional advisor, offering detailed comments and invaluable advice. I am most grateful to Jim and Nick for the time they have invested in this project that continued throughout the most strenuous parts of the COVID-19 pandemic.

Book projects simply don't emerge without funding. I am grateful to the Arts and Humanities Research Council, who funded a PhD studentship (award AH/1791919) based in the Department of History and Philosophy of Science at the University of Cambridge in addition to overseas research trips, fellowships, conferences, and support toward an affiliation with the Natural History Museum in London. Churchill College, Cambridge, also sponsored part of this award in addition to several conference and research trips. I owe the Linda Hall Library in Kansas City and the Lewis Walpole Library at Yale University my gratitude for offering funded fellowships to support my visits to and research on their collections. I thank the British Society for the History of Science, the History of Science Society, the European Society for Environmental History, the Royal Historical Society, and the Department of History and Philosophy of Science at the University of Cambridge for providing grants toward conference participation.

My thanks extend further to Cambridge University Library and Darwin College for electing me as Munby Fellow between 2020 and 2021, time used to write and revise large sections of this book. Darwin College went on to support the project "The Darwin Family and Cambridge: Science, Art and Nature" between 2021 and 2022 and elected me as advanced research fellow in 2022. Since then, my research has received generous funding from the AHRC, who support the grant "Natural History in the Age of Revolutions, 1776–1848" (reference AH/W007193/1) in the department of History and Philosophy of Science. This research funding has proved vital for the revision of large parts of this manuscript, making it the book it is today. As this book shows, images are far more expensive to produce than plain text and it remains so in the twenty-first century. I am very grateful to the Department of History and Philosophy of Science for funding the image production costs. I must offer my erstwhile thanks to Graham Copecoga, who has generously given time to assist with formatting the images.

Many people read and advised on parts of this book. Janet Browne brought a series of new perspectives and approaches while welcoming me as visiting fellow in the History of Science Department at Harvard University in 2018. Others who read and advised on drafted sections include Emma Spary, Simon Schaffer, Anne Secord, Dániel Margócsy, Roger Gaskell, Patricia Fara, Staffan Müller-Wille, and the Longest Nineteenth Century Reading Group, including Jules Skotnes-Brown, Sarah Qidwai, Henry-James Mearing, Eoin Carter, and Laura Brassington. I thank many other scholars for sharing ideas, including Boris Jardine, Anna Marie Roos, Anne Salmond, Anita Herle, Scott Mandelbrote, Daniel Simpson, Sebastian Kroupa, Sujit Sivasundaram, Melissa Calaresu, James Raven, Gianamar Giovannetti-Singh, Chris Preston, Alex Csiszar, Aileen Fyfe, Noah Moxham, Rebekah Higgitt, Kim Sloan, Paul Evans, Michael Bycroft, James Poskett, Geoff Bil, Isabelle Charmantier, Alexandra Cook, Susannah Gibson, Gordon McOuat, Charlie Jarvis, Mark Spencer, Sachiko Kusukawa, David Mabberley, Richard Sorrenson, Hamish Spencer, Hugh Slotten, Dominik Hünniger, and others.

I thank Mary-Ann Constantine, Nigel Leask, and the teams of the two AHRC-funded projects, "Curious Travellers: Thomas Pennant and the Welsh and Scottish Tour," for their continued advice and opportunities to present in London and Glasgow. I thank Simon Werrett, Jordan Goodman, and the AHRC-funded network project "Joseph Banks, Science, Culture and the Remaking of the Indo-Pacific World" for offering advice and an opportunity to present during the first stages of this project in 2016. I thank those who have invited me to speak at, attend, and contribute to numerous conferences, workshops, seminars, and public talks in the United Kingdom, Switzerland, Germany, the Netherlands, Madeira, the United States, Canada, Australia, and New Zealand. These talk and the questions I received from audience members undoubtedly shaped this research for the better.

The foundation of this book in material culture means I am especially grateful to the librarians, archivists, curators, and museum staff who have provided me with access to and invaluable advice on the collections they curate. Special thanks go to Mark Carine, Paul Cooper, Andrea Hart, Hellen Pethers, Consuelo Sendino, Andreia Salvador, Hein van Grouw, Jacek Wajer and others in the Life Sciences, Earth Sciences, and Library and Archives departments of the Natural History Museum in London. I thank the staff at the British Library, the British Museum, the National Library of Wales/Llyfrgell Genedlaethol Cymru, Linda Hall Library in Kansas City (especially Benjamin Gross and Cynthia Rogers), Houghton Library at Harvard University, Ernest Mayr Library at Harvard University, the Warwickshire Records Office in Warwick (especially Sharon Forman), Uppsala University Library, State Library of New South Wales in Sydney, the National Library of Australia in Canberra, the Buckinghamshire Archives in Aylesbury, the Lewis Walpole Library at Yale University, the Beinecke Library at Yale University, Cambridge University Library, and the Whipple Library at Cambridge. Some of these institutions have even purchased items I recommended and several are cited in this book. I must also extend my thanks to the owners of private collections for giving me access, especially Graham Arader and Alison Petretti of Arader Galleries in New York, who have both been generous hosts and allowed me to reproduce images from the Arader Galleries collection.

At the University of Pittsburgh Press I extend a special thanks to Abby (Collier) McAllister and Bernie Lightman. Abby and Bernie found two enthusiastic, engaged, and knowledgeable referees whose advice and close reading of the manuscript has led to the thorough revision and improvement of this work—a level of engagement represented by one referee's thirteen-page report. I add my thanks to all of those at the press who have been extraordinarily attentive during the publication process.

In the Department of History and Philosophy of Science at the University of Cambridge, my thanks go to all who have supported this endeavor. I especially thank the speakers and attendees of the Cabinet of Natural History seminar group that I had the pleasure of organizing between 2016 and 2017 and continue to attend. I thank the staff of the Whipple Library and Museum at the University of Cambridge, especially Anna Jones, Jack Dixon, Dawn Kingham, Laura Burgazzi, and Josh Nall, the History and Philosophy of Science office staff, and the strong community of postdoctoral researchers and graduate students. I must also extend my thanks to Darwin College for providing a friendly and supportive setting in which to embark on this project, within the rooms of which many pages of this book were written. Finally, I thank my family for their interest and unwavering support in this work, especially my parents, John and Heather Rose; my brother, Ben; grandmother Marjorie and my nan Annie who always placed academic achievement in the highest esteem. I have always been impressed at my parents' ability to tolerate a son who never left university.

Reading the World

Introduction

Natural History and the History of Books

On March 19, 1766, the young naturalist Joseph Banks called on Thomas Pennant in St James's Street, London, to present a scarce book by the sixteenth-century ornithologist William Turner.[1] Pennant later considered this volume "as a valuable proof of his esteem," and it initiated a major correspondence and research collaboration.[2] Banks's gift exemplifies the deep involvement of books in the enterprise of natural history. Books formed one side of a tripart relationship with manuscripts and physical specimens, all of which were sorted, described, and annotated to formulate, consolidate, and shape natural knowledge. Books, manuscripts, and specimens variously tied to natural historians' social standing and facilitated networks of exchange. Within three years of this event, Banks and Pennant were both participating in major voyages. Pennant traveled across the wilds of northern Scotland, taking a ship to explore the outer islands. Banks travelled to the Pacific Ocean on James Cook's first circumnavigation (1768–1771), collecting numerous species previously unknown to Europeans. Journeys through Europe and the Pacific built collections used to solidify these naturalists' reputations as authorities over knowledge of the natural world.

This book reconstructs the processes associated with the practices of obtaining, organizing, and distributing information on natural history as the British Empire witnessed a period of unprecedented contraction in the Americas and expansion into Asia and the Pacific. This involves a close study of the material culture of natural history, examining the processes of integrating manuscripts, books, illustrations, and specimens with the philosophical systems developed to apply a systematic structure to nature. These all integrated with a production line geared toward introducing knowledge in the form of natural history books to an increasingly prevalent global society. Books facilitated the processes of accumulating and synthesizing knowledge in ever-expanding repositories. Mid-eighteenth-century notions of production lines for natural knowledge originated from the earlier accounts set out by the philosopher and lord chancellor of England Francis Bacon in his posthumous *New Atlantis*, published as part of *Sylva Sylvarum: Or a Natural History in Ten Centuries* (1670). Bacon's work included a description of the fictional institution "Solomon's House" that outlined how the practical aspects of knowledge production embodied in a mechanized workshop were intimately connected to philosophical ideas. Bacon initiated a division between natural history and other philosophical pursuits, notably natural philosophy, establishing an intent to develop standardized approaches for gathering and systematizing empirical facts on nature.[3]

The past twenty years have seen an increased interest from historians in the systems employed to gather and organize information during the seventeenth and eighteenth centuries. The question remains, however, as to what exactly happened to this knowledge? How did information sourced, compiled, and organized across global networks reach its intended audience? And to what extent was it used? Central to this is integrating our understanding of natural history with the history of books. Although there has been a recent surge in literature on the rise of scientific publishing, this concentrates on journals while often overlooking the production of more substantial monographs.[4] Much previous work examines seventeenth-century processes of synthesizing information formed from both manuscript and print to construct books on topography, antiquities, and natural history. Here, I extend scholarship by Elizabeth Yale and others into the global world of the eighteenth century, where the disciplinary divides initiated by Bacon become defined while solidifying the relationship between natural history and imperial expansion.[5] Studies examining the practices of managing textual illustration and scientific illustrations, which dominated the practices of natural history from the late seventeenth century, have remained disconnected with scholars often confining analysis to image or text while overlooking the inherent connections between these two media.[6] There is also a substantial divide between studies of systems of classification and the material practices of natural history. Many accounts of the development of philosophical systems overlook the practical collaborative approaches naturalists used to produce printed books that remained central for the simultaneous standardization of working practices and philosophical approaches across continents.[7]

This book asks the question as to what books did to stimulate the formation of new global natural histories. It combines the processes of producing printed books with the practices of natural history. Understanding these processes, from the initial collection of specimens through to the use of the finished volumes that stimulated further global natural history enterprises, moves attention away from the existing focus on private and public museum collections as the main sites for knowledge production.[8] Rather, though integrating a study of natural history with the history of books, this analysis will show, in Robert Darnton's terms, how "the history of books must be international in scale and interdisciplinary in nature."[9] Natural history books were not only constructed to disseminate knowledge, but shaped practices of collecting on both philosophical and practical levels. This necessitates a detailed exploration of the use and construction of books alongside practices of natural history and the networks that governed knowledge exchange.

Visions of Natural History

The years between 1760 and 1820 are a particularly productive period through which to address the practices of natural history. These decades coincided with a huge expansion in British trade and industry after the financial ruin that followed the successive wars in the Americas, including the Seven Years' War (1756–1763) and the American Revolutionary War (1775–1783). The unprecedented economic recovery in Great Britain and its empire has led C. A. Bayly to refer to this period as the most "dramatic example of national resurgence" and the beginning of global "modernity." As a result of American independence, Britain began to look to Asia and the Pacific for trade and resources.[10] British expansionism was central to the changes made to the daily workings of empires and states, through which natural history was connected as a discipline that relied on the identification, illustration, and publication of species discovered on a global scale. The increased geographical range of species entering Britain connected natural historians to the wider world, stimulating naturalists' interests in distant regions and their attempts to shape the research practices employed when collecting information on different continents. The rapid commercial and colonial expansion of Britain into Asia and the Pacific facilitated the emergence of standardized practices used by individuals across the globe, fuelling the development of collecting practices and increasing the number of new publications to extend an era of uninterrupted progress in natural history.[11] Natural historians received and distributed material throughout Europe, the Americas, Asia, and the Pacific, a feat accelerated by the growth of the British book and periodical trade. To contextualize these issues, the main concentration here will be on working practices exhibited across the networks established by particular naturalists. These include Joseph Banks (1743–1820) and Thomas Pennant (1726–1798), in addition to figures such as Gilbert White (1720–1793), whose brother was the notable Fleet Street natural history publisher Benjamin White (1725–1794). Despite their different levels of personal wealth brought about through private landownership, it is possible to place these individuals within the group of "gentlemanly" or "genteel" naturalists, a subset that dominated British natural history and natural philosophy between 1760 and 1820.[12]

Although the eighteenth-century pursuit of natural history has been widely studied, scholarship has tended to concentrate on its integration with the centralizing attempts of states in Continental Europe. In France, in the lands governed by the Hapsburgs, and in other regions such as the Dutch Republic and Scandinavian states, there were consistent attempts to combine practices of natural history and natural philosophy with emergent national agendas. This developed more professionalized disciplines that were integrated into a state bureaucracy and initiated the standardization of approaches to organizing information. Practices of recordkeeping were embodied through cameralism, or the science of administration and recordkeeping, a philosophic approach aimed at the improvement of all levels of society, from state bureaucracies to household management. As David F. Lindenfeld has shown, cameralism was central for aligning these practices, which included natural history and natural philosophy, with the objectives of emergent European states during the eighteenth-century.[13] Interests in communicating, analysing, and manipulating vast quantities of information inspired practices of reducing nature to numerical values, aspects that became essential for defining standards, facilitating imperial expansion, and solidifying the power of states by the nineteenth century.[14] In France institutions such as the Jardin du Roi and Le Muséum national d'Histoire naturelle attracted state funding in an attempt to create central sites for producing natural-historical knowledge. As several historical works have shown, European governments made the roles of natural historians' official paid state appointments with the remit to accompany voyages of discovery due to the professed economic advantage of the new species they collected. Naturalists such as Carl Linnaeus (1707–1778) advocated cameralism, aligning natural history with economic ideas of import substitution to avoid the drain in state bullion reserves brought about by international trade.[15]

As Roy Porter emphasized, however, the British state did not play this interventionist role. The British, unlike the continental powers, were not concerned with the depletion of bullion. Indeed, a major source of their revenue came from the resale of Oriental goods on continental markets.[16] By the mid-eighteenth century, Britain did not have the same bullion shortages as its continental neighbours. This reduced the perceived need to align scientific practices with the state. Natural historians rarely received official appointments on voyages of discovery. If they wished to participate, it was expected for naturalists to fund themselves and contribute to the overall costs of the enterprise. Emergent institutions, such as the British Museum, remained insignificant and underfunded. A lack of intervention from the state allowed natural history in Britain to develop as a field of enquiry dominated by private landowners ranging from aristocrats to rural vicars. Although, as Richard Drayton has suggested, many British initiatives such as the foundation of the Royal Botanic Gardens, Kew, were stimulated by Bourbon reforms and continental cameralist outlooks, there was far less governmental control and the management of these enterprises was left to the independently wealthy. For example, Kew was not considered a "public" garden until the 1840s. Practitioners used their wealth from landed estates, emergent industries, and international trade to fund private research programs, using these to practice and patronize natural history to maintain and advance their positions in society.[17]

The dispersed state of natural history has correlations with British industry, which thrived from an abundance of natural resources, private wealth and investment, access to the largest free trade area in the world, and a lack of state regulation. This marks the emergence of a laissez-faire economic attitude in the face of centralized political structures and administration bought about through the Jacobite risings. Economic freedom and a lack of state regulation has certain correlations with the fragmented nature of scientific development in Britain. The different doctrines associated with various branches of natural history influenced the systematic approaches naturalists used to classify nature. These research programs, independent of state influence, reflected typical British, or more particularly English, enlightened attitudes. It was a very different case in France, arguably Britain's greatest rival of the period, where philosophes such as Voltaire encouraged a unified effort to remove despotic autocrats and the corrupt Catholic Church. This resulted in an increase in interventionist policies used by the state to align scientific practice with a growing national agenda. In comparison, the Glorious Revolution of 1688 had imposed certain limitations on the power of the British monarchy and some religious tolerations had been introduced for nonconformist Protestants in 1689. Combined with economic independence, this gave individuals the ability to define themselves as independent gentlemen, a major symbol of the Enlightenment in Britain, which intertwined natural history with concepts of freedom brought about through private property ownership, wealth, and education. To emphasize a connection between landed wealth and natural history, many sought to distance themselves from other emergent naturalists who made a living by selling their manuscripts to publishers. These included figures such as John Hill and Oliver Goldsmith, regarded by many as little more than literary hacks who exploited the content of recent natural history publications to earn a living from growing commercial publishing markets.[18]

Genteel Naturalists

The decentralized character of British natural history and the relative freedom of individual practitioners from religious persecution, press censorship, and stamp duties levied on political publications and newspapers—together with attempts by the state to align itself with private research programs—led to the emergence of a wide array of different practices. These were dominated by private individuals who maintained independent incomes secured by private property ownership.[19] Often defined as "genteel" or "gentlemen," a title ascribed to those who interacted with and built positive reputations in the learned circles of "polite society," individuals within this group ranged between aristocrats, local gentry, wealthy yeomen farmers, and professional figures (such as physicians), through to parochial clergy. Natural history gave the lower and middle ranks of genteel society an opportunity to define themselves as connoisseurs of a particular subject and integrate themselves with aristocrats, upper gentry, and landowners through distributing useful information based on their observations and natural history collections. Practices of accumulating useful knowledge on nature joined these groups through their shared interests in improving property across the wider empire, emergent nations, country estates, individual parishes, and gardens. The distribution of useful knowledge within polite society provided an opportunity for social advancement, showing how this was not tied to wealth but relied on the perceived novelty and usefulness of the information disseminated throughout genteel networks.[20] This was in spite of the fact that many genteel naturalists had a lower annual income than emergent industrial elites, who were often excluded from more traditional networks since the maintenance of businesses was seen to leave insufficient leisure time for the genteel pursuits of polite society. Despite the difficulties obtaining entry to this group, the flexibility within British genteel society was very different to ancien régime France, where social mobility between the three rigidly defined estates remained infrequent and advancement had to be ratified by official state appointments.[21]

The ability to define oneself as *genteel* was important for practitioners of natural history and natural philosophy since it validated their claims on the natural world and associated these with the circulation of truthful information. Many had received a university education and were deeply committed to notions of improvement in all areas of society through the application of natural philosophy and natural history.[22] As Steven Shapin has suggested, the supply of truthful information was a central pillar of gentlemanly etiquette and played into colonial communication networks that, since the late seventeenth century, had been dominated by this group, many of whom invested their surplus income into joint stock companies, international trade ventures, and plantations. Some defined their genteel standing through positions in government offices, emergent banks, and the church, appointments many combined with an interest in natural history that was quickly integrated into their colonial endeavors.[23] Genteel positions in society were brought about through wealth and the communication of "useful knowledge," an essential attribute that connected those defined as *genteel* throughout the world. Genteel naturalists aligned their own standing with that of Indigenous informants, ranging from parochial clergy in northern Scotland to local elders and priests in the South Pacific, whose social position and knowledge of their surroundings validated the information they supplied. This was essential for establishing relationships with correspondents and individuals referred to as "go-betweens," maintaining a continual supply of information from correspondence networks and through the people naturalists met on expeditions.[24] The relative flexibility of British notions of gentility facilitated communication among different social, economic, and cultural groups, encouraging a diverse supply of useful information while facilitating the collaborative collection, synthesis, and redistribution of information on natural history.

My concentration on three main individuals is designed to facilitate engagement with the practices employed by the full range of naturalists encompassed by genteel society. Out of these, two naturalists dominated the practice of natural history in Britain through their ownership of extensive private natural history collections and publication programs. One is Joseph Banks, a botanist and president of the Royal Society of London from 1778 until his death in 1820. Banks represents the highest ranks of gentlemanly naturalists and assembled a huge natural history collection and library in his London mansion at 32 Soho Square. This was maintained by numerous staff whilst Banks stimulated a vast global correspondence network. Banks's activities were funded by private wealth generated by estates in Lincolnshire and the Midlands, which by 1820 brought him an annual income of £16,000.[25] The second is Thomas Pennant, a member of the Welsh country gentry, who maintained a large natural history collection and library at Downing Hall, the house on his estate in Flintshire, North Wales. When compared to Banks's landholdings, Pennant's estate was more modest since he was descended from wealthy yeomen farmers, placing Pennant toward the middle tier of genteel society.

The lower rungs of the genteel naturalists include figures such as Gilbert White, curate of the rural Hampshire parish of Selborne, who kept what he described as "a little shelf of natural history" in his home at the Wakes.[26] White had a modest income and did not have the resources to indulge in the trappings of emergent consumer society.

Indeed, by the time of his death in 1793, the total value of White's personal possessions did not exceed £30.[27] White's family were well-known to polite society, although their lower economic status ensured that he maintained connections with a growing mercantile elite whose lack of leisure time often excluded them from traditional genteel circles. A typical example is White's younger brother, Benjamin, who was far better known to contemporaries for his prominent bookselling and publishing business. From 1766 this was based under the sign of Horace's Head at 63 Fleet Street and specialized in works of natural history. Benjamin White was responsible for publishing Pennant's and his brother's works. Although few archival materials survive concerning White's business, it is clear that this featured as central node in natural history circles and was well-known to Banks and Pennant. The relationship between Gilbert and Benjamin White is a typical example of a younger sibling who moved to London to establish a business designed to cater for polite society, representing the crossover and relative confusion between these ranks that emerged toward the mid-eighteenth century.

The different levels of wealth encompassed by genteel circles ensured Banks, Pennant, and White all knew one another and exchanged information. Landowners of varying degrees represented by Banks, Pennant, and White characterize the largest and most established component of genteel society.[28] Elite landowning naturalists held distinct notions of who was eligible to engage with natural history. This is in sharp contrast to physicians such as Erasmus Darwin, John Hunter, and Robert Thornton, who often published to supplement their incomes and promote their medical practices to wealthy clients. For example, Erasmus Darwin (1731–1802) was paid "ten shillings a line" for his *Loves of the Plants*, part 1 of his immensely popular *Economy of Vegetation* (1791). It is evident that Darwin and other physician-authors viewed these payments for their time and expertise as having correlations with their medical fees.[29] In comparison, Banks, Pennant, and White each developed a diverse natural history collection. These emanated from research programs across the fields of botany and zoology that were facilitated and extended by communication throughout genteel circles.

Unlike other well-known naturalists of the period who are remembered for developing new systems of classification, Banks, Pennant, White, and many other genteel naturalists did not make any significant philosophical contributions to natural history. They did not develop new systems of classification. Rather, understanding British naturalists' approaches to managing an array of information through paper are central for connecting their working practices to diverse interests in the wider world. These individuals used philosophical systems, such as those developed by Carl Linnaeus (1707–1778), John Ray (1627–1705), and Antoine Laurent de Jussieu (1748–1836), to intertwine natural history with global expansion, influencing approaches to ordering information across research programs. Through examining everyday working practices, this book decenters correspondence as the primary method of circulating information. Instead, it concentrates on the intellectual and practical processes associated with scientific research which, as Erika Milam and Robert Nye have suggested, depended on gendered rules of knowledge production, the embodiment of which was essential for intellectual reproduction and dissemination.[30] The concentration on these three individuals will cast new light on the gendered spaces they worked in and traveled through, connecting them with diverse communities in settings ranging from global expeditions to book production and distribution, communication, and the organization of collections.

Banks, Pennant, and White have been subjected to several studies over the last thirty years. Banks has been well examined since the mid-1990s by John Gascoigne and David Miller in relation to his role in Enlightenment culture and various imperial endeavors. Since then, Richard Drayton and Julian Hoppit have built on this material and examined the resonances between Banks's scientific interests, imperial program, and meticulous management of his Lincolnshire landholdings.[31] However, the practical approaches to natural history and its integration within an intricate network have remained an important understudied aspect of Banks's life that interrelates with his global ambitions, relationship with the British state, and estate management. Understanding how Banks's broader program shaped approaches to organizing information reveals the extent of his agency when managing a global network and specimen collection. Banks relied on others when producing natural knowledge in addition to linking his activities in Britain with the vast global programs he is remembered for.[32]

In comparison to Banks, scholarship on the scientific work of Pennant and White has remained relatively sparse. Pennant is best remembered as a travel writer, a point exemplified by recent studies of his tours in Wales and Scotland. To this day Pennant has not received any substantial published biography in spite of being referred to by historians such as Averil Lysaght as "the most able British zoologist between John Ray and Charles Darwin." Despite the concentration on the literary construction of Pennant's tours, the vast majority of Pennant's published output, surviving library, manuscript, and specimen collections relate to his work on natural history. These Pennant gathered as part of the global collection he assembled at Downing Hall, Flintshire, North Wales. In addition to collecting on his tours around Britain, most notably in Wales, Scotland, Ireland, and England, Pennant built connections with naturalists who had traveled to or resided in Asia, the Pacific, Continental Europe, Africa, the Americas, and the polar regions. Pennant's tours were a by-product of his interests in natural history, especially zoology, publishing numerous works in several editions that include *British Zoology, Indian Zoology, History of Quadrupeds,* and *Arctic Zoology,* among others.[33] This book gives the first account of how Pennant worked to gather and organize information to construct and distribute this vast published output. The main concentration here, however, is on Pennant's seminal work, *British Zoology,* published in numerous editions and formats between 1766 and 1812. Pennant's publishing played a central role in his collecting: the books were designed to cover the expenses incurred on his tours and the publishing process, serving as an important stimulant for gathering, ordering, and publishing new information.

Despite being recognised as one of the most prominent naturalists of the age by contemporaries, Pennant has been placed in Gilbert White's shadow by historians and literary scholars as one of the latter's main correspondents.[34] White remained obscure until the publication of *The Natural History and Antiquities of Selborne* (1789), a book that only achieved widespread readership in the nineteenth century. White's position as a peripheral figure is evidenced through the letters his other main correspondent—the judge, antiquary, and naturalist Daines Barrington (1727/8–1800)—sent to Pennant throughout the 1770s. In the fifteen surviving letters White is only mentioned once in relation to the printing of his letters on swallows in the Royal Society's *Philosophical Transactions.*[35] Most previous scholarship on White examines the literary composition of *The Natural History and Antiquities of Selborne.* Although there have been attempts to analyze White's approach to keeping the records used to formulate *The Natural History and Antiquities of Selborne*, there have been few examinations of how the letters and other famous manuscripts—such as his *Naturalist's Journal*—connected to the annotated printed books, lists, and physical objects in White's collection. For example, White's *Naturalist's Journal* was used to record changes in nature alongside momentous political and parochial events, causing his notes to spill over the spaces allocated for new information. These notes relate to White's interests in the wider world and allowed him to formulate ideas on distant regions.[36] The examination of annotated printed books, manuscript lists, physical objects, in addition to the materials used to construct books, integrates the broader practice of natural history with the burgeoning global society of late eighteenth-century Britain. These sources show unified approaches to recording and ordering information, drawing the activities of naturalists in metropolitan settings and the field together, developing methods of thinking across cultural, geographical, and chronological divides.

Paper and the Practice of Natural History

Global outlooks on natural history became intertwined with emergent notions of geographical scale as practices of gathering and organizing information became embedded in imperial endeavors. Natural historians had to develop means for coping with the flood of specimens, many representing new species, being discovered in diverse parts of the globe. These ranged from their own back gardens to remote islands in the South Pacific that were only just being explored by Europeans. It therefore became necessary to standardize diverse information, arriving at a common approach for thinking across a series of diverse objects with a variety of geographical provenances and collected at different times. To overcome similar problems, eighteenth-century naturalists developed paper systems to organize and collect information obtained over broad geographical areas.[37]

Naturalists' use of paper connects the development of a global vision with everyday working practices. Practical approaches to managing information were more important to contemporary actors than the systematic classification of species. For example, when writing to the trustees of the British Museum on June 29, 1765, Daniel Solander (1733–1782), the first librarian of the museum's Department of Natural and Artificial Curiosities and former student of Linnaeus, commented that he had developed a cataloguing system that allowed any naturalist to arrange the names and descriptions of species "according to his own faivorite [sic] system."[38] The cataloguing system Solander referred to was his "Manuscript Slip Catalogue," an approach that took a similar format to index cards, which could be ordered according to whatever system natural historians deemed to be the most appropriate for classifying each branch of nature. All of Solander's slips contained references to the relevant botanical, zoological, and mineralogical literature, allowing users to check the synonyms while also providing a basis through which one might rearrange the order of the slips. As Bettina Dietz has suggested, the inclusion of page numbers in references made this process quicker and easier for readers of botanical books and Solander's use of them in this context at the world's first public museum was designed to have a similar effect.[39] This range of material was of particular importance to individuals such as Solander, who had been charged with the task of ensuring that the British Museum's collection was accessible to natural historians. It also allowed Solander's successors to revise the slips alongside changes to the Linnaean system and completely reorder them if Linnaeus's aim of finding a "natural system" of classification was ever achieved.[40] The most commonly used systems of classification in Britain between 1760 and 1820 were those devised by Linnaeus in the 1730s, which became more widely accepted after the publication of the 1753 edition of *Species Plantarum*, and the late seventeenth-century system of John Ray. Divergent classificatory approaches encouraged naturalists to organize information according to a scheme that could be used regardless of philosophical agendas and adapt to cope with the addition of new species and the subtle revision of systems.

Over the last two decades the term *information overload* has been used by historians who have examined the different paper-based approaches taken to arrange empirical facts from the sixteenth to eighteenth centuries, the ordering of which was designed to increase efficiency and extend the understanding of nature. More recently, there has been an additional interest in the practicalities of ordering and understanding information through paper, bringing the different users of paper to the foreground.[41] Naturalists assembled advanced paper systems from annotated and interleaved printed books, index cards, bound notebooks, specimen labels, images, and lists. These were all designed to bring information collected over different geographical and time scales together as a unified repository. As Staffan Müller-Wille and Isabelle Charmantier have shown, figures such as Linnaeus relied on several "paper technologies" to order, accumulate, and inspire the creation of information added to new editions of *Species Plantarum* and *Systema Naturae*.[42] However, previous scholarship on the development of paper-based systems for managing information has often attributed these to a certain philosophical or classificatory framework devised by the individual user of this material. This was not the case in Britain, where many natural historians used alternate systems to classify the different kingdoms of nature. Rather, paper systems were developed as a means for moving between different practical and theoretical approaches.

The word *bridging* is used in this context as an analytical construct to situate how paper was used to transfer, organize, communicate, and standardize information on natural history at a practical and theoretical level. The practicalities of natural history collecting often depended several different scales in time and geography governed by the size of an area surveyed that influenced the time an individual or team had to undertake observations.[43] For example, those on global voyages often only had a few hours to observe and record species, whereas others could undertake observations of the place they inhabited over several decades. The need to overcome these problems and mediate between different collecting and recording strategies resulted in the emergence of unified systems for managing information though paper. The physical arrangement of paper records allowed naturalists to arrive at a common standard and move between the different quantities of information accompanying each specimen. Paper created a physical bridge between methods of collecting in the field and the different systems of classification naturalists used. In some cases, naturalists alternated between systems for different branches of nature within their own collections, meaning that paper was crucial for creating a unified cataloguing structure designed to bridge between different kingdoms of nature. It also bridged the physical gulf between the two central pillars that made natural history collections in the period: that of printed books and physical objects. In others, paper created a common standard for communication between naturalists, allowing them to exchange specimens, names, and descriptions to transfer information and objects between collections compiled by individuals with conflicting philosophical approaches. Paper created physical and intellectual bridges between collections, building genteel networks of exchange and creating standards across different scales of time and geography that governed certain individuals' practices of collecting and the range of philosophical systems naturalists employed. Thus, the term *bridging* refers to the physical bridges paper created both within the practice of collecting and through facilitating the communication and exchange of physical objects.

The term *paper technologies* is used to represent the wide variety of distinctive, albeit connected, means for recording and transferring information through notebooks, paper slips, annotated printed books, images, specific specimens, labels, and other things. Although it might be considered somewhat anachronistic, the obviousness of paper technologies as a construction by historians has earned it a relatively secure position in the historiography of science.[44] Initially used by Anke te Heesen when describing the advantages of using bound notebooks alongside a variety of loose paper slips to add, organize, and retrieve information, historians such as Ann Blair, Richard Yeo, James Delbourgo, Elaine Leong, and others have since applied this terminology when discussing the scribal techniques used to record, organize, and keep track of observations. However, the overwhelming emphasis of previous research is on bound notebooks which, as te Heesen suggested, have the unavoidable effect of locking information into a certain structure that cannot be revised without removing the binding or cutting out pages.[45] Matthew Eddy's recent analysis has successfully shown how these problems were commonplace among those who kept notebooks, dividing this into three linked aspects examining notebooks as artifacts constructed over time, the practices of note-taking, and the specific educational community in which these practices developed.[46]

The term *flexible paper technologies* has been employed to describe materials used to arrange and transfer information outside the confines of a bound notebook, ledger, and museum catalogue. These practices rose to prominence from the mid-eighteenth century with the expansion of colonial empires and state bureaucracies, bringing the need to describe, catalogue, classify, and reorganize thousands of new species that began to enter natural history collections to the foreground.[47] Paper technologies became connected with the development of collections alongside colonial expansion, meaning the "paper empires" described by Müller-Wille as collections and books developed by peripheral European naturalists due to the relative ease in accessing information within the Linnaean system, began to spill over into colonial contexts. As a result, the term *paper technologies* is used in this book to represent different paper-based mediums, such as notebooks, paper slips, interleaved books, images, and copperplate illustrations. These materials were used to transfer, organize, and refine the same information, incorporating and revising records during processes that ranged from collecting in the field through to the printing and distribution of books.[48]

In comparison to previous scholarship on the everyday use of paper by natural historians—which tends to concentrate on single objects such as a notebook or a range of paper technologies compiled by a specific individual—the emphasis here is on how diverse groupings of flexible paper systems facilitated collaborative research practices. Collaboration between individuals is apparent throughout the major processes of natural history outlined in this book. These range from collecting in the field to the production, distribution, and use of books. For example, naturalists on larger expeditions rarely traveled alone. They were accompanied by a team including other naturalists, secretaries, artists, translators, servants, among others, all of whom obtained information on new species from a wide variety of sources. Such information often drew on material provided by Indigenous peoples who, in addition to those employed to participate in these enterprises, were essential for shaping the means for obtaining and recording information while translating between different cultures.[49]

Local knowledge was crucial if naturalists were to formulate names for different species according to a predetermined system of classification. Many classificatory systems were designed to integrate vestiges of vernacular names and hints toward the economic and medicinal uses for species, practices that show significant continuities between global, national, and local expeditions. The examination of collaborative approaches to obtaining and incorporating information into a variety of paper technologies adds a new slant to recent scholarship on Indigenous engagement with imperial programs by emphasizing the influence informants ranging from European farmers to Polynesian priests had over the process of recording, naming, and ordering knowledge on new species.[50] Collaborative practices continued during the construction of a publication and evidence for these is preserved in the paper systems assembled alongside the collection. Teams of amanuenses share certain similarities with the "invisible technicians" of late seventeenth-century England discussed by Steven Shapin.[51] However, when compared to laboratory technicians, it seems those who worked alongside natural historians had far more independent authority when describing, ordering, collecting, and depicting nature.

Although practices of collaborating and using paper technologies to manage, accumulate, and disseminate information on eighteenth-century natural history have received attention over the last two decades, such studies of the practical processes of managing information and the different participants have remained disconnected. For example, Bettina Dietz's *Das System der Natur* and several articles have cast new light on the various collaborative practices of eighteenth-century botany. Dietz's analysis concentrates on botanical correspondence and how textual descriptions of species were formulated according to standard conventions to allow several individuals to contribute a botanical work. This resulted in a vast increase in botanical correspondence from the eighteenth century and the consistent use and citation of standard reference catalogues. The transfer of information went toward what Dietz has described as "iterative books," collaborative compilations brought together through correspondence and bibliographical exchange that emerged through the circulation of presentation copies and private library catalogues. Although published under the name of a single author, these volumes contained the collaborative findings of numerous research projects, ranging from global voyages to metropolitan attempts to reduce synonymy in botanical nomenclature. However, Dietz's analysis only goes so far as to examine "the European botanists of the eighteenth-century," and when analyzing the Linnaean apostle Pehr Osbeck's (1723–1805) travels in China between 1750 and 1752 suggests "the question of whether Osbeck had help from indigenous or local informants when *collecting* natural history specimens in China, and if so how, will not be discussed here," attributing this to a perceived lack of source material.[52]

A certain disconnect remains between collaborative practices of knowledge production in botanical publishing and accounts of the paper technologies naturalists exchanged, developed, and used both in libraries, private studies, and on voyages of exploration.[53] Through exploring the practical collaborative processes that went into formulating natural history collections, this current study not only reduces the emphasis on botany through examining several zoological collections, it combines the collaborative processes of collecting, organizing, printing, distributing, and interpreting information, moving beyond letters to explore the diverse range of communicative strategies revealed in diverse groupings of paper technologies. This book explores the range of paper naturalists assembled to expose numerous contributions from Indigenous groups to move away from earlier research that examines the textual conventions used to produce and translate published accounts. This serves to reveal the diverse contributors of information and expertise to these enterprises and the large source bases that reveal the input of Indigenous groups throughout Britain, its empire, and the wider world.

Moving beyond current scholarship on the eighteenth-century Republic of Letters, which overemphasizes correspondence networks as a medium for exchange and collaboration between naturalists, the analysis of diverse collaborative uses of paper will give a new insight into the practicalities of knowledge production. Although letters did play an important role, they were not the primary means for circulating information through what has been referred to as the "public sphere." Loosely defined as a center for discourse between propertied individuals, the public sphere has often been seen as a means of formalized communication though letters, print, and conversations undertaken in coffeehouses and theaters.[54] The overall notion of a public sphere crowds out more geographically diverse and less formal collaborative approaches to communicating and gathering information through diverse paper technologies. Examples include annotated printed books, the pages of which were added to and read by different individuals; manuscript index cards designed to accompany specimens given to correspondents, creating a catalogue compiled by individuals dispersed throughout the globe; in addition to notebooks that were lent, exchanged, and added to. Each state of recordkeeping retained a definitive structure of users and compilers; different tasks represent distinct social ranks in a team of natural historians, amanuenses, and artists, mirroring the hierarchic ranks of British society.

Systems of Classification

Underlying collaborative paper-based approaches to managing information were specific systems of classification. These emphasize the continual connections between the materiality of natural history and philosophical systems, although, as previously stated, systems of classification were not central to the formulation and construction of paper technologies. From the 1760s onward, the two main systems of classification used in Britain were those developed by Carl Linnaeus from the 1730s and the system developed by John Ray in the late seventeenth century. In many cases this mixed reception mirrors attitudes across Europe, where naturalists continued to debate the merits and drawbacks of the Linnaean system.[55]

The Linnaean approach to classifying species was based on a narrow range of physical characters used to define the main groups. These were arranged according to a hierarchic method, starting with kingdoms and moving down to classes, orders, genera, and species. In botany, Linnaeus devised the sexual system of classification based on the number and distribution of pistils and stamens in the flowers of plants. These were compiled from the number of stamens in each flower used to define the twenty-four classes; the quantity of stigmas used to subdivide the first thirteen classes of flowering plants into orders; before moving onto genera, which were assigned by comparing the parts of fructification; before arriving at individual species. These were ascribed a binomial name, compiled from that of the genus placed before an arbitrary or trivial name, to give the plant a unique identifier. Linnaeus admitted the artificiality of the sexual system, a result of it being based on a limited array of physical characters that tended to separate species with a high degree of overall similarity. In *Systema Naturae* (1735), Linnaeus stated that "no natural system of plants has been constructed up till now . . . nor do I maintain that this system is in some way natural."[56] Nevertheless, Linnaeus always strived to produce a natural system based on morphological continuities between species, although this system remained fragmentary and subject to revision throughout his lifetime. Linnaeus maintained that artificial systems were necessary in the absence of a natural system.[57] The usefulness of Linnaeus's sexual system of plant classification was stressed by James Edward Smith, who stated that it was "professedly artificial . . . its sole aim is to learn the name and history of an unknown plant in the most easy and certain manner."[58] This created a set of defined standards that could be used by British naturalists to secure an elite network.

The Linnaean system of classification and nomenclature gained immense popularity with botanists throughout Britain and its empire from the mid-1750s. This was the main reason for why the British

Museum hired Daniel Solander in 1763 with the physician William Watson (1715–1787), a trustee of the British Museum, describing Linnaeus's *Species Plantarum* as a work that "no doubt, will be thankfully received by botanists, . . . as the master-piece of the most complete naturalist the world has ever seen."[59] A year before Solander started at the museum, Thomas Martyn (1735–1825) introduced a course of lectures at the University of Cambridge that "were the first public notices of the Linnaean system."[60] This represents the rapid spread of Linnaean botanical practices in Britain after the publication of *Species Plantarum* in 1753. In comparison to earlier renditions of Linnaeus's system for classifying plants, *Species Plantarum* was the first book to align botanical naming practices with the sexual system of classification. For example, the book was broken into twenty-four sections, representing each Linnaean class starting with Monandria, containing plants with only one stamen, a section subdivided in up to eight orders, the first being Monogynia, represented by the presence of a single stigma in the flower. The last class was Cryptogamia, which included all plants with concealed reproductive parts. The orders were then subdivided into genera, groupings many critics of the Linnaean system viewed as somewhat arbitrary. Individual species were listed under each genus. The last two names, of the genera and species, formed a unique binomial name for each individual species. The relative ease of using this approach when organizing and communicating information on botanical species encouraged naturalists to use the system and nomenclature outlined in *Species Plantarum* to arrange herbarium collections, structure botanical books, and organize university courses.

In contrast, the system developed by John Ray between the 1660s and 1690s relied on a series of general physical features and their combination with social and behavioral differences, such as the environment a particular species inhabited and the sounds they made.[61] In his classification of plants published in the monumental *Historia Plantarum* (1686–1704), the first attempt to produce an all-encompassing account of known plants, Ray grouped plants into three broad categories of herbs, shrubs, and trees. In his classification, Ray distinguished between monocotyledons and dicotyledons, noting differences between the open or enclosed nature of the seeds in addition to the overall size of the plants. Perhaps most importantly Ray adopted an idea not addressed at any length since the time of Aristotle and revived by his contemporaries such as Nehemiah Grew (1641–1712) on the sexual reproduction of plants.[62] The differences between the stigmas and stamens in the flowers were of great importance when thinking about modes of sexual reproduction as a means for plant classification—a central part of Linnaeus's system from the 1730s. Ray's religious interests were central for his interest in compiling *Hisotria Plantarum* and many of his other natural history books. In *Catalogus Plantarum circa Cantabrigium Nascentium* (1660) Ray commented that "we know of no occupation either worthier or more delightful for an honourable man than to concentrate than to contemplate on the glorious works of *nature* and so to honour both the infinite wisdom & goodness of the Divine creator."[63] Linnaeus emphasizes a similar theological view, drawing on the work of Ray and the natural theologian William Derham (1657–1735), whose work had appeared in Swedish translation in 1737. When opening the tenth edition of *Systema Naturae* (1758), the first to consistently use binomial names, Linnaeus used the same quotation from the book of Psalms to that Ray used to open his *Wisdom of God Manifested in the Works of Creation* (1691).[64] Armed with its brevity, consistency, and firm foundation in a distinct Anglican natural theology, the Linnaean system quickly eclipsed Ray's earlier system in British botany from the 1750s.

By contrast, Ray's approaches to classifying the different branches of zoology were not replaced by the Linnaean system for many British naturalists. This resulted from Linnaeus's controversial reliance on specific physical characters to define species. Disputes and problems with the Linnaean system resulted from naturalists' divergent methods of observing living animals, for which they relied on other senses in addition to sight. When Linnaeus proposed the class Mammalia and the abandonment of the Aristotelian Quadrupedia, his terminology sparked controversy among many naturalists for its violation of traditional classificatory boundaries and emphasis on female sexual organs. Mammalia included human beings and infused the systematic classification of nature with European notions of sexuality. Linnaeus's downgrading of human beings to mere animals caused Pennant to note that "my vanity will not suffer me to rank mankind with *Apes*," adding that there was a series of fundamental flaws when grouping species under Mammalia, "which have paps and suckle their young."[65]

The Linnaean system was not universally accepted in Britain. Although it was often used for the classification of insects, crustaceans, molluscs, and fish, the observation of which relied on the physical differences between species, many naturalists could not accept the Linnaean system for the major groups of birds and quadrupeds. This often came down to Linnaeus's removal of any description of the social attributes of species—such as the sounds they made, their habitation of land or water, whether they were diurnal or nocturnal, feeding habits, their ability to fly, and methods of collecting food—all of which were taken into account by Ray. Gilbert White commented that "Lin. is too general in some of his assertions: too many exceptions occur under his general rules."[66] Others, such as Oliver Goldsmith, suggested Linnaeus's work gave "the dry and distinguishing air of a dictionary," adding that his *History of the Earth and Animated Nature* (1774) had been compiled from "great obvious distinctions that she [nature] herself deems to have made." Many ideas contained within Goldsmith's book had been compiled from the work of genteel naturalists such as Pennant and Georges-Louis Leclerc, Comte de Buffon.[67]

Ray's zoological classifications, most notably of quadrupeds, birds, fish, and insects, take far more physical features, preferred environments, social habits, and feeding practices into account than his classification of plants. For example, in the ornithological classifications Ray developed with Francis Willoughby (1635–1672) in the 1670s and published as *The Ornithology* (1678) and *Synopsis Methodica Avium at Piscium* (1713), birds were initially ordered according to habitat, descending from land to water birds, and then keyed the specific preferred environmental conditions for each species onto their physical features, starting with rapacious birds that have hooked beaks and talons, while gradually descending to water birds, the last of which have a bill, webbed feet, and inhabit ponds and lakes. The main purpose of this approach was to identify features to distinguish species while also showing the connections between them.[68] For example, Ray's suggested purpose of *The Ornithology* was to help readers identify birds in the system through comparing specimens and the plates to trace "the characteristic notes of the *genus's* from the highest or first downward will easily guide him to the lowest *genus*."[69] There was also sufficient freedom for those who used Ray's system to include details on the specific economic uses of species. This feature became increasingly important to naturalists as the eighteenth century progressed, giving them the ability to combine notions of improvement with systematic classification. Ray outlined a similar classification for quadrupeds in his *Synopsis Methodica Animalium Quadrupedium et Serpentini Generis* (1693), where the classification he devised was designed to correlate the physical shape of animals' feet with the environments they inhabited, their size, and overall anatomy.[70] Despite the major differences between Ray's and Linnaeus's systems, a main Linnaean innovation that became universal by the mid-eighteenth century, regardless of the system a naturalist used, was binomial nomenclature. Binomials induced stability when applying a definitive name, reducing synonymy while incorporating the concepts of genus and species.

Natural History and the History of Books

The relationships between paper, specimen collections, and systems of classification are manifested in the construction and distribution of natural history books. Books formed part of the collaborative structures embodied through the paper technologies naturalists constructed to undertake research. They also served to unite networks, inspiring the loyalty of those who received presentation copies. Many books were the combined products of overseas voyages and paper information management systems, uniting diverse sets of objects with a series of printed volumes and manuscripts designed to refine and order information obtained from observations and specific specimens.[71]

Unlike previous scholarship on the history of books in the eighteenth century, which has concentrated on publications designed for emergent consumer markets such as almanacs, novels, ballads, poetry, and the *Encyclopédie*, natural history books do not conform to the more general publishing models proposed by Robert Darnton, James Raven, and others. To study the production, distribution, and use of natural history books, this book moves away from accounts of these processes from the perspectives of publishers and booksellers to show how specific authors went about producing, distributing, and integrating publications within the broader practices of natural history. Unlike the authors of popular books, who have been neglected due to their relative obscurity,[72] naturalists such as Joseph Banks, Thomas Pennant, and Gilbert White were not obscure individuals in late eighteenth-century society. Rather, they are representative of the three main tiers of a genteel elite who combined natural history publishing with concepts of wealth and leisure to secure their reputations as genteel naturalists who did not need to engage in commerce for their incomes.

Many of the diverse materials naturalists and their teams of associates used to produce publications survive, examples being account books, original illustrations, copper printing plates, proof sheets, specimens, letters, and paper technologies, in addition to the copies of these publications they circulated. Books interacted with and embodied the content of physical collections of specimens and the paper systems that surrounded them, often acting as substitutes for objects by creating a portable version of a specific collection.[73] Many served to solidify standards in practices of collecting and description to aid communication and grow networks. Numerous surviving annotated volumes show how books were designed to be used by naturalists, who tended to incorporate them into a wider series of paper technologies designed to govern a physical collection.

The examination of the processes that went into the production and distribution of natural history books compiled by Banks, Pennant, and White shows how authors influenced the physical construction and readership of these publications. Natural history authors paid meticulous attention to printing techniques and had a relative lack of interest in commercial publication. Naturalists' major ambitions were not to make profits, but to ensure a limited quantity of their work was distributed among scholars to solidify patronage networks, influence imperial voyages, and stimulate the discovery and classification of new species. The lack of financial gains associated with book production reflects how many of these individuals considered natural history a devotional activity. Taxonomy and the correlations between the various parts of organisms revealed the extent and variety of God's creation. Anatomy, and the study of functions of parts revealed the beneficence of God's design, reinforcing, according to Pennant, "the theory of religion and the practice of morality."[74] The relevance of natural history for improving society and its connections to a distinct theological agenda became increasingly apparent with the increased prevalence of natural theology toward the end of the eighteenth century. For example, in *Natural Theology* (1802), the Cambridge cleric William Paley echoed Pennant's remarks: "There is no subject in which the tendency to dwell upon select or single topics is so usual, because there is no subject, of which, in its full extent, the latitude is so great, as that of natural history applied to proof of an intelligent Creator."[75] Natural history, then, was essential for providing evidence for the divine actions of the deity, a practice that intertwined the act

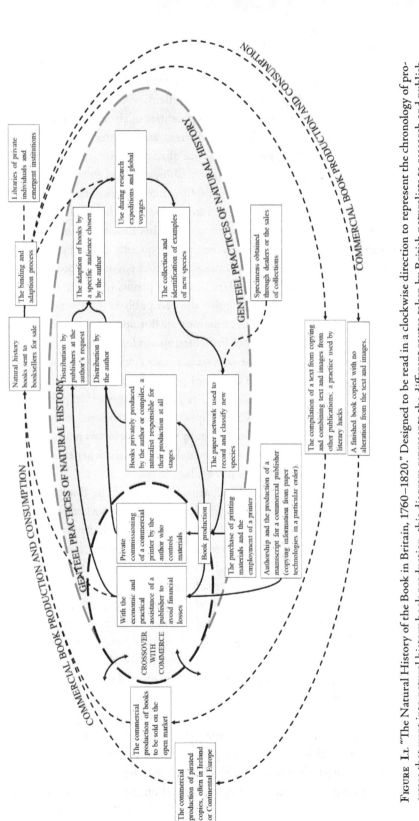

FIGURE 1.1. "The Natural History of the Book in Britain, 1760–1820." Designed to be read in a clockwise direction to represent the chronology of processes that went into natural history book production, this diagram represents the different routes taken by British naturalists to research and publish natural histories. Genteel practices were central to these processes and are emphasized by the central sections of this diagram that fans out into the more commercial sections with which these individuals sometimes interacted. At certain points naturalists initiated a certain "Crossover with Commerce," represented by the checked sections.

Labels within the figure:

Libraries of private individuals and emergent institutions

The binding and adaption process

Natural history books sent to booksellers for sale

Use during research expeditions and global voyages

The adaption of books by a specific audience chosen by the author

The collection and identification of examples of new species

Specimens obtained through dealers or the sales of collections

GENTEEL PRACTICES OF NATURAL HISTORY

COMMERCIAL BOOK PRODUCTION AND CONSUMPTION

Distribution by publishers at the author's request

Distribution by the author

Books privately produced by the author or compiler, a naturalist responsible for their production at all stages

The paper network used to record and classify new species

The compilation of a text from copying and combining text and images from other publications; a practice used by literary hacks

A finished book copied with no alteration from the text and images.

GENTEEL PRACTICES OF NATURAL HISTORY

Private commissioning of a commercial printer by the author who controls materials

Book production

The purchase of printing materials and the employment of a printer

Authorship and the production of a manuscript for a commercial publisher (copying information from paper technologies in a particular order).

COMMERCIAL BOOK PRODUCTION AND CONSUMPTION

With the economic and practical assistance of a publisher to avoid financial losses

CROSSOVER WITH COMMERCE

The commercial production of books to be sold on the open market

The commercial production of pirated copies, often in Ireland or Continental Europe of reading with general notions of moral and economic improvement that became immensely popular from the 1770s.[76]

A relative disinterest in the financial potential of the commercial publishing industry is reflected by the different ways genteel naturalists' integrated publishing into the practice of natural history (figure I.1). Some used emergent commercial publishing markets to produce and circulate a certain percentage of complimentary copies of the total print run of a book, while others privately produced their publications and distributed copies to a select group. This imitates an older aristocratic literary culture, perhaps the most frequently studied proponent of which is Voltaire, who famously rejected payments for his manuscripts and relied on booksellers' desire for profits to ensure the wide circulation of his books. Voltaire allowed publishers to retain his literary profits and circulated numerous author approved pirated editions to stimulate the spread of Enlightenment ideas. Unlike Voltaire, British naturalists did not face state censors when publishing their works, giving them a high degree of personal power and influence when disseminating knowledge.[77] The growing commercial publishing sector was seen as somewhat peripheral when compared to the broader processes of natural history—many practitioners saw it as an industry that could be exploited so they could meet their other theological, social, economic, and political goals.

The diagram in figure I.1, "The Natural History of the Book in Britain, 1760–1820," is designed to visualise the interrelation between the practices of natural history and the processes of constructing, distributing, and using books in relation to the different genteel social attitudes that governed these. Designed to be read in a clockwise direction, the content outlines the broader paths taken by the main chapters of this book. Such means for visualizing the processes and practices of knowledge and book production have been in use since the 1970s, a famous example being Robert Darnton's "communication circuit," designed to bring the processes associated with producing and distributing books together. This has been revisited several times by historians of the book such as Thomas R. Adams and Nicholas Barker, who have produced an inverted diagram showing the cycle of book production in the center with the indirect social, economic, intellectual, and political forces looking inward. The sociologist Pierre Bourdieu has used diagrams to outline the relationships between the genera of a book, the social standing of the author, the audience they intended to engage, and the relationship with profits.[78] In the history of science, Martin Rudwick has attempted to encapsulate the "visual language of geology" as a diagram, bringing together the diverse relationships between the economic and technical changes that went into producing visual materials, the theoretical meaning of images, the construction of a visual language used to read and interpret images, while dividing this between different "cognate zones" defining the different activities individuals and groups commenced. Still more recently Eddy has utilized a diagram in *Media and the Mind* to connect intellectual processes with the physical practices of note-taking.[79]

Building on renditions by scholars in book history and the history of science, the diagram in figure I.1 brings together these branches of scholarship to combine the practices of natural history with the process of book production. Central to this are the notions of naturalists' genteel practices of natural history that integrate with the practices of producing and distributing books. These influenced the variety of routes naturalists took to accumulate collections and the process of transferring these into assemblages of printed images and letterpress designed to be distributed that conformed to their different interpretations of genteel practices of natural history. As such, the central gray shaded area represents undisputed genteel practices of natural history, a key defining attribute of which is their isolation from the emergent commercial world of the late eighteenth century. However, some naturalists did initiate a variety of "crossover with commerce," emphasized by the checkered area of this diagram, and their motivations for doing so are outlined as this book progresses.

Commercial crossover often meant naturalists had to collaborate with different sectors, such as publishers and booksellers, in addition to a range of artisans, including letterpress and intaglio printers, engravers, and artists, aspects that undermined their attempts to reserve knowledge for a genteel group. Finally, the most peripheral area of this diagram is the surrounding unshaded area. This shows processes viewed as commercial, such as the production of books for profit and the purchase of materials. It is designed to show the integrated practices of natural history and book production in the late eighteenth century, the genteel social constraints surrounding which subvert many earlier accounts of commercial book production.

The desire of natural historians and natural philosophers to define themselves as gentlemen who did not depend on commercial profits and government grants was central for shaping the production and distribution of natural history books. Many naturalists, philosophers, and aristocrats detested the wealth commercial publishing brought to authors such as Samuel Johnson and Alexander Pope.[80] These common views were expressed in an unpublished pamphlet by the physician William Heberden (1710–1801), who stated that a philosopher lost any established reputation "as soon as he ventures to give anything to the public," adding that harmony in oneself and a certain level of intellectual prestige was obtained by maintaining "a view to publishing, though without ever doing so."[81] This was a model adhered to by many aristocratic natural historians and natural philosophers, perhaps the most famous examples in Britain being Joseph Banks and the chemist Henry Cavendish.

Although many did not go quite so far as Heberden, natural historians made efforts to restrict the audience of their works. Books were used to fashion naturalists' reputations as benevolent gentlemen who gifted natural knowledge to a select group of elites, in so doing fashioning their "empires" of natural history and patronage networks. There has been little effort to understand how distributing books was central to Banks's and other naturalists' efforts to develop what David Miller termed "the Banksian Learned Empire" in relation to natural history.[82] Not only did books forge links with other naturalists, they were the product of, and promoted, the specific working practices and philosophical frameworks naturalists used to collect and organize specimens throughout the world. These involved standard approaches to description, image production, and the communication of information through paper technologies.

Distributing books had been central to the building of networks amongst members and associates of princely courts since the sixteenth century. Books were exchanged to secure political allegiances and the patronage of aristocrats and princes. Elite networks became far more sophisticated by the early eighteenth century when Continental European scholars began to distribute complimentary copies of their books to their peers.[83] Others, such as those discussed by Sebastian Kühn, sought to distribute printed works to raise the profile of their respected academies, societies, and institutions, justifying the use of collective funds to foot publication expenses. Some natural philosophers relied on direct state sponsorship to fund their publishing programs and broader scientific enterprise, combining these with the interests of the state.[84] In comparison, the majority of British naturalists who published between 1760 and 1820 relied on their personal income, or that of a wealthy patron, to fund the entire production process. This allowed naturalists to utilise the expansion of the commercial publishing industry, exerting high levels of control to influence what Roger Chartier has described as the "triangular relationship" between texts, books, and readers.[85] Authors could ensure these publications were being produced and used according to their directions, drawing together the physical construction of books, the author, and the reader, which have been portrayed as fairly separate entities in previous scholarship. Naturalists' high level of control over their publications consigned the activities of commercial publishers to the periphery. Publishers often followed natural historians' instructions when distributing individual books and would not make editorial changes without first consulting the author. In many cases limited print runs gave authors the opportunity to check the quality and content of every single copy.

Examining the steps individuals such as Banks, Pennant, and White took during the production and distribution of natural history books connects to their global ambitions, showing how they integrated publishing with the broader practices and social connotations of natural history collecting. However, natural historians designed to keep their collections of books and objects out of the economic circuit, shaping their reactions to what has been described as a "print revolution."[86] The high levels of control wealthy landowners maintained over research expeditions and the publication of natural history books shows how they utilised the growth of imperial trade and publishing industries to fit their broader aims.

Practices of book production reveal how British naturalists reacted to the growth of the commercial book trade, suggesting that contemporary society was somewhat less "revolutionary" in terms of commerce and politics than has been previously assumed.[87] Rather, most books were regarded as exclusive objects distributed to a landowning elite. In comparison to the gentlemen-scientists of the 1830s and 1840s, who established laboratories in their homes and founded the British Association for the Advancement of Science, British naturalists of the Enlightenment intertwined their means for distributing books with a limited vision for improvement concentrating on their own landholdings and personal interests in the wider world.[88] As a result, the emphasis here is on how enlightened knowledge on natural history remained within the confines of elite knowledge networks, serving the more prominent national and imperial agendas of these individuals.

Many naturalists had firm views on aristocratic privilege, some of which became embedded in their natural historical work and publishing practices. Figures such as Banks and Pennant viewed the communication, dissemination, and general investigation of the natural world from within the parameters of traditional views on blood, lineage, social hierarchy, and religion. Independent gentlemen were seen as being qualified to create unbiased accounts of natural history due to their education and leisure. In a letter sent to Samuel Barker in 1781, White commented that "you do very right, I think, in looking into the study of history, which is a very gentleman-like study."[89] Considering these specific views on social class and its relationship with the practice of natural history, I show how individual practitioners shaped economic and imperial expansion alongside natural knowledge production to conform to their interests and beliefs.

This book is divided into five chapters, each focusing on a stage in the process of making natural-historical knowledge, ranging from the collection of information, book production, the distribution, and use of these materials. Each chapter has been built around specific case studies outlining a major stage in the working practices of Joseph Banks, Thomas Pennant, and their assistants while comparing these with other naturalists such as Gilbert White and William Jackson Hooker. The aim is to show the interactions between these individuals and a variety of physical materials, ideas, and information that fed into the growing global culture of natural history. Major themes outline the processes of producing natural-historical knowledge from collecting specimens on expeditions, ranging from parochial surveys through to global voyages, to the construction of natural history books, before ending with the different strategies used to distribute printed information and stimulate engagement with natural history. Together, the analysis demonstrates how some of the fundamental questions posed by historians of science cannot be understood by examining isolated aspects of the practice of natural history or a select geographical area.

Chapter 1 examines natural historians' approaches to collecting and organizing information during research expeditions undertaken within Britain itself, ranging from parochial surveys, journeys through the mountains and islands of Wales and Scotland, through to visiting the global collections compiled by East India Company bureaucrats. National expeditions encompassing different geographical scales set the precedent for standardized working practices employed on the global voyages explored in chapter 2. Perhaps the most well-known global voyage, at least so far as the British Empire was concerned, is James Cook's first voyage to the Pacific (1768–1771). Joseph Banks, Daniel Solander, and their team of field assistants accompanied this expedition, employing numerous paper technologies. These records infused Western understandings nature with those of people naturalists encountered in the Pacific, laying the foundations for the swift transferal of these materials into printed formats after expeditions returned to Europe. Chapter 3 reassesses how the diverse materials collected on expeditions were connected to construct illustrated natural history books. Concentrating on Banks's, Pennant's, and to a lesser extent White's different approaches to producing expensive copperplate books, chapter 3 shows how these do not conform to more general commercial publishing models. Books were designed to embody both the physical structure of natural history collections and the interests of their compilers and funders, reflecting on the diversity of classification systems British naturalists employed. This chapter draws out the relationship between natural historians and a host of artists, engravers, papermakers, and other artisans, casting new light on these individuals' agency in the production process.

Chapters 4 and 5 explore the distribution and reception of the books produced by Banks, Pennant, and other naturalists, suggesting these were central for defining British naturalists as independent gentlemen whose incomes were not derived from commercial markets. Books were strategically distributed to build global networks of naturalists. These connections facilitated the transfer of information compiled to form new editions while influencing voyages of discovery and global trade ventures, which became more frequent as the eighteenth century progressed. My approach serves to connect notions of gentility to scientific publishing, showing how naturalists only had a limited interest in the commercial viability of their books. The concluding chapter examines the practical use of these books both on expeditions and in the library. Many naturalists used print to develop unified networks extending from the South Pacific to European libraries while standardizing practices of communicating and organizing information. Books also caused their genteel authors, patrons, and distributors significant problems. Some escaped their select networks to be engaged with by a whole range of different peoples. Books and collections were designed to continually assimilate natural knowledge—in a time when more new species were being described than ever before or since. Despite these issues, genteel natural history authors did not consider their works commercial objects, believing books and their associated collections should generate new information, serve emergent national and imperial interests, while stimulating the cyclical process of collecting, publishing, distributing, and using natural knowledge.

Chapter 1

From Parish to Nation

Thomas Pennant on Tour

After visiting Joseph Banks at his home in New Burlington Street, London, in May 1768, the Reverend Gilbert White wrote to the Welsh naturalist Thomas Pennant. White commented that "even Mr Banks (notwithstanding that he was soon to leave the kingdom, & undertake his immense voyage) afforded me some hours of his conversation at his new house, where I met Dr. Solander."[1] White's visit is representative of the standard practice of leaving a local parish to converse with other naturalists and view collections, examples being the specimens Banks had collected in Newfoundland, while discussions centred on Banks's preparations for traveling to the South Pacific.

Travel was integral when undertaking natural historical research in Britain itself. Expeditions ranged from local parochial surveys resulting in publications such as White's *Natural History and Antiquities of Selborne* (1789) through to broader national journeys. Examples include Pennant's tour of Scotland in 1769 and voyage to the Hebrides in 1772. Journeys to survey the natural history of Britain were undertaken by a wide range of genteel naturalists; examples include Gilbert White, Joseph Banks, Thomas Pennant, Anna Blackburne, Benjamin Stillingfleet, William Borlase, Alexander Catcott, Thomas Falconer, Hugh Davies, Henry Jenner, and others. Given the quantity of individuals who traveled within Britain and the numerous archival resources they produced, this chapter analyzes the journeys taken by Pennant, White, and Banks, naturalists who traveled to observe, collect, and record specimens. These naturalists encountered objects when researching a specific area, during national expeditions and when visiting private natural history collections compiled during international voyages. This resulted in the emergence of works that ranged from parochial accounts through to global natural histories.

The main analysis explores the development of naturalists' use of paper alongside different scales of time, classificatory systems, and the geopolitical areas they covered. The initial concentration is on the integration of paper into natural history research practices, exploring how this was formulated and presented to facilitate the processes of recording and collecting information over different periods of time and geographical scales. This progresses to an analysis of approaches to collecting information on natural history research trips that covered specific geographical areas. Local trips to the same places were repeated over several decades and naturalists rarely carried large quantities of books and collecting equipment on these expeditions.

Through building on smaller-scale collecting enterprises naturalists started to embark on more extensive national journeys, which include the two tours of Scotland Pennant commenced in 1769 and 1772. National trips often took several months, required considerable preparation, and show how Pennant communicated with local populations in the Outer Hebrides in addition to working and traveling with a team of fellow naturalists, field assistants, and a train of porters who carried the equipment. This casts new light on contemporary understandings of indigeneity within Britain, showing how naturalists who traveled in remote areas collected information on the native flora and fauna while working with local peoples who communicated in several different languages. The notion of "indigenous" became increasingly common in accounts of British natural history in the late eighteenth century, a prominent example being Colin Milne and Alexander Gordon's *Indigenous Botany*, published in 1793. This has certain correlations with Alix Cooper's exploration of local collecting in early modern Europe, arguing that Carl Linnaeus's systematic innovations initiated important changes in local natural histories.[2] The final section examines how natural historians undertook global research during journeys across Britain, by traveling to the increasingly diverse collections owned by other naturalists and the menageries and art collections of the aristocracy. These collections were used to produce accounts Pennant referred to as "general" natural histories designed to encompass a global array of flora and fauna, emphasizing the connections between national travels and naturalists' interests in the wider world.[3]

Interest in natural history grew throughout the eighteenth century, often inspired by the rulers of emergent European nation-states who sought to centralize administrative structures and catalogue the resources of specific regions. Individual catalogues accommodated different geographical scales and often concentrated on specific areas to consolidate regional power and assess their economic value. In Britain the state was far less centralized than the administration of absolutist states in Continental Europe, although geographical scale and the economic importance of specific areas were of particular importance to British landowning elites when it came to the management and general improvement of their property.

The decentralized administration and increased importance of specific areas shaped the natural history enterprises administered by Pennant, Banks, and White, all of whom had economic and political interests in improving the regions they studied. Geographical scale and personal interests molded the nature of expeditions, naturalists' approaches to recording information, and their main outputs of correspondence and publications. In Britain natural history books published since the 1760s tended to take the form of parochial, county, and national natural histories and floras. Localized accounts compared records of species with those compiled in other parishes, nations, and on global expeditions. The concept of Great Britain as a "nation" was still new during this period. Wales had been unified with England since 1536, Scotland since 1707, and Ireland was not incorporated into a political union until 1801. The British Isles became more politically distinct from the Continent after the 1750s due to frequent wars, which resulted in the construction and definition of a maritime border alongside the solidification of internal administrative units. The defined nation-state inspired naturalists to conduct research on particular areas and compare these with potential rivals.[4]

The geographical scales British naturalists worked across were determined by internal boundaries that solidified administrative entities. Each constituent country was divided into counties under the jurisdiction of a sheriff, who answered to the monarch and central government. Counties consisted of parishes, usually defined as an area administered by a member of the clergy under the jurisdiction of the Church of England. From the 1750s, these political and religious administrative entities became more geographically distinct, developing into important units of official administration that often defined the parameters of an area used by natural historians for their research. By 1800 parishes emerged as essential administrative and social units toward which individuals developed significant personal connections.[5] As a result of landlord absenteeism, the local clergy often became the most important people in the parish. The very nature of their work, of traveling through the parish to converse with and record the births, marriages, and deaths of parishioners, gave them an insight into more general paper-based recordkeeping practices and— if so inclined—the opportunity to pursue work on the natural history of their surroundings.[6]

The Emergence of Traveling "Paper Technologies"

Practices of recordkeeping on the natural world had to be adapted to a range of different information sources, physical localities in which they were used, and movement between geographical areas, ranging from surveys of specific areas taking place over decades to more general accounts covering larger geographical distances over shorter periods of time. From the late seventeenth century, it became a routine practice for naturalists and antiquarians to assemble paper in a variety of different formats for use on expeditions. These included notebooks, separate bound volumes that contained specific information pertaining to a single expedition, geographical region, or kingdom of nature; annotated printed books, often compiled from octavo and duodecimo volumes that were frequently interleaved with blank pages so information could be incorporated into a system of classification; and loose notes that could be reordered, added to, inserted between the pages of books, and posted back to a naturalist's home after new information was acquired. Many of these practices adhered to a programme set out by Francis Bacon and promoted by the Royal Society of London, which encouraged the systematic recording of empirical facts in journals and commonplace books that could be collated at a later date.[7]

Examples include the emergence of natural histories defined by a specific county or constituent kingdom pioneered by figures such Robert Plot, who wrote accounts of Staffordshire and Oxfordshire, and Edward Lhuyd's work on Wales.[8] Naturalists held the Baconian belief that information gathering depended on "the united labours of many," collaborating with groups ranging from local people to employed assistants while consistently comparing these views with published works and their own observations to construct detailed records that allowed this information to be received and adapted by different people. As Staffan Müller-Wille has suggested, these Baconian ideas inspired Linnaeus's own approaches to developing technologies of writing and classification of the sciences.[9] A product of the popularity of Baconian philosophy and the Linnaean system for certain branches of natural history, these two aspects led to the emergence of a variety of different collecting practices used by naturalists throughout Britain, which led to the emergence of systems designed to exchange and compare vast quantities of information.

Thomas Pennant's interests emerged from within this tradition, collaborating across cultures to assimilate information recorded in a variety of diverse paper repositories. For example, Pennant described how he "kept a regular journal," which took a similar form to a traditional commonplace book, on his expeditions around Britain, which he wanted to remain unpublished "as they contain inaccuracies." However, Pennant's journals were made available to the select group who visited his private library at Downing Hall "as they contain many descriptions of buildings, and accounts of places in the state they were at the time they were made."[10] These processes of collecting information and visiting collections were all governed by strict social hierarchies between traveling gentlemen-naturalists, the staff they employed, traveling companions, and those they encountered on the expedition. One legacy of the seventeenth century that becomes clear throughout Pennant's working practices is his annotation of printed books. Many of the annotated books Pennant used as he traveled were printed in the seventeenth century. Examples include copies of Francis Willoughby and John Ray's *Ornithology* (1678) and Christopher Merrett's *Pinax rerum naturalium Britannicarum* (1667). The annotations Pennant added to these volumes are designed to locate species he observed and collected within a prescribed systematic order, standardizing information as it moved between Pennant's own field notes, illustrations correspondence, and printed text. For example, in his copy of Merrett, Pennant inscribed on the front flyleaf "such as are found in Flintshire I have marked thus — ," adding a dash to the margin next to the description of every species found in this specific geographical area to compare this distribution with the more generalist survey.[11] This reflects on how Pennant consistently integrated information when formulating his zoological works and travel accounts, adding and extending entries in printed books.

Paper, in a variety of different formats, became increasingly available from the 1760s with the general expansion of paper production and printing. This allowed naturalists to complexify their different means for recording information, using interleaved books, almanacs, instruction manuals, and ledgers, objects designed to be carried in pockets and encased in wallet-style bindings to protect them from hostile weather conditions.[12] These books became central for unifying times and dates across the country and were used by naturalists such as Banks, White, and Pennant, who incorporated them into their wider systems for managing information.

Natural Histories of the County and Parish

The mid-eighteenth century saw an increase of natural histories defined by a county or parish, studies that served as an intricate jigsaw puzzle to be slotted together by naturalists who aspired to national projects. For example, when gathering detailed information on an entire country, Thomas Pennant circulated questionnaires titled "Queries, addressed to the Gentlemen and Clergy of North-Britain, respecting the Antiquities and Natural history of their respective Parishes, with a view of exciting than to favour the World with a fuller more satisfactory account of their Country, than it is the Power of a Stranger and the transient Visitant to give."[13] These questionnaires indicate the necessity of parochial knowledge for a broader survey, showing how localized accounts could fit within a national framework. Pennant's questionnaire, which contained twenty-five questions relating to natural history, was inspired by his correspondence with William Borlase (1696–1772), who circulated questionnaires to every parish in Cornwall when compiling information for his *Natural History of Cornwall* (1758). Borlase was motivated by a long tradition of circulating questionnaires established in the seventeenth century by figures such as Edward Lhwyd, Robert Plot, Gerard Boate, and Robert Boyle, who used the Royal Society's journal, *Philosophical Transactions*, to distribute lists of questions titled "Articles of Inquiries Touching Mines" and "Other Inquiries Concerning Sea." Many early questionnaires remained general and did not seek to extract geographically specific information. The early eighteenth century saw the emergence of surveys of counties such as Leicestershire, Rutland, Norfolk, Dorset, Oxfordshire, Staffordshire, County Down, Durham, and others. Pennant's questionnaires were circulated with the hope that "parochial *Geniuses* will arise and favour the Publick with what is much wanted, LOCAL HISTORIES."[14]

Pennant intended to inspire similar working practices to those he used to record the natural history of specific areas. For example, when collecting information for his *Of London*, Pennant described: "I often walked around several parts of London, with my notebook in hand, that I could not help forming considerable collections of materials."[15] These new projects inspired lengthy correspondences on particular geographical areas, some of which Pennant published. One typical example is Charles Cordiner's *Antiquities & Scenery of the North of Scotland* (1780). After realizing that he would not be able to explore the northernmost parts of Britain in any detail, Pennant financed Cordiner's research and publication. Similarly, at the recommendation of Banks, Pennant engaged the Reverend George Low of Birsa to explore and produce a manuscript on the Orkney and Shetland Islands. Although this manuscript remained unpublished until 1813, Pennant integrated Low's observations into his manuscripts and publications.[16] Pennant's patronage of figures such as Low and Cordiner shows how he emerged as a significant supporter for parochial natural histories.

Perhaps the most famous parochial natural history is Gilbert White's *Natural History and Antiquities of Selborne* (1789), a book compiled from detailed surveys of a specific area repeated over several decades. In compiling these records, White relied on his copies of Daines Barrington's *Naturalist's Journal* (1767) and his garden *Kalendar*, books used to record the gradual changes that occurred in specific places. White often went beyond the traditional scope of garden and natural history records, adding a wealth of detail on the seasonal changes to specific species, their interactions with one another, and supplementary information on wider global events. White's interests in the wider world reflects the numerous conversations he had in London. In addition to visiting Banks, he conversed with Pennant and Barrington, whose correspondence reveals their integration with a global network ranging from those who reported sighting James Cook's ships off Kamchatka Peninsula in the late 1770s through to the American Revolutionary War and an explanation for the delays in printing "[Gilbert] White's letters on swallows" in the Royal Society's *Philosophical Transactions*.[17] White hoped that *Selborne* would encourage "stationary men" to "pay some attention to the districts in which they reside," forming a model that could be followed by those creating parochial accounts across the country.[18]

To produce surveys defined by the natural borders of a single county or parish, naturalists combined their own observations with information recorded in printed books with a wide geographical outlook. Naturalists saw the improvement of specific local areas, which often included a substantial proportion of their own landholdings, as particularly important to their work and personal connections to regional politics. For example, Banks held the position of sheriff of Lincolnshire from 1792 to 1793, Pennant served as high sheriff of Flintshire from 1762 to 1763, and White was responsible for the spiritual well-being of the parishioners in Selborne. As parishes grew in prominence, many began to play a greater role in supplying poor relief and maintaining common land, much of which was removed to fulfill ideas of national improvement after the passing of successive Enclosure Acts in 1769 and 1773.[19] The parish, as an increasingly significant focal point, gave a greater call for naturalists to undertake surveys of these areas, many of which did not have specific works of natural history associated with them.

County natural histories by figures such as Robert Plot and John Morton remained the most geographically specific natural history books of the seventeenth century. In addition to the emergence of general inventories of counties, the late seventeenth century saw systematic works that concentrated on a specific aspect of the flora or fauna of the entire globe, such as Francis Willoughby and John Ray's *Ornithology* (1678) and Ray's *Historia Plantarum* (1686–1704).[20] From the 1750s, the combination of the Linnaean reform and the solidification of national political boundaries in Europe stimulated the emergence of national floras and faunas that conveyed lists of a single kingdom of nature from a specific geopolitical area, giving a sense of uniqueness and topographical limits to a population. As Janet Browne has suggested, the construction of Linnaean national floras and faunas in Britain tended to follow the examples set by Linnaeus's apostles in the 1740s and 1750s. By the 1780s there was a full range of works available on the open market that represented the flora or fauna of different nations. Typical examples included John Hill's *Flora Britannica* (1760), Johann Georg Gmelin's *Flora Sibirica* (1747), Georg Christian Oeder's *Flora Danica* (1766–1789), and John Lightfoot's *Flora Scotica* (1777).[21]

Geographical specialization influenced naturalists' means for using books on research trips that presented global inventories of a single branch of nature. Books were used to identify and associate certain species with a specific area, connecting these regions with the wider world whilst emphasising their uniqueness. For example, Pennant commented that to undertake his zoological work, he combined information "from the works of general naturalists, from the Fauna of different countries, and from my own observations."[22] An example of a "general" book is Pennant's copy of Willoughby and Ray's *Ornithology* (1678). This gave Pennant "a great love for natural history in general" from the age of twelve when he received a copy from his uncle, John Salisbury. In his working copy of *Ornithology*, Pennant added annotations that relate to species he observed and collected in his local county of Flintshire. On the first endpaper Pennant wrote, "N. B. The Birds marked thus ✳ are found in Flintshire."[23] This symbol has been added next to every entry in this work that concerns a species Pennant found in his local county. Pennant traveled throughout Flintshire to observe birds, some of which he viewed in the grounds of Downing Hall, where his library overlooked a drive, extensive grass lawn, and wooded area. Pennant and his son, David, both encouraged birds to come near to the house. The regular purchase of "Seed for the Birds to Downing" is recorded in their 1803–1807 household account book.[24]

A species Pennant observed and collected in Flintshire was the bittern, as is apparent from an annotation in his copy of Willoughby and Ray's *Ornithology* (figure 1.1). The close observation of the bittern allowed Pennant to build on Willoughby and Ray's description, giving additional information on its anatomy, weight, geographical distribution, character, and calls. In *British Zoology*, Pennant described the bittern as "a very retired bird, concealing itself in the midst of reeds and rushes in marshy places."[25] This was an improvement on Ray's brief description, which only described its call and nesting habits, and fulfilled Pennant's ideal of conducting fieldwork that aligned with the model initiated by Willoughby and Ray a century before: "In the prosecution of our plan, we shall to avoid the perplexity arising from forming a new system, adopt (as far as relates to the *Quadrupeds* and *Birds*) that of the inestimable *Ray*, who advanced the study of nature far beyond all that went before him."[26]

Pennant believed his detailed descriptions in *British Zoology*

FIGURE I.I. Left: page 15 from Thomas Pennant's copy of Francis Willough-by and John Ray's *Ornithology* (1678). By permission of Llyfrgell Genedlaethol Cymru/The National Library of Wales. Right: Pennant's specimen of a bittern that he used to formulate the description for *British Zoology*. © Jonathan Jackson, Natural History Museum, London.

would prove useful for others with interests in the natural history of their local parishes. This is apparent from the various parochial users of Pennant's *British Zoology* (1776–1777). For example, John Blackburne of Orford Hall, near Warrington, used his copy to record species he observed in Orford, inscribing, "Those marked with the letters a. b. are in my Collection. 1800. J. Blackburne." Blackburne recorded details of the birds he shot and added to the collection originally compiled by his sister, Anna Blackburne, whose name is undoubtedly represented by the "A. B." initials. The Kingfisher is representative of a new addition to the collection that Blackburne recorded as being "Shot at Orford Janury. 18th 1803" (figure 1.2).[27] Similarly, White used his copy of *British Zoology* to record observations of choughs in Sussex and the nesting habits of sand martins.[28] Pennant's *British Zoology* became an essential resource for naturalists when surveying local

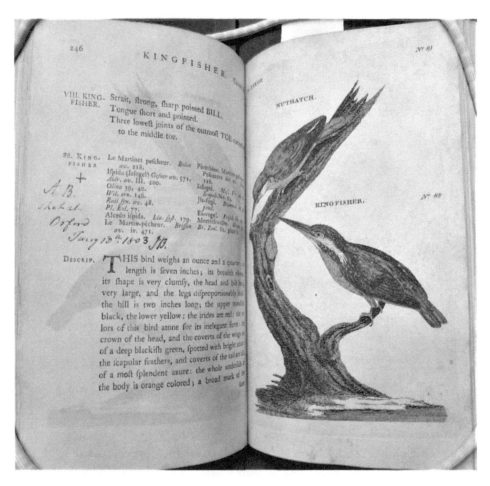

FIGURE 1.2. Annotations next to the description of the kingfisher in John Blackburne's copy of Thomas Pennant's *British Zoology* (1776–1777). By kind permission of the Trustees of the Natural History Museum, London.

parishes, who then communicated their observations to Pennant, who was compiling a new edition.

The use of national natural histories to identify and account for species in a specific area is apparent from Pennant's and White's use of national floras to produce inventories of plants that could be found in their local parishes. For example, in his copy of William Hudson's *Flora Anglica* (1762), White inscribed, "The Plants marked thus × have all been found within the parish of Selborne in the County of Southampton," marking a total of 439 species (figure 1.3).[29] Similarly, in his copy of John Lightfoot's *Flora Scotica* (1777), Pennant noted

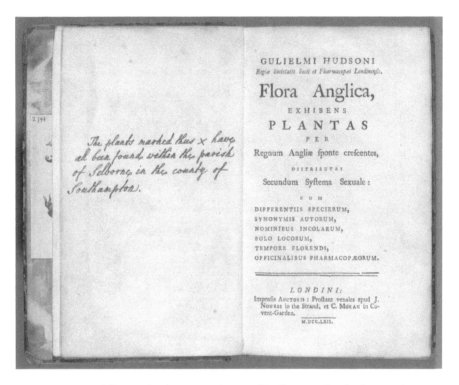

FIGURE 1.3. Gilbert White's annotated copy of William Hudson's *Flora Anglica* (1762). Houghton Library, Harvard University.

that all the species he marked with "+ [are] In whiteford parish."[30] These national botanical works were crucial for identifying and isolating the flora of a specific area while arranging it according to the Linnaean system. This practice allowed naturalists to integrate species observed in a specific place with a broader program, comparing specific areas to develop a national picture. For example, marginal notes were used by Pennant and White to construct publications on their local parishes. In his *History of the Parishes of Whiteford and Holywell* (1796), Pennant listed plants found in the parish and ordered these according to the Linnaean system. Many were collected and identified by Lightfoot, who had visited Downing Hall when touring Wales in 1773.[31] White also published an inventory of the botanical species he encountered in Selborne, providing descriptions of the local conditions responsible for the high diversity of species in the parish.[32] All of these plants are ordered and named according to the synonyms given in Hudson's *Flora Anglica*.

Annotated natural history books recorded and codified species from a specific area, structuring these under a definitive system of classification to follow the Baconian ideal of collecting, stabilizing, and ordering empirical factual information. For Pennant and White, the system they used depended on the branch of natural history. For plants, fish, shells, and insects, they preferred the Linnaean system. For quadrupeds and birds, they leant toward the system used by Ray. For example, White's annotations in his copy of Ray's *Synopsis Methodica Avium & Piscium* (1713) relate the printed descriptions to his observations of specific ornithological species in Selborne and specimens communicated by his brother, John White (1727–1780), from Gibraltar.[33] Similarly, Hudson's *Flora Anglica* was central to White's approach for using the Linnaean system when assessing the diversity of species in Selborne. These small books could be carried inside a coat pocket or saddlebag and would be taken on frequent parochial or even countywide expeditions. Many were adapted to accommodate notes. Examples include Thomas Martyn's copy of *Methodus Plantarum circa Cantabrigiam Nascentium* (1727). Martyn annotated his interleaved copy with additional notes on the species mentioned in the printed text that he observed during his travels around Cambridge.[34]

Annotations and descriptions in printed books were often associated with loose pieces of paper to aid with the incorporation and accessibility of information. For example, White kept small folded over sheets of paper alongside his copy of Ray's *Synopsis Methodica Stripium Britannicarum* (1724).[35] The Oxford botany professor Johann Jacob Dillenius (1684–1747), whom White had met during his time at the university, had edited this book, and it remained an essential tool for White's research before it was superseded by Hudson's *Flora Anglica*. White's supplementary notes were used alongside his copy of Dillenius's edition of Ray's work and provide an index for the main classificatory divisions that allowed him to locate specific sections, speeding up the process of identifying species in the field. Although White suggested that "to enumerate all the plants that have been discovered within our limits would be a needless work," he emulated Pennant by marking every species he observed within the natural parochial boundaries of Selborne in *Flora Anglica*. This facilitated the comparison of similar records kept by naturalists across the country who kept annotated copies of *Flora Anglica*, including Richard Pulteney, James Edward Smith, and Humphrey Sibthorp.[36]

FIGURE 1.4. The image, specimen, and annotation on the interleaved page in the hand of David Pennant in the Downing Hall copy of August Johann Rösel von Rosenhof's *Insecten-Belustigung* (1746–1761). By kind permission of the Trustees of the Natural History Museum, London.

Sometimes naturalists employed pre-Linnaean publications to record local observations and classify species according to the Linnaean system. For example, Pennant's copy of August Johann Rösel von Rosenhof's *Insecten-Belustigung* (1746–1761) was interleaved with blank pages designed for annotation.[37] Compiled from a series of copperplate images that display the metamorphoses of insects, this work is characteristic of what Pennant described as a "general" natural history that aimed to create a global inventory of entomological species. The interleaved pages in Pennant's copy have been annotated by his son, David, who also added specimens and watercolor images that describe and depict insects he collected in the local parish. For example, on February 12, 1812, David Pennant caught a small tortoiseshell butterfly at Downing Hall. He then took this specimen back to the library and pressed it between the pages of Rösel's work. Pennant used the interleaved page to note the Linnaean binomial, referring it to *Systema Naturae*, adding the local temperature, the time it was collected, and the original locality (figure 1.4).[38]

Pennant's meticulous recording of the temperature and date show the close relationships between practices of recording natural history, the weather, and atmosphere by the early nineteenth century. The annotations in Pennant's copy of Rösel serve to unify these two sets of observations, locating them in a specific time and place, situating localized observations within a global insect classification. Simultaneous observations of the local weather and natural history were central for connecting a specific area with a broader national picture. For example, between 1780 and 1835, David Pennant kept daily weather readings at Downing Hall. Following a similar format to the tables of annual weather reports published in *Philosophical Transactions*, his notebooks show that he checked the weather two or three times each day, noting the date, hour, minute, and temperature, taking readings from a thermometer, barometer, and weather vane.[39] His notebook has a white parchment wallet-style binding to protect the paper interior from the weather—for David Pennant would have to go outside to take readings from various instruments, record the direction of the wind, and assess the quantity of rain, taking a very similar structure and purpose to the notebooks used by Thomas Barker (1722–1809), brother-in-law to Gilbert White, who kept meticulous weather records from his home at Lyndon Hall, Rutland, between 1738 and 1798.[40] David Pennant's use of different instruments reflects how they permeated into wealthy households with the general expansion of luxury consumer goods.[41] Instruments imposed standard measures on the unpredictable North Wales weather, readings recorded in notebooks, and interleaved books designed to standardize information on local flora and fauna and the weather that governed it.

Interconnected approaches to recording the weather and natural history was essential for applying structure to nature, allowing for the accurate prediction of the annual life cycles of certain species. For example, next to the weather records for March 13 and 14, 1793, David Pennant added a description of how the plants "Ficaria verna has been in flower some days; on the same walks I saw the anemone memorosa, & by the pond a tuft of the Caltha palustris." Pennant's records of the cyclical and seasonal changes species went through in his immediate locality are similar to those White added to the *Naturalist's Journal*. White placed his descriptions of the natural world alongside momentous political events, descriptions that quickly spilled over the edges of the printed forms. In comparison, Pennant's notebook gave

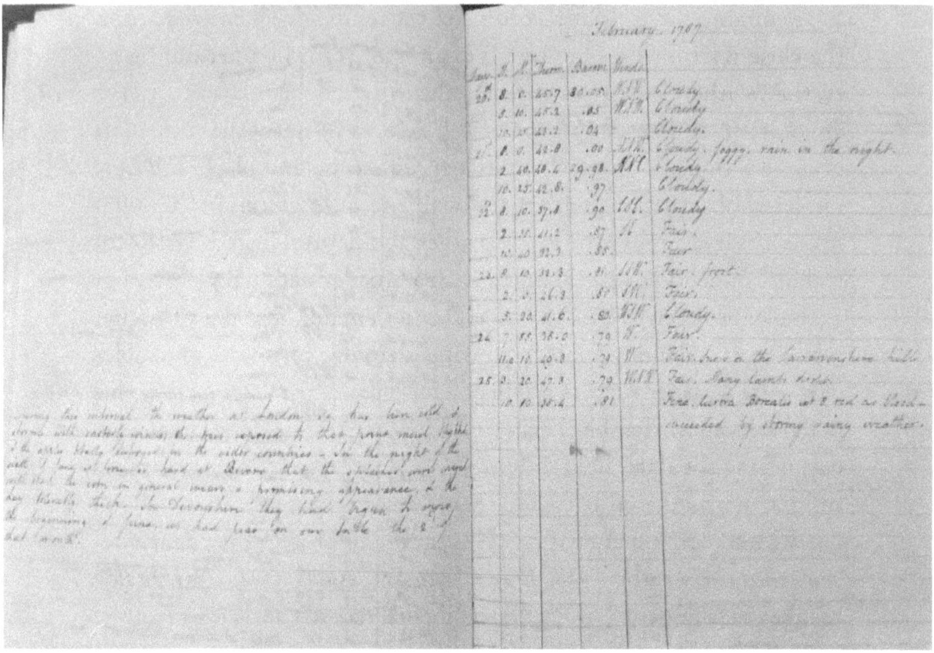

FIGURE 1.5. David Pennant's weather records and related observations for the end of February 1787. By permission of Llyfrgell Genedlaethol Cymru/The National Library of Wales.

him the flexibility to relate information on local events to news circulated throughout the wider nation. This came in the form of letters, newspapers, and visitors to Downing Hall. For example, on February 25, 1787, Pennant described seeing aurora borealis "red as blood" at Downing Hall and added information on more general events from across the country on the blank left-hand page of the notebook (figure 1.5). These include the "cold & stormy" weather in London, the impact of this on apple trees in the cider-producing counties of the Midlands, and its effects on the harvest in Devonshire.[42]

The relationship between parochial records, national natural-historical, and political events was essential for inspiring collaborations between naturalists. Examples include that between White and Daines Barrington, whose handwriting appears throughout several copies of the *Naturalist's Journal* from White's collection. Barrington often added information that relates White's specific observations of Selborne to a broader national picture.[43] These notes relate to Barrington's interest in creating "a General Natural History of Great

Britain," which was to result from the combination of "many such journals kept in different parts of the kingdom."[44] Barrington and Pennant had similar reasons for distributing the *Naturalist's Journal* and questionnaires: to become overarching patrons of parochial natural histories as exemplified by the central role of their correspondence in White's *Selborne*. However, the joining up of parochial accounts only went part of the way toward creating national natural histories and many believed more extensive travel was required to lend authority to these accounts.

Collecting on a National Expedition

In comparison to journeys around local counties and parishes, longer expeditions naturalists took to survey major constituent parts of Britain, such as Scotland, Wales, and Ireland, required more preparation. By the 1760s, Thomas Pennant was regarded as a main authority on the natural history of these regions. The geographical extent of Pennant's journeys, or "tours," had a significant impact on his approach to recording and ordering information on the natural history of constituent kingdoms. The main purpose of these trips was to survey, observe, describe, collect, and enumerate species from around Britain, thereby giving Pennant licence to write an expanded edition of *British Zoology* published between 1776 and 1777—the page count of this edition is more than double that of the edition published between 1768 and 1770. Providing a numerical analysis was of great importance to Pennant, who stated that an "enumeration of the species of certain classes of the animal kingdom would be equally agreeable and serviceable to the travelling Zoologist."[45] Many contemporaries reaffirmed Pennant's qualification to study Britain in its entirety. For example, in a letter to his brother, John, Gilbert White remarked that Pennant "has now taken great pains to investigate Great Britain and its Islands, and will be well qualified to put the last hand to the 'British Zoology' in a quarto edition." In *The Scientific Tourist through England, Wales, Scotland* (1818), Thomas Walford commented that "those tours that come from the pens of scientific travellers are not only most pleasing, but always the most instructive," identifying Pennant as a known authority and reprinting the itineraries of his national "tours."[46]

Pennant traveled on two main tours of Scotland in 1769 and 1772. The latter included a voyage to the Hebrides. Pennant's second journey was the most extensive and he allowed John Lightfoot,

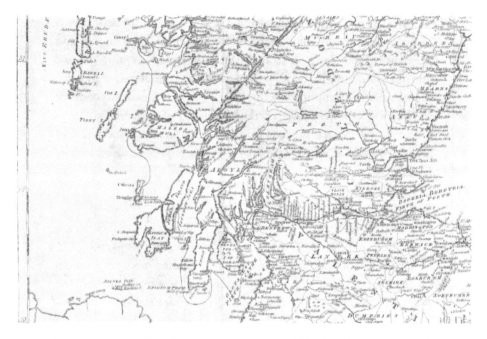

FIGURE 1.6. Detail from Thomas Pennant's map of Scotland (1774). The black line around the islands marks the route of Pennant's voyage in 1772. Reproduced by kind permission of the Trustees of the National Library of Scotland.

the Linnaean botanist and curator of Margaret Bentinck, the Duchess of Portland's private natural history collection, to accompany his party. This planned trip was inspired by Joseph Banks's voyage to the Pacific aboard the *Endeavour*, Pennant's previous tour of Scotland in 1769, and his *Tour on the Continent* in 1765. Pennant either traveled on horseback or paid for passage on postal coaches during overland trips. For example, after his visit to George-Louis Leclerc, Comte de Buffon's (1707–1788) estate at Montbard, Pennant described how "M. de Buffon lent me horses to convey me to the nearest post . . . took the post horses at Maison neuves."[47] Pennant privately charted ships to commence sea voyages. Examples include the cutter *Lady Frederick Campbell*, which Pennant boarded in Glasgow on June 17, 1772, and traveled on when surveying the Hebrides (figure 1.6). In addition to Lightfoot, Pennant was accompanied by staff, including Archibald Thompson, the ship's master; Dr. John Stuart of Luss, a Gaelic expert and Linnaean botanist; Moses Griffith, an artist; Louis Gold, a French valet; an unnamed landscape painter; a groom; and a hawker.[48] The expenses Pennant incurred from this "journey and voyage from May 18th to my return [in] sept.r." totalled £296, a sum he offset against sales of the published account of his *Tour in Scotland*.[49] Pennant's large team presents a very different mode of travel when compared to others who visited this region. For example, when writing about Samuel Johnson's 1773 tour, James Boswell wrote, "Dr J thought it unnecessary to put himself to the additional expense of bringing with him Francis Barber, his faithful black servant; so we were attended only by my man, Joseph Ritter, a Bohemian; a fine stately fellow above six feet high, who had been over a great part of Europe and spoke many languages."[50] This reflects on their very different outlooks on funding these tours. For example, Pennant made clear that he only ever intended to break even on his expenses through sales of his published work. By contrast, Johnson relied on profiting from the published account for his income.

The structures of authority Pennant established with fellow travelers resemble the workings of his country estate at Downing Hall. Most of those who accompanied Pennant were employed and treated as servants. For example, Pennant described the artist Moses Griffith, who lived in a cottage on Pennant's estate for over thirty years, as having "distinguished himself as a good and faithful servant, and able artist." During the tours and Griffith's early career, Pennant regarded the latter's artistic capabilities as his personal property, describing how "in the spring of this year [1769] I acquired that treasure, Moses Griffith," who was "descended from very poor parents." Similarly, Pennant described his valet Louis Gold as a "servant and friend," reflecting their good relationship for much of Pennant's life.[51]

Pennant utilized the talents of his servants as trained amanuenses and artists, employing these skills alongside more general tasks associated with cooking, cleaning, and carrying baggage. The hierarchies and practices employed on Pennant's tours were nearly all established in his library, where Griffith, several secretaries, and family members had specific defined roles in the process of managing and recording incoming information, reflecting the hierarchic structure of the Downing estate. Each stage of Pennant's information-management system utilized the skills of an individual and defined their specific assigned role, practices later transposed onto the materials and people that accompanied Pennant on his tours.[52] Other figures who accompanied Pennant, such as Lightfoot, received similar treatment to the honored guests who visited Downing. As a result, Pennant allowed his guests to employ the skills of his natural history staff and servants, although he continued to maintain authority over the use of this material in publications.

Pennant exercised his private wealth to employ the crew of the *Lady Frederick Campbell* for the purpose of surveying the natural history and topography of the Hebrides, thus reducing the potential for conflicts between Pennant's party and the seamen, and they parted on good terms. Pennant noted how Thompson's "obliging conduct throughout, and skill in his profession, demand my warmest acknowledgements."[53] This is very different from the relationship between many other traveling naturalists and their employers. For example, in 1778 the Earl of Sandwich described Johann Reinhold Forster, who had traveled on James Cook's second voyage of exploration, as "a person who could not keep a friend for any length of time, his behaviour to me, who did my utmost to serve him, was a plain proof of the truth of this affliction."[54] Daines Barrington requested Pennant "to continue silent with regard to matters between myself and Dr: Forster leaving that ungrateful madman to me as I shall know how to deal with him."[55] Barrington's and the Earl of Sandwich's comments reflect on the frequent conflicts between the naval administration, captains, and naturalists on ships chartered by the government where natural history remained a secondary concern—a difference in opinion that caused Banks to back out of Cook's second voyage.[56]

However, Pennant's and Thompson's relationship was not without disputes. On July 11, 1772, when the *Lady Frederik Campbell* was approaching the Isle of Staffa, the rough weather compromised the ship's safety. Thompson refused Pennant's request to dock at the island. Staffa was a recent discovery for naturalists of the late eighteenth century, and Banks formulated the first account of the island after visiting on his way through the Hebrides to Iceland on August 12 and 13, 1772. Pennant was interested in the geological formation of the tall hexagonal basaltic columns and wished for the ship to approach the rocky foreshore of the island, commenting that "I wished to make a nearer approach, but the prudence of Mr. *Thompson*, who was unwilling to venture in these rocky seas, prevented my farther search of this wondrous isle: I could do no more than cause an accurate view to be taken of its Eastern side, and those of the other picturesque islands then in sight."[57] Rather than producing his own account, Pennant relied on that provided by Banks whose artist, John Cleveley

(1747–1786), had produced an illustration of Fingal's Cave. A large cave formed from basaltic columns, Pennant reproduced Cleverley's image in his *Tour in Scotland and Voyage to the Hebrides, 1772* (1774).[58]

Although Pennant was dismayed at not being able to land, he regarded the publication of Banks's description of Staffa to be "a great consolation," allowing him to "lay before the public a most accurate account."[59] Pennant was reassured by his fellow traveler John Stuart, who suggested shortly after the voyage that "I think it does not hitherto appear that Mr. Banks has made any new considerable discoveries in his late voyage. As for the Island of Staffa, by passing near it on a fine day I doubt not but you had as good an opportunity of observing the general appearance of it's curious columnar rocks as he could have had by landing there."[60] Despite Stuart's comments, the inclusion of Banks's account emphasizes the collaborative nature of this book. As the editions progressed, Pennant added dozens of descriptions concerning various Scottish regions communicated by correspondents and respondents to questionnaires in a voluminous appendix.

The specific roles assigned to those who accompanied Pennant on his 1772 tour of Scotland mirrors the collaborative nature of the main published products of this journey. Lightfoot was responsible for botanical matters since Pennant had "quit all thoughts of Botany" by 1767.[61] Pennant regarded Lightfoot as having equal social status, similar to Samuel Johnson and James Boswell, who traveled through the Hebrides in 1773, and used Lightfoot's notes for his botanical descriptions in *A Tour in Scotland and Voyage to the Hebrides*. Pennant also intended for Lightfoot to publish *Flora Scotica*, to which he contributed an introduction on Scottish zoology.

A major difference between the workings of the library at Downing and hierarchies exhibited on Pennant's tour was the employment of Stuart for his combination of botanical knowledge and expertise in the Gaelic or Erse language. Stuart acted as a translator for Pennant and his companions in rural areas. Pennant remarked in the preface to his *Tour in Scotland* that he was indebted to Stuart "for a variety of hints, relating to customs of the natives of the highlands, and of the islands, which by reason of my ignorance of the *Erse* or *Galic* language, must have escaped my notice."[62] Lightfoot thanked Stuart for "a great portion of *Highland* botany, for *many* of the *medical* and *oeconomical*, and all the *superstitious* uses of plants" in addition to "the supply of their *Erse* and *Gaulic* names."[63] Stuart was essential for providing

Pennant and Lightfoot with Erse plant and animal names, which they placed alongside the Latin and English descriptions in *Flora Scotica* (1777), *A Tour in Scotland and Voyage to the Hebrides* (1774), and *British Zoology* (1776–1777). He was also used to translating and publishing information on the Erse language, publishing a revised version of the Gaelic Bible in 1767. Stuart's knowledge of the economic potential of species and his ability to translate Erse plant names into Latin and English was indispensable when Pennant and Lightfoot came to formulate descriptions of species, combining information on Indigenous uses with their systematic accounts.

Pennant and Lightfoot followed Linnaean practices when integrating Indigenous names and uses for new botanical species into Latin names and diagnoses, providing clues in the binomial name or generic description that could lead the reader to Indigenous Scottish uses of a species while reducing synonymy.[64] The desire to understand the original etymological root followed a long tradition that had been common among British naturalists since the seventeenth century. Naturalists such as Ray frequently recorded the vernacular names in books such as *Historia Plantarum* and compiled works based on local names, terms, and sayings, such as *A Collection of English Proverbs* (1670).[65] As Alix Cooper has suggested, interests in making local floras bilingual or even trilingual was a means for moving beyond groups of classically educated scholars to integrate knowledge from local communities into botanical collecting.[66] Thus, Pennant's interests in aligning Indigenous British names, contemporary English names, and those ascribed by Linnaeus were central for placing these species in their historical context and allowed him to improve on earlier descriptions.

Pennant made efforts to obtain Indigenous names for plants and animals during his travels around Britain and devised paper tools to record this information. For example, when it came to England and Wales, Pennant published lists of Indigenous names at the end of *British Zoology*. These were sourced from the people he visited, correspondents, and his team of field assistants. Pennant obtained the majority of "British" names, derived from the Welsh language, from William Morris of Anglesey.[67] The process of accumulating these multilingual lists can be found in the loose papers Pennant tipped into his copy of Willoughby and Ray's *Ornithology*, which Pennant and Morris used to tabulate comparisons of "English," "British," and the "Translation

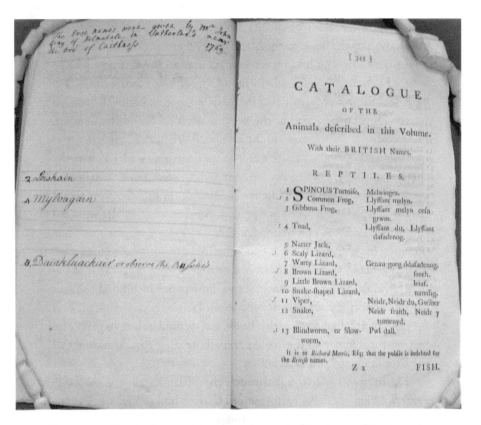

FIGURE 1.7. Thomas Pennant's interleaved copy of a "Catalogue of the Animals Described in This Volume with their British Names" extracted from the third volume of *British Zoology* (1769). By permission of Llyfrgell Genedlaethol Cymru/The National Library of Wales.

of the British" names, the last column being in Morris's hand representing their collaborative working practice.[68] To collect Indigenous names on his broader national journeys to Scotland, Pennant had the printed appendices of English and Welsh names for quadrupeds, birds, and fish from the 1768–1770 edition of *British Zoology* interleaved and bound as a separate notebook.[69] Pennant and the people he encountered then annotated the interleaved pages during his trip. Some annotators added Erse names and keyed these to the number in the printed text (figure 1.7).[70] This practice of using numbers to align printed descriptions and annotations on interleaved pages compares with Johann Reinhold Forster's use of an interleaved copy of *A Catalogue of British Insects* (1770). As Staffan Müller-Wille has suggested, Forster produced this work to supplement Pennant's *British Zoology* and added the names of species he acquired or observed since the publication of the book to the blank pages. Pennant carried his notebook-style indices on his journey to Scotland in 1769, during which he sourced Erse names from a "Mr. John Gray of Helmesdale in Sutherland near the ord of Caithness." Pennant stopped in Helmsdale on his way to Duncansby Head at the northeastern corner of the Scottish mainland.[71] The unification of the English, Welsh, and Erse languages in this interleaved pocketbook embodies Pennant's wish to unite the three major languages of Britain, giving him enough information to produce a national natural history. Knowledge of different vernacular names for species throughout England, Scotland, and Wales was central for the creation of a national account that assessed the economic potential for species, and, in a similar manner to a Linnaean binomial, offered a roadmap to any Indigenous uses. The physical status of the light, flexible, interleaved book shows how paper facilitated the collaborative accumulation of information. Not only did Pennant and his team of amanuenses use this book, but it could be lent to people he met on the journey so they could contribute their own knowledge of the Erse language and local fauna.

During his 1772 voyage through the Hebrides aboard the *Lady Frederick Campbell*, Pennant ensured that he had several notebooks, interleaved books, and printed volumes at his disposal. Some of these he brought from Downing Hall; others were obtained from stationers in various Scottish towns. By the late eighteenth century, stationers had become more widespread in remote areas and it was even possible to purchase supplies in the islands of the Hebrides. For example, Samuel Johnson, who traveled throughout the Hebrides with James Boswell in 1773, noted that when "Mr. Boswell's journal was filled" he purchased some paper from the one standing shop on the island of Col.[72] As Mary Poovey has suggested, Johnson viewed the emergence of shops and the ability to purchase supplies for writing as part of a chain of material conditions that facilitated knowledge production and the collection of information.[73] The increased connections between remote rural areas and the cities of southern Scotland after the Jacobite Rising of 1745 facilitated detailed studies of these regions and their integration into a centralizing nation-state.

During the Hebridean voyage, Thompson acquiesced with the majority of Pennant's requests, defining the relationship Pennant established with his employees to enable his survey of the natural history

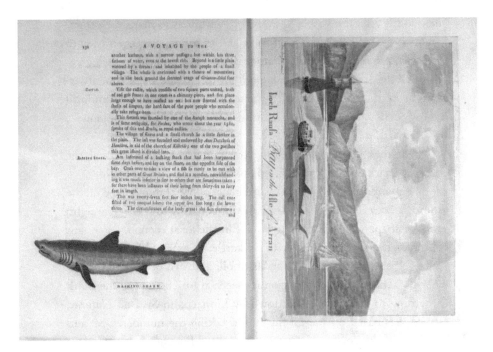

FIGURE 1.8. Thomas Pennant's extra-illustrated copy of *A Tour in Scotland and Voyage to the Hebrides* (1790) showing Moses Griffith's illustration of the capture of a basking shark in Loch Ranza. Mounted under Pennant's description is the related print from *British Zoology*, vol. 3 (1776). By permission of Llyfrgell Genedlaethol Cymru/The National Library of Wales.

of the Western Isles. This becomes apparent from the use of the ship for natural historical pursuits. An example can be found in Pennant's descriptions and Moses Griffith's illustrations of the basking shark, a species they observed at Loch Ranza on the Isle of Arran on June 20, 1770.[74] When Pennant visited this site he commented that basking sharks "were so tame as to suffer themselves to be stroked" from the boat, and Griffith produced a pen and ink wash image that shows a cutter, possibly the *Lady Frederick Campbell*, in close proximity to a crew from another ship who are represented as attempting to harpoon the shark in a method similar to that described by Pennant (figure 1.8).[75] In *British Zoology* (1776–1777) and his *Tour in Scotland and Voyage to the Hebrides*, Pennant commented that he observed living and dead specimens of the basking shark: "They will permit a boat to follow them, without accelerating their motion, till it comes almost within contact; when a harpooner strikes his weapon into them as near to the gills as possible."[76] Pennant's authority to make such statements is most likely a product of his ability to instruct Thompson to sail with pursuits of natural history in mind, approaching such creatures as the *Lady Frederick Campbell* navigated the Hebrides, observations he then combined with the accounts he obtained from local clergy such as "Mr. Lindsay the minister," who led him on a journey overland from Ranza in June 1772.[77]

Pennant's interest in carrying out detailed observations of the basking shark reflects its perceived economic potential. In *British Zoology*, Pennant described "the measurements of one, I found dead on the shore of *Loch Ranza* in the isle of *Arran*" equating for its size, weight, and fin shape while describing its use in the local economy. Pennant observed how the liver was boiled in kettles to extract oil, adding that "a large fish will yield eight barrels of oil; and two of worthless sediment."[78] This precise description is essential for classifying the shark according to the system laid down by Linnaeus in *Systema Naturae*, which was based on the fin ray count, taking the number, size, and position of the dorsal, anal, pectoral, and tail fins into account before moving onto more general physical characters such as size. Unlike quadrupeds and birds, Pennant believed the Linnaean system provided the best means for describing and classifying fish, commenting, "I should be very disingenuous, if I did own my obligations in this respect to the works of ARTEDI, Dr GRONOVIUS, and LINNÆUS."[79] A copy of the engraving of the basking shark from *British Zoology* has been pasted under the description in Pennant's copy of his *Tour in Scotland and Voyage to the Hebrides*, showing the close relationship between these two sets of descriptions and images.[80] In this way, Pennant extrapolated different sorts of information from his field notes and tailored these to specific works. Material on the events that transpired when hunting the basking shark and the romantic backdrop of Loch Ranza were set aside for his *Tour in Scotland and Voyage to the Hebrides*, and the systematic description was reserved for *British Zoology*.

Another product of the collaborative practices employed on this voyage was Lightfoot's *Flora Scotica*. Pennant funded this book and wrote the introduction on Scottish zoology to give a general overview of the interactions between plants and animals. Pennant's financial outlay for *Flora Scotica* was made clear in a letter Lightfoot sent on October 21, 1777: "I am sorry you gave yourself the trouble to particularise your Expenses in the Publication of the *Fl: Scotica*. I hear

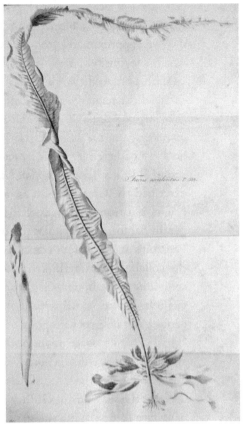

FIGURE 1.9. Left: an image of the Crag of Ailsa from Thomas Pennant's *A Tour in Scotland and Voyage to the Hebrides* (1774). Private collection. Right: the copperplate image of *Fucus esculentus* from John Lightfoot's *Flora Scotica* (1777). By kind permission of the Trustees of the Natural History Museum, London.

they must be great. But if the Public should intertain as favourable an opinion of the Work as you are pleased to do, I hope you will be repaid with interest."[81] Pennant's bill for the for the publication expenses of *Flora Scotica* was a colossal £471/10/2. In perspective, this cost considerably more than Pennant's entire journey, at £296, a sum that equated to the value of a considerable landholding. Pennant added publication expenses to the "expenses on Mr. Lightfoot's acct. in the voyage," which totalled £30/0/1.[82]

Pennant's *A Tour in Scotland and Voyage to the Hebrides* and Lightfoot's *Flora Scotica* are connected through descriptions of species and geographical localities. One locality is the Crag of Ailsa (now referred to as Ailsa Craig), a bleak inhospitable rock at the mouth of the Firth of Clyde approximately ten miles off the western coast of Scotland, which they visited on June 25, 1772. Whilst on the island, Moses Griffith produced two illustrations. One depicts a species Lightfoot ascribed the name *Fucus esculentus* in *Flora Scotica*. The other shows the Crag of Ailsa and was published in Pennant's *A Tour in Scotland and Voyage to the Hebrides* (1774) (figure 1.9). Lightfoot described how "we gathered it [*Fucus esculentus*] at Ailsa Craig, on the western shore." [83] In his *Tour in Scotland and Voyage to the Hebrides*, Pennant described the Crag of Ailsa as "a perpendicular rock of an amazing height, but from the edges of the precipice, the mountain assumes a pyramidal form: the whole circumference of the base is two miles." [84] Pennant's and Lightfoot's descriptions of the Crag of Ailsa emphasize the collaborative interlinked process of collecting information and compiling these books. Griffith's images show how Pennant shared Griffith's artistic skills with Lightfoot to complement their simultaneous natural history programs.

In comparison to Lightfoot, who concentrated on producing a systematic survey of the botany of Scotland, Pennant was interested in the zoology of the Crag of Ailsa. This is emphasized by the prominence of the sea birds in Griffith's image and the time and space Pennant gave to describing the spatial distribution of each species: "The birds that nestle on the precipices are numerous as swarms of bees; and not unlike them in their flight to and from the crag. On the verge of the precipice dwell the gannets and shags. Beneath are guillemots, and the razor bills: and under them the grey bills and kittiwakes, helped by their cry to full the chorus. The puffins made themselves burroughs above: the sea pies found a scanty place for their eggs near the base." [85] These observations reflect Pennant's interest in local animal populations, observing an integrated community of different species that all inhabited a range of climates over a specific geographical area. This observational approach was essential for classifying birds according to Ray's system that relied on the preferred habitats and social interactions between species.

The listing of species present in a small, but diverse, geographical locality reflects Pennant's desire to chart the differentiation between them and emphasize "everything I thought would be of service to the country." [86] For the Crag of Ailsa, Pennant described the "people who come here to take the young gannets for the table and other birds for their feathers," an industry that earned the Earl of Cassilis an annual rental income of £33. The contemporary and historical economic value of species is apparent throughout Pennant's *British Zoology*. In the appendix to volume 2, Pennant even provided a published account of the financial costs of species caught for the dining table in the early sixteenth century, combining his interests in natural history, the economic potential of species, and the historical value of nature.[87]

The ability to convert diverse information into a numerical account was essential for connecting a national natural history of Britain to similar enterprises undertaken across Europe and the globe. Approaches to creating what Pennant described as a "general" natural history became more intertwined with biogeographical research by the close of the eighteenth century and became dependant on comparing the distribution of species across broad geographical areas.[88] For example, Pennant's combination of accounts communicated by correspondents allowed him to map the flora of Britain. In his *Supplement to the Arctic Zoology* (1787), Pennant stated:

> In about *lat.* 53, I may draw a line from the *North Sea* to the opposite
> part of the kingdom, which will comprehend a small part of the north
> of *Norfolk*, the greater part of *Lincolnshire, Nottinghamshire, Derbyshire,*
> the moor-lands of *Staffordshire,* all *Cheshire, Denbighshire, Flintshire,*
> *Caernarvonshire,* and *Anglesey.* Beyond this line nature hath allotted
> to the northern part of these kingdoms certain plants, of which I am
> about to make an enumeration, which are rarely or never found to
> transgress that line to the south. Those which are nearest the south
> shall be first taken notice of.[89]

Pennant's mapping of British plants was facilitated by the solidification of internal boundaries, which allowed him to piece together information sourced from specific parishes and counties to give an overall national picture. The process of enumerating species diversity becomes apparent from the notes Pennant tipped into his copy of Lightfoot's *Flora Scotica* where he recorded the total numbers of species found in each constituent country. For example, when it came to zoology, Pennant noted the presence of forty-one species of quadrupeds in England, thirty-seven in Scotland, and sixteen on the Orkney and Shetland Islands. At the end of his listing of birds and plants from Scotland and Orkney, Pennant added that "the enumeration of the

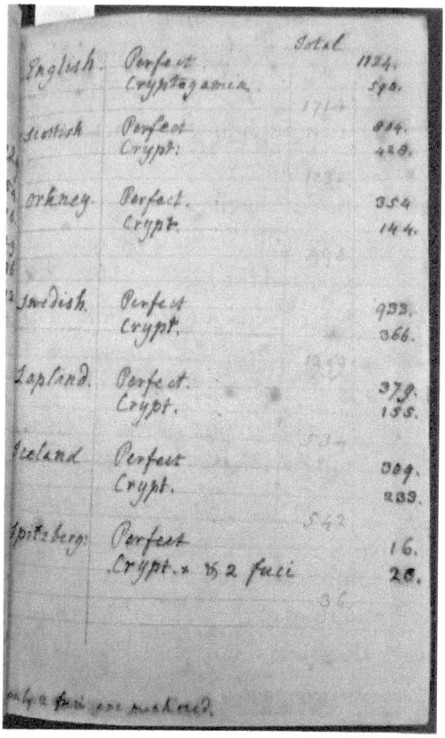

		Total	
English	Perfect		1124.
	Cryptogamia	1714	590.
Scottish	Perfect		814.
	Crypt:	1232	418.
Orkney	Perfect.		354
	Crypt.	498	144.
Swedish.	Perfect		933.
	Crypt.	1299	366.
Lapland.	Perfect.		379.
	Crypt.	534	155.
Iceland	Perfect		309.
	Crypt.	542	233.
Spitzberg:	Perfect		16.
	Crypt.* & 2 fuci	36	20.

only 2 fuci are mentioned.

Figure 1.10. Thomas Pennant's enumeration of botanical species in his copy of *Flora Scotica* (1777). The totals for England, Scotland, and Orkney are compared those of Sweden, Lapland, Iceland, and Spitzbergen. By kind permission of the Trustees of the Natural History Museum, London.

Orkney plants are by the Revd. Mr Low of Birsa possibly imperfect."[90] Pennant gave similar enumerations for birds and plants before comparing the numbers of British species with the total enumerations for Sweden, Lapland, Iceland, and Spitzbergen. Pennant derived these comparative national enumerations from Linnaean national floras and Banks's journals after his voyage to Iceland in 1772 (figure 1.10).

The consolidation of the total numbers of species available in several nation-states was central for assessing the diversity of each region and its potential for improvement. In the list of species Pennant tipped into his copy of *Flora Scotica*, England has more species of animals and plants than any of the other countries listed. Pennant's approach shows how he used a mixture of his own research, other national floras, and information from correspondents. Examples include George Low, who supplied the enumerations of plants and birds from Orkney. This allowed Pennant to rank the British Isles against Continental European competitors, proving the economic superiority and potential of the emergent nation-state. Pennant was familiar with these practices of managing information and converting textual descriptions into numerical formats since they have a distinct correlation with the methods of double-entry bookkeeping used to manage his estate and accounts he drew up when publishing books. This literal accounting for species was a common feature in Linnaean natural history and was used to link references across field notes, publications, correspondence, and collections. Pennant's conversion of systematic lists into a numerical format was essential for comparing the potential to improve emergent nation-states. Books, such as Lightfoot's *Flora Scotica*, could be ranked alongside similar works such as Hudson's *Flora Anglica* and Linnaeus's *Flora Lapponica* (1737) to assess the natural-historical importance of Britain and the potential for national improvement on the global stage.

Travel and the Development of "General" Natural Histories

National tours of Britain were not limited to the observation and collection of indigenous flora and fauna. From the mid-eighteenth century, a main motivation for naturalists to travel was to examine the collections compiled by their peers, institutions, and those who had embarked on intercontinental trips. For example, on a journey across England and Wales between 1767 and 1768 Joseph Banks visited "a small collection of rarities hung up museum with nothing uncommon except one monkey" at Cheatham's Library in Manchester and the collection of one "Mr Newton" in Lichfield, who had "lately returned from the East Indies" and accumulated "heaps of shells."[91] Other collections took the form of menageries. Examples include the private menageries of the Duchess of Portland, whom Thomas Pennant visited in 1774; King George III and Queen Charlotte; the Duke of Norfolk, who acquired a reindeer for his menagerie at Greystoke Castle, Cumberland, in 1799; and Joseph Banks, who in 1800 had some emus imported from New South Wales that were released in Kew Gardens.[92]

The opening of the British Museum in 1759 presented a direct motivation for naturalists to travel from the provinces to London and view Hans Sloane's collection. Another collection that came to

London in 1775 was that of Sir Ashton Lever, which remained in the city until its sale in 1806.[93] In London it became easy to view both dead and living exotic animals in zoos, menageries, parks, or in the shops of "animal merchants" such as Brookes of Holborn. For example, poet and naturalist Thomas Gray described seeing a myna bird, cassowary, macaw, leopard, and armadillo at Charing Cross in 1766.[94] By 1805 the American chemist Benjamin Silliman described how he had examined "the lion and lioness, royal tiger of Bengal, panther, hyena, tiger cat, leopard, orang-utan, elephant, rhinoceros, hippopotamus, great white bear of Greenland, the bison, elk or moose deer, the zebra" in a single afternoon in the city, adding that "most of these were living."[95] Pennant examined numerous collections of living and dead animals when on his tour across Continental Europe in 1765 and during his travels around Britain. These included Anna Blackburne's collection at Orford Hall near Warrington, who had "formed a Museum from the other side of the Atlantic, as pleasing as it is instructive." This was compiled from specimens sent by her brother, Ashton Blackburne, who had immigrated to New York in the 1760s.[96]

The gathering of information on a global variety of species was not limited to natural history collections. Interests in examining and copying exotic animals from paintings in the private collections of the British genteel elite or those exhibited at the Royal Academy of Arts was another major motivation for Pennant to undertake numerous tours around Britain. The growth of animal painting was especially useful to Pennant when compiling publications such as his *Synopsis of Quadrupeds* (1771).[97] This book was Pennant's first attempt to create a "general" natural history designed to include all known quadrupeds from across the globe. While Pennant was preparing this work, George Stubbs (1724–1806), emerged as one of the main pioneers of the genre of animal painting, producing numerous images of exotic animals by the 1760s. Examples include Queen Charlotte's zebra (1763); the Duke of Richmond's moose (1770); the kangaroo for Banks, based on a preserved skin and skeleton from New Holland (1772); and an image commissioned in 1765 by the governor general of Madras, Sir George Pigot, which depicts a hunting cheetah, a stag, and two Indian servants (plate 1). All of these images found their way into Pennant's publications, such as the moose, published as the frontispiece for *Arctic Zoology* (1785) and the kangaroo, published in his *History of Quadrupeds* (1781).[98]

To view valuable artworks, Pennant followed an approach similar to that used when collecting portraits for his Welsh and Scottish tours, many of which were in the possession of aristocrats and fellow collectors.[99] This involved going to see the owners of these artworks, as evidenced by the inscription on the verso of the watercolor copy of the cheetah written by Pennant's artist, probably Peter Paillou (c. 1720–1790): "The Chittah or Hunting Tyger taken from lord Pigot's painting & with his leave, it is a very exact drawing, the tail except which is a little thick."[100] After obtaining permission from Pigot, Pennant instructed his artist to copy the cheetah from the painting. However, Pennant had his artist omit the surrounding context so a single image of the cheetah could be inserted into a systematic work of natural history. For this to take place, Pennant's artist had to adapt the original image by removing the crimson linen strips the Indian servants are using to restrain the beast, although its shape, general posture, and number of spots remained the same.

These alterations are reflected in Pennant's description in his *History of Quadrupeds*, which pays particular attention to the physical characters of the cheetah, such as the shape of the head, legs, and feet, in addition to its hunting habits and how it "is tamed and trained for the chase of antelopes."[101] Pennant and Stubbs had a close relationship, and both believed the images they produced and published presented faithful representations of the natural world. Both prided themselves on observing animals in person, such as the cheetah in Pigot's painting, an animal brought back from India and presented to King George III. It could be viewed in Windsor Great Park.[102] Stubbs's personal observation made his painting of the cheetah and stag with two keepers, a reliable source. Paintings were of great importance to Pennant, who in 1768 commented that "painting is an imitation of nature in the representation of objects," adding that an accurate depiction was impossible "with out consulting the original."[103] A major part of Pennant's national tours was taken up by visiting paintings such as that Pigot commissioned from Stubbs, which, after its display at the Royal Academy of Arts in 1765, was kept at Pigot's Staffordshire residence at Patshull Hall. Pennant almost certainly visited the Royal Academy of Arts exhibition and Pigot's estate when traveling from Chester to London.[104] Pennant's use of these images in his publications linked his books to collections compiled by the aristocracy and allowed the owners of these artworks to become patrons of natural history—building a network to promote the accumulation of useful knowledge. Images extracted from works of art were published in Pennant's *History of Quadrupeds* (1781), *Indian Zoology* (1790), and the first volumes of *Outlines of the Globe* (1798–1800), books that sought to define the biogeographical natural history of major global regions and cement Pennant's natural history patronage network.[105]

North American natural history collections became common during the 1760s. Examples include the specimens Banks brought back from Newfoundland in 1767. Shortly before Banks's departure for Newfoundland in 1766, Pennant gave him a notebook titled "These queries I drew up for Mr. Banks during his voyage to newfoundland april 1766."[106] A small book enclosed in tough card wrappers, its general lightness made Pennant's queries easy to carry as Banks traveled over the Atlantic to Newfoundland and across England and Wales to visit Pennant at Downing Hall. At the end of the notebook, Pennant invited Banks to Downing—"Mr Pennant will be happy to receive Mr Banks's orders addressed to him, at Downing, Flintshire." A year later Banks took Pennant up on his offer, arriving at Downing on November 21, 1767, when he gave Pennant several Newfoundland bird specimens and drawings, examined Pennant's natural history collection, explored the local area, and returned the notebook. Banks noted that they observed "a very strange Phenomenon called the Burnt rock" on the local coastline and spent most of their time "almost intirely at home in reviewing a collection of English seashells & crabs."[107] Descriptions of this collection were published in the final volume of Pennant's *British Zoology* (1777). Later on, Pennant added comparative descriptions of Newfoundland species to his interleaved "Catalogue of British Birds" that traveled with him to Scotland in 1769.[108]

The notebook of "queries" Pennant gave Banks before his Newfoundland voyage represents the integral role of paper for stimulating collaborative information gathering. It shows how by the late 1760s the geographical scale of questionnaires expanded beyond parochial and national boundaries. This notebook contains questions in Pennant's hand and Banks's responses on the verso of the previous page, supplying Pennant with information he incorporated into *Arctic Zoology* (1785–1787) (figure 1.11). For example, Pennant asked Banks, "What does Charlevoix mean by white porpoises which he says are found in the river St Lawrence"? Banks responded on the facing page:

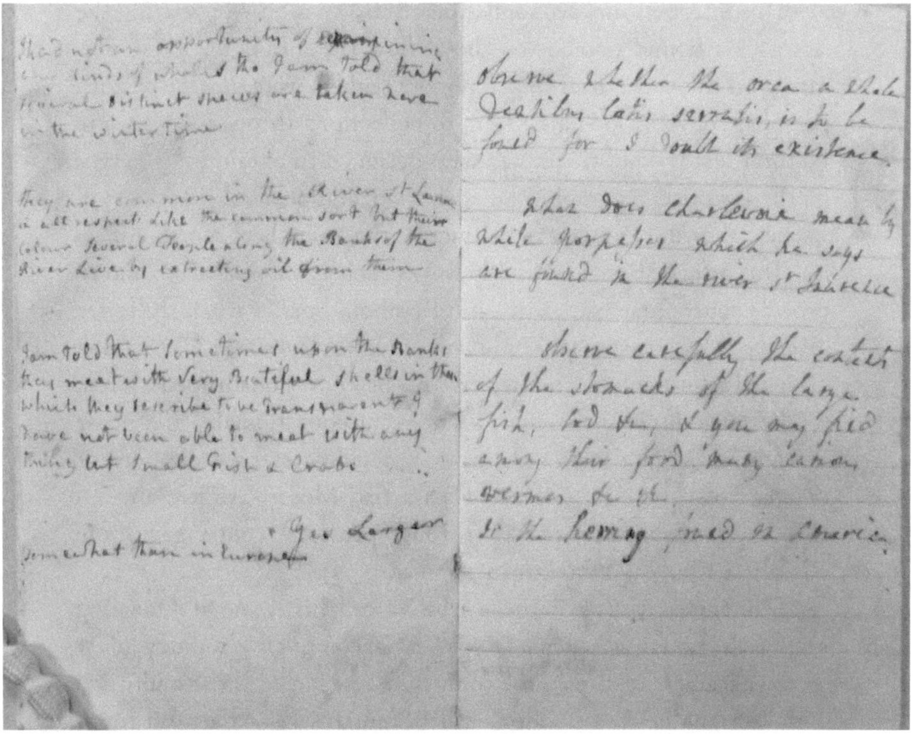

FIGURE. I.II. Thomas Pennant's notebook in which he poses questions to Joseph Banks about the fauna of Newfoundland (right), who then responds on the verso of the opposite page (left). Reproduced by kind permission of the Warwickshire County Records Office.

"They are common in the River St Lawrence in all respects like the common sort but their colour Several People along the Banks of the River Live by extracting oil from them."[109] Pennant then used Banks's information on Newfoundland cetaceans in *Arctic Zoology*: "They [porpoises] are numerous in the gulph of St. *Lawrence*; and go with the tide as high as Quebec. There are fisheries for them, and the common *Porpoesse*, in that river. A considerable quantity of oil is extracted."[110] Pennant's lack of citation of Banks's observations likely contributed to the fierce debates over intellectual property that contributed to a notable dispute between these individuals that lasted for most of the 1780s. Despite this, knowledge on the extraction of oil was of great importance to Pennant, who believed understanding natural history in relation to "all its particular uses in common life" was essential for national improvement.[111]

This interpersonal accumulation of information—by sharing the leaves of a bound notebook—shows how the means for communicating on the natural history of remote regions were integrated with personal conversations and visits to collections throughout Britain. Natural history questionnaires were designed to obtain information on the quantity, geographical distribution, and physical characters of species, allowing Pennant to enumerate the animals of North America. This information was used in works such as *Arctic Zoology* and to provide source material for Johann Reinhold Forster when the latter produced *A Catalogue of the Animals of North America* (1771), a book designed to present "an enumeration of the known quadrupeds, birds, reptiles, fish, insects, crustaceans and testaceous animals."[112] Similar to Lightfoot's *Flora Scotica*, Pennant supplied Forster with important manuscripts; shared the skills of his artist, Moses Griffith; and assisted in the publication process through providing contacts with the relevant artisans and financial assistance.

Pennant reciprocated those who visited his home at Downing Hall with visits to other collections held in comparable country houses, towns, and cities across the country. These included the multitude of collections held in London, which Pennant visited several times in the 1760s and 1770s. Other collections included Banks's library, which was noted for the specimens and descriptions brought back from James Cook's first voyage to the South Seas. Banks wrote to Pennant on the day he returned to London in 1771: "Our Collections will, I hope, satisfy you: very few quadrupeds; one mouse, however, (Gerbua) weighing 80 Ib weight. I long for nothing so much as to see you, but must delay that pleasure for some time."[113] Later that year, Pennant "took a journey to *London*, to see sir *Joseph Banks* and doctor *Solander*, on their arrival from their circumnavigation." When recording information on Banks's specimens Pennant used a notebook titled "Quadrupeds and Birds." Pennant outlined that the notes contained within relate to creatures "observed and collected by Joseph Banks esq. & Doctor Solander in the voyage round the world begun august 25th 1768 ended July 12th 1771."[114]

Pennant's notebook "Quadrupeds and Birds" is representative of the standard bound notebooks purchased during his journeys and has been encased within a sturdy binding of green vellum. These were combined with Pennant's manuscript slips and stored in a card envelope inscribed "Birds from the South Sea Mr Banks's Voyage."[115]

Pennant started to use paper slips to manage information after using the system Daniel Solander established at the British Museum in the 1760s and later used for managing Banks's collection. The use of paper slips to manage information was a common practice when compiling lists for taxonomic works, encyclopedias, and dictionaries. For example, Pennant's contemporary, Samuel Johnson, relied on paper slips when revising new editions of his *Dictionary*. These were placed between the pages of the last edition, adding new words alongside their respective definitions.[116] Information from the slips was then transferred into an interleaved copy of the *Dictionary*, to which Johnson and his amanuenses added annotated descriptions. This reflects many naturalist's approaches to duplicating names and descriptions across a variety of paper technologies when collecting, compiling, and publishing information.

Pennant's use of Solander's system for managing information on the zoological material becomes apparent from the descriptions of specific genera and species in Pennant's notebook. An example is the genus Solander named *Nectris*, of which Banks collected representative specimens from the South Pacific on December 15, 1769. Banks shot this specimen from a small boat he kept on the *Endeavour*, recording: "Calm this morn. Went in the boat & Killed Procellaria velex Nectris munda & fulginosa, which two last are a new genus between Procella & Diemendia this we [Banks and Solander] rekon a great acquisition to our bird collection. My stay out today was much shortened by a breeze of wind which brought me abroad by 11 o clock & before night blew very fresh."[117] In addition to Banks's journal entry, Solander composed a description on a separate manuscript slip and instructed Herman Spöring, a Finnish naturalist and amanuensis who accompanied them on the voyage, to transfer information onto the relevant interleaved pages in their copy of *Systema Naturae* (1766).[118] These descriptions were transcribed into another manuscript after Banks and Solander returned to London, which provides a systematic classification for all the new zoological species they discovered.[119]

When he visited Banks in 1771, Pennant was given supervised access to the manuscripts, illustrations, and specimens Banks and Solander compiled during this voyage. Pennant almost certainly copied these ornithological descriptions from Solander's "Fair Copy of the Descriptions of Animals" before combining this with content from Banks's journal and information obtained through conversations with

Banks and Solander. For example, when Solander derived the name *Nectris* from the Greek word *Nukteris*, the literal translation for which is "Night Bird," he followed a similar practice to Linnaeus by relating the origin of the name to a behavioral trait common to all species of this genus. In addition to using Solander's name, Pennant described how the species *Nectris carbonaria* "fly in flocks innumerable at once dip under water all together disappear & then rise as suddenly. These birds with various sorts of *Procellaria* are the common birds of the s. sea as auks are of the north."[120] Pennant believed the additional information on social attributes was essential for classifying species according to Ray's system, which took into account features such as birds' habitation of land or water, their feeding and sleeping habits, in addition to their physical appearance.

Flexible paper tools were central for bridging between the different systems of classification used by British naturalists. For example, Solander had designed his manuscript slips to facilitate the movement of information across a broad range of theoretical frameworks. As mentioned earlier, when reporting to the trustees of the British Museum in 1765, Solander stated that he had "taken care to describe all those [new species] so minutely, that any Botanist whatsoever, may range them [manuscript slips] according to his own faivorite system."[121] Solander's system also applied to the British Museum's zoological collections, solidifying these practices of accessing information for the full range of naturalists regardless of their preferred classificatory system. It also allowed for the information to be restructured as classificatory systems developed.

The use of slips to mediate between systems is apparent from those written by Solander that Pennant integrated into his collection. This is exemplified by the slip concerning the "Natter Jack Toad." Solander gave this paper slip to Pennant, along with a specimen of the toad, prior to his departure aboard the *Endeavour* (figure 1.12). Pennant kept these slips in small folders made from thick card and his insertion of Solander's slips shows how these groupings were never compiled by a single actor. Rather, descriptions and slips are the products of collaborative productions involving multiple individuals, reflecting Bettina Dietz's point on the inherently collaborative nature of Linnaean natural history.[122] However, Pennant's slips show that these practices extended much further than the realm of Linnaean botany through his use of them to accumulate descriptions and arrange species published

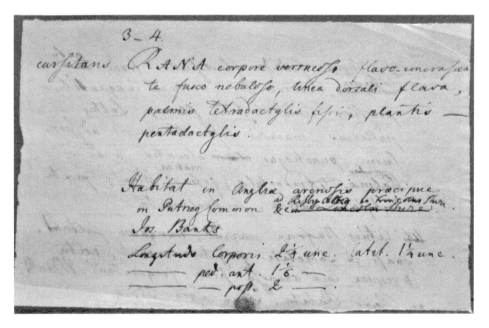

FIGURE 1.12. The slip Daniel Solander gave Thomas Pennant concerning the natterjack toad. By permission of Llyfrgell Genedlaethol Cymru/The National Library of Wales.

in the 1768–1770 edition of *British Zoology*—a book arranged in accordance with the earlier system of John Ray. In his description of the natterjack toad, Pennant added that "it is found on Putney Common, and also near Revesby Abbey, Lincolnshire." These two localities are places with which Banks was familiar—being his country estate and a popular London botanizing location—and match those outlined on Solander's slip, while Pennant has added "we are indebted to *Joseph Banks*, esq; for this account" in his published description.[123]

Flexible manuscript slips were essential for managing and transferring information between naturalists, regardless of systems of classification. This becomes apparent in the case of *Mosacilla*, a genus containing wagtails and flycatchers, Pennant observed at Banks's home in 1771. After examining the bird skins, Pennant translated the content of Solander's manuscript slips into English. Slips were then placed into what would soon be known as Solander boxes, each of which contained slips relating to one Linnaean order. At the top of each slip, Solander gave the page reference to Linnaeus's *Systema Naturae* (1766), under which he listed earlier published descriptions and depictions of

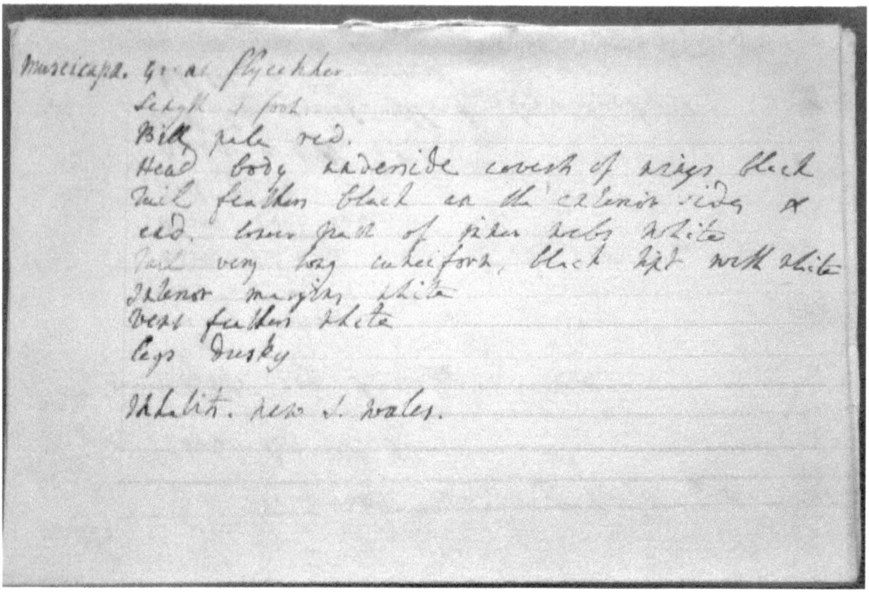
Muscicapa. great flycatcher
Bill pale red.
Head body underside coverts of wings black
Tail feathers black on the exterior sides &
end. lower part of inner webs white
Tail very long cuneiform, black tipt with white
Exterior margins white
Vent feathers white
Legs dusky
Inhabits. new S. wales.

FIGURE 1.13. The last slip in the gathering that relates to the genus *Muscicapa*, which Thomas Pennant used to describe the great flycatcher Joseph Banks and Daniel Solander collected from New South Wales. Signs of previous binding can be seen along the top edge. By permission of Llyfrgell Genedlaethol Cymru/ The National Library of Wales.

the species before describing its physical characteristics. In comparison, Pennant used card envelopes to group slips that relate to a specific class of animals and then arranged these according to geographical locality, emphasizing his biogeographical interests. Pennant's slips on the South Seas have been extracted from a notebook; each gathering of leaves was used to describe specimens from a specific genus and many have gilt edges. This is similar to Solander's production of manuscript slips, many of which were cut from the pages of his zoological notebooks, reflecting the transferral of information from a static repository into flexible paper technologies. In the case of the flycatcher, Pennant added a title page to the gathering of slips, on which he noted the genera described within, these being "Muscicapa. Motacilla." Pennant described four new species of flycatcher in this gathering; all of these came from the South Seas, and the descriptions were laid out in a similar manner to those in his notebook (figure 1.13).

Pennant recorded the genus for this bird on the top left of the slip, next to which he gave the vernacular name for the species, the

"Great flycatcher." This was followed by a description of the bird's physical features, starting with its general size and moving onto specific characters. At the end of his description Pennant gave the bird's geographical distribution, "Inhabits. New S. wales."[124] The partially bound nature of Pennant's slips reflects the movement from the use of bound notebooks in natural history collecting to more flexible paper technologies, allowing for the addition and rearrangement of information. The more static nature of Pennant's manuscript slips and their arrangement by geographical locality, and then by genera, was a result of his long descriptions of physical features, social interactions, and intention to publish this information to present an enumeration of species from defined geographical regions. Slips that relate to species Banks observed in the Pacific could be placed alongside records Pennant obtained from other sources, such as the art collections of the aristocracy, and combined in a published account to create a "general" natural history. The diversity of sources and the need to travel to view collections reflects how Pennant's global publications were not the product of a sedentary deskbound scholar, but the result of a lifetime of travels across the country.

Through exploring the different kinds of expeditions naturalists took throughout Britain to gather information on a global fauna and flora, we see the development of new practices used to record, order, and standardize information. This was gathered from a diverse array of sources in different settings and across a range of scales of geography and time, extending from explorations of local parishes over decades to more extensive tours taken over a shorter timeframe. The practices explored in the previous pages are representative of those in the right-hand portion of the "genteel" zone of the diagram depicted in the introduction (figure I.1). All the books, notebooks, and paper slips discussed represent objects adapted for use in various natural history enterprises. Several were designed or adapted to facilitate the accumulation of information. Typical examples include interleaving to accommodate notes, while many notebooks were bound in a tough wallet-style binding to protect them from the weather. Other books—including Thomas Pennant's *British Zoology*, William Hudson's *Flora Anglica*, and John Lightfoot's *Flora Scotica*—were designed by their authors to accommodate notes. Many were distributed as gifts or by specialist booksellers to naturalists who planned to use these items to assess the diversity of plants and animals in their own collections or across a specific geographical locality. Annotated books interacted with a range of notebooks and manuscript slips, many of which were purchased with the intent or adapted to be taken on expeditions. Examples include Pennant's manuscript slips, the thick card envelope being specifically designed to keep them together and protect the precious contents and the small notebook Banks took to Newfoundland—an item designed to travel since it could fit in a jacket pocket.

Interleaved and annotated books were integrated with a broader framework of notebooks, manuscript slips, and images ranging from romantic views to taxonomic representations of species and specimens collected, observed, and extracted from artworks. The processes figures such as Pennant, Gilbert White, and others used to tabulate information both when traveling and in the library was similar to those used by Samuel Johnson, who employed numerous interleaved copies of his dictionary to rearrange words and definitions. Johnson's practices reflect on Baconian approaches to synthesizing and ordering factual information, initiating collaborative working structures with teams of secretaries and family members to edit the dictionary. Collaborative working practices are also apparent through Johnson's notable travels with James Boswell and his bohemian servant Joseph Ritter through the Western Isles, a journey that relied on Johnson's and Boswell's continual exchange of information, aspects of which extended into their literary works. However, as John Radner has suggested, the relationship between Johnson and Boswell remained unequal, with Johnson maintaining control over the public narrative of the trip and straining their relationship in later years.[125]

Similar practices for managing information were used by literate cultures on a global scale. For example, collections of manuscript slips were not only created by naturalists but became a well-established practice for managing information across a range of scholarly disciplines extending into literary cultures outside of Europe. Within Europe, these practices became essential to a range of emergent disciplines. For example, Elisabeth Décultot has explored the "excerpt collections" assembled by the art historian Johann Joachim Winckelmann (1717–1768), suggesting how they form a link between the practical world and book-related knowledge. However, many of examinations of these approaches to managing information have explored them from the perspectives of assimilating and organizing knowledge or organizing the content of published works from within the confines of a library or study.[126]

In comparison, the current exploration of British natural history practices has shown how such paper technologies became the practical tools of natural history. They accompanied naturalists on journeys ranging from local surveys to national tours, assimilating diverse information while facilitating collaborations between teams of naturalists and with the people they encountered. Many served to structure information ranging from individual images extracted from paintings through to specimens found on the shores of the Crag of Ailsa. They initiated the development of standards for recording information both as illustrations and textual descriptions that brought a broad range of different knowledge bases together. A standardization of practices of collecting and recording information became all the more important as the eighteenth century progressed, witnessing an age when more species defined as new to natural history were discovered before or since, a direct consequence of an increase in global voyages of discovery such as that discussed in the next chapter.

Chapter 2

A New World for
Natural History

You are also carefully to observe the nature of the Soil, & the
Products of thereof, the Beasts & Fowls that inhabit or frequent it,
the Fishes that are to be found in the Rivers or upon the coast &
in what plenty & in case you find any mines, minerals, or valuable
Stones you are to bring home Specimens of each as also such Specimens of the seeds of the Trees, Fruits & Grains as you may be able
to collect & Transmit them to our Secretary that We may cause
proper Examination & Experiments to be made of them. You are
likewise to observe the Genius, Temper, Disposition & number of
the natives.

—British Admiralty Instructions to
Lieutenant James Cook, 1768

James Cook's three voyages were the main sources of natural history collections from the Pacific entering Great Britain during the late
eighteenth century (figure 2.1). These collections were compiled due
to the development of standardized practices for collecting, describing, and classifying information, processes stimulated through an increased interest in surveying the full extent of God's creation alongside
its perceived economic benefits. This inspired independent naturalists
and scientific societies to plan voyages of exploration. Many aspired to
create what Thomas Pennant referred to as a "general" natural history,
to classify and describe all species from each kingdom of nature.[1] A result of these expeditions, the two decades before Carl Linnaeus's death
in 1778 saw an upsurge in publications reflecting the multitude of new

Figure 2.1. Map showing the route of the *Endeavour* voyage, 1768–1771.

species entering Europe. Many were collected by Linnaean students or "apostles" who traveled the globe, sometimes on state-sponsored voyages of discovery inspired by the potential of commercial opportunities.[2] As Bettina Dietz has suggested, Linnaean apostles utilized the standards being developed by Linnaeus and his contemporaries to name, describe, and communicate information. For example, writing to Pehr Osbeck (1723–1805), who explored the Canton region of China, Linnaeus praised his early adoption of standard practices of naming and describing species, suggesting, "I seem myself to have travelled with you, and to have examined every object you saw with my own eyes."[3]

Exploring the working practices of Joseph Banks, the Linnaean "apostle" Daniel Solander and their team of field assistants, this chapter investigates the processes of collecting and recording information on the natural world during James Cook's first voyage of discovery between 1768 and 1771. Under commands to sail to the South Pacific, this expedition was jointly funded by the Royal Society of London and British Admiralty with the remit to observe the transit of Venus from Tahiti that had first been visited by Samuel Wallis, who named it King George the Third's Island, in 1767. Cook's remit was to chart, survey and collect specimens that could benefit European society from unexplored territories of the *Terra Australis Incognita*. Although much scholarship has concentrated on Cook's instructions to chart new territories and build relations with Native peoples, a significant portion of the "additional instructions" concentrated on the natural history of the Pacific.[4] Banks and his team were essential for fulfilling this section of Cook's instructions. However, unlike the state-employed naturalists who participated in voyages sponsored by Continental European powers, Banks self-funded his party so that everything they collected and produced remained his personal property.[5]

Practices developed on expeditions and in collections throughout Britain had to be adapted to cope with the variety of new timeframes and geographical areas covered on a global circumnavigation. For example, scales of geography ranged from traveling across oceans and rapid surveys of continental coastlines to detailed surveys of small islands. Time scales shaped the nature of the natural-historical observations undertaken in different regions that were influenced by the number of days the naturalists spent on shore, the size of the area to be covered, and the diversity of the species they encountered. An unprecedented volume of new species and specimens caused a crisis within the paper structures designed to cope with the rapid accumulation of new information. The objective to visit previously unknown regions presents a very different scenario when compared to other examples of intercontinental travel when naturalists maintained limited communication with Europe.[6] In contrast, after the *Endeavour* rounded Cape Horn and until its arrival in the Dutch enclave of Batavia, the naturalists were completely cut off from European trade networks, relying on the materials they had aboard the ship and information obtained from Indigenous peoples to identify, record, classify, and store numerous new species encountered on an hourly basis (map 2.1).

The rapid expansion of the quantity and diversity of material Banks and Solander added to their specimen collection contrasted with their relatively static means for recording and classifying this information. This became most apparent after they passed through the Strait of Magellan, when consistent changes in geography, time, and the quantity of information obtained conflicted with their means for managing these different scales through paper technologies ranging from bound notebooks to paper slips. It was no longer possible to purchase paper in the Pacific and expand the capacity for record-keeping. However, this was overcome through Banks's and Solander's expectation to encounter and record thousands of new species and representative specimens. In a letter to Thomas Falconer, Banks stated: "From the Scarce Intelligible accounts of Travelers That almost Every production of Nature here [the South Seas] is very different from what we see at this end of the globe."[7] Solander expressed similar views to the trustees of the British Museum, stating that he "may be of great utility to the British Museum, in collecting Natural Curiosities for that repository, from countries that perhaps never before were investigated by any curious men."[8] A main preparation for the assumed abundance of new species was to leave London with a huge quantity of blank paper. This included a range of bound notebooks, paper Solander used for manuscript slips, ledgers used for systematic lists of species, large sheets of paper for illustrations, a selection of printed books that were adapted to accommodate manuscript notes, and a large quantity of waste printed pages used to dry botanical specimens. This emphasizes an important transition outlined by the chart figured in the introduction to this book: the adaption of books as tools for systematizing and accumulating information on global voyages.

Assumed differences between the natural history of the Pacific and that of Europe reflects emergent concepts of regional integrity involving the study of species in geographically defined sets based on their attachment and adaption to various climatic zones. The continuous accumulation of new material meant Banks and Solander embraced the emergent flexible paper technologies and solidified their use in natural history fieldwork.[9] To collect new species Banks assembled a team that consisted of himself, Solander, the artists Sydney Parkinson and Alexander Buchan, the assistant naturalist Herman Spöring, in addition to four servants and field assistants. The rounding of Cape Horn in January 1769 increased the epistemic value of the books and writing paper aboard the *Endeavour* and it was now impossible to replace these. The value of books on such expeditions was emphasized by Georg Forster: "Books are very dear, but a naturalist without them, is like an artist without tools."[10] In comparison to previous scholarship on practices of collecting and recording information on global expeditions that categorizes books and manuscripts as separate entities,[11] the main emphasis here is on how paper technologies and the people who managed them formed a unified, flexible system for recording, accumulating, collecting, and classifying information.

Objects, people, and paper formed an integrated system on global voyages. Through tracing the hierarchic organisation of Banks's team, the analysis uncovers how these roles were combined with the physical materials used to manage information. These practices emerged with the assessment, recording, and classification of previously discovered species before manifesting themselves in the processes of incorporating new species into the Linnaean system. Collaborative frameworks of managing information proved essential in the assessment of new species' usefulness to European commerce and approaches to recording novelty.

Organizing Labour to Classify Nature

On October 3, 1769, just four days before the *Endeavour* arrived in Poverty Bay, New Zealand, Joseph Banks described how he and Daniel Solander divided the process of recording, classifying, and describing species: "Dr Solander setts at the Cabbin table describing, myself at my Bureau Journalizing, between us hangs a large bunch of sea weed, upon the table lays the wood and barnacles; they would see that notwithstanding our different occupations our lips move very often, and without being conjurors might guess that we were talking about what we should see upon the land which there is now no doubt we shall see very soon."[12] This quote represents Banks's and Solander's strict daily routine of writing in their journals while using their shipboard library to describe, classify, and catalogue species. Banks recalled how they commenced this process "each day from about 8 a. m. to 2 p.m., and after the smell of food had disappeared, from 4 or 5 pm until dark," timings influenced by James Cook's daily naval routine.[13] This description presents Banks as an overall manager and coordinator of collecting, describing, classifying, illustrating, and preserving natural history specimens, while recording daily events in his journal.[14] Solander was in charge of writing Linnaean descriptions and transferring these through a range of manuscripts.

Activities of describing and classifying species were at the pinnacle of a rigid hierarchy Banks established for the collection, ordering, and description of natural history materials. As shown in figure 2.2, this top-down hierarchy formed a pyramidal structure with internal feedback as specimens were collected, fed back to Banks and Solander, who then initiated the recording process in the various manuscripts and supervised the artists to create visual representations of species.

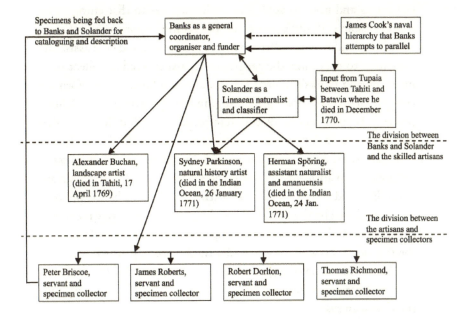

FIGURE 2.2. Diagram showing the top-down social hierarchy and working structure Joseph Banks established between himself and his team of natural history staff aboard the *Endeavour*.

In addition to Banks and Solander, this process involved an amanuensis, two artists, and four field assistants. The employment of staff and collecting apparatus inspired the naturalist John Ellis to comment that "No people went to sea better fitted out for the purpose of natural history."[15] The skilled artisans Banks employed occupied a middling position in this social hierarchy and included Herman Spöring (1733–1771), a Finnish watchmaker, surgeon, engraver, and scribe who had been employed by Solander as an assistant and secretary at the British Museum since 1766; the artist Sydney Parkinson (1745–1771), whose task was to produce illustrations of specimens in their living state; and Alexander Buchan (d. 1769), an artist who specialized in landscapes. The four servants and field assistants included Thomas Richmond and George Dorlton, both of African descent, who Banks employed in London; in addition to Peter Briscoe and James Roberts, who had been employed as servants at Revesby Abbey, Banks's Lincolnshire estate. These individuals occupied the lowliest position in Banks's regime, forming the final layer of the pyramid emphasized in

figure 2.2 and serving as a constant reminder of Banks's higher social status and need to apply similar hierarchies to that enforced on
country estates. Banks's properties placed him among the 250 or so
wealthiest landowners in Britain by 1770. Servants and field assistants
were used to maximize the geographical area covered to collect specimens during Banks's limited time ashore in addition to performing
the general tasks of cooking, cleaning, and carrying baggage.[16] Defined
places in Banks's hierarchy were reinforced through different contributions to the various paper structures and each individuals' place of
abode aboard the ship. For example, Banks's cabin had direct access to
the great cabin adjacent to Cook's; the space occupied by the assistant
naturalists and artists was aligned with that occupied by the philosophers employed by the Admiralty and higher-ranking crew members;
the servants and field assistants occupied hammocks belowdecks with
the rest of the crew. Banks viewed a rigid hierarchy, which formed a
parallel to Cook's naval regime, as essential for ensuring the swift collection, description, illustration, and incorporation of new species into
the Linnaean system.

Banks's hierarchy enforced a mechanical collecting strategy, facilitating the acquisition of around 30,400 plant specimens. Among
these around 1,400 species were previously unknown to European
naturalists while approximately 3,600 species had already been recorded.[17] To keep track of this rapid accumulation of material Banks
and Solander employed advanced systems for classifying and cataloguing the botanical specimens, emulating the social hierarchy set
out by Carl Linnaeus in *Philosophia Botanica* (1751), who suggested
botanists should work alongside skilled artisans to take accurate records.[18] Banks, Solander, Spöring, and Parkinson were the main actors in this process and relied on bound notebooks, paper slips, images, interleaved books, and a large library of natural history and travel
literature to record, describe, and classify species represented in their
ever-growing specimen collection. The delegation of these materials
between different individuals shows how their working practices and
relationship adapted after they entered uncharted territories and collected huge quantities of information. Solander had tested the capability of these frameworks alongside Linnaeus when he worked on the
collections of the Swedish nobility during the 1750s, during trips to
Lapland and at the British Museum—approaches that shaped the information management methods used aboard the *Endeavour*.

When cataloguing specimens and classifying species, Banks's role was to record the collection date and geographical distribution in his journal. This bound notebook served as a narrative account used to incorporate information from different sources. Alongside his journal, Banks kept a catalogue of species represented in the specimen collection. For much of the voyage, this took the form of Banks's "A Catalogue of Plants Collected at Madeira, Brazil, Tierra del Fuego and the Society Islands, Arranged for Each Locality According to Linnaeus' Species Plantarum." Compiled from a list of species divided by geographical locality and ordered according to the Linnaean system, this manuscript list records the quantity of specimens of each species collected while referencing their localities within the stacks of printed sheets Banks referred to as "books" used to dry botanical specimens.[19]

Banks's enumeration reflects his position as a director and coordinator of this inventory of natural history. This position gave him sufficient authority to use this information to make general claims about the economic potential of each region. For example, in 1779 Banks emphasized the suitability of Botany Bay for the establishment of a penal colony before a House of Commons select committee. Banks based this claim on observations of the good climate and vast quantity of new species he and Solander collected from the eastern coast of New Holland and associated with the potential for "improving" this region alongside European notions of agriculture and commerce.[20] This mirrors Pennant's assertions on specific parts of Britain, although a major problem Banks experienced in the Pacific, which fueled disagreements with Cook, was the limited time spent in each place. Banks hinted at these conflicts decades later when relaying to the botanist Robert Brown, after the latter complained that Matthew Flinders, the captain of HMS *Investigator*, never gave him enough time on land, that "Had Cooke paid the same attention to the Naturalists as he [Flinders] seems to have done, we should have done much more at that time."[21] Constraints on time ensured Banks did not have the opportunity to observe the seasonal development of plants, necessitating further voyages of exploration.

Banks and Solander collected a total of 906 species they deemed new to the Linnaean system from eastern New Holland, a region Cook named New South Wales (figure 2.3). This rapid accumulation of material tested the capability of their system for classifying

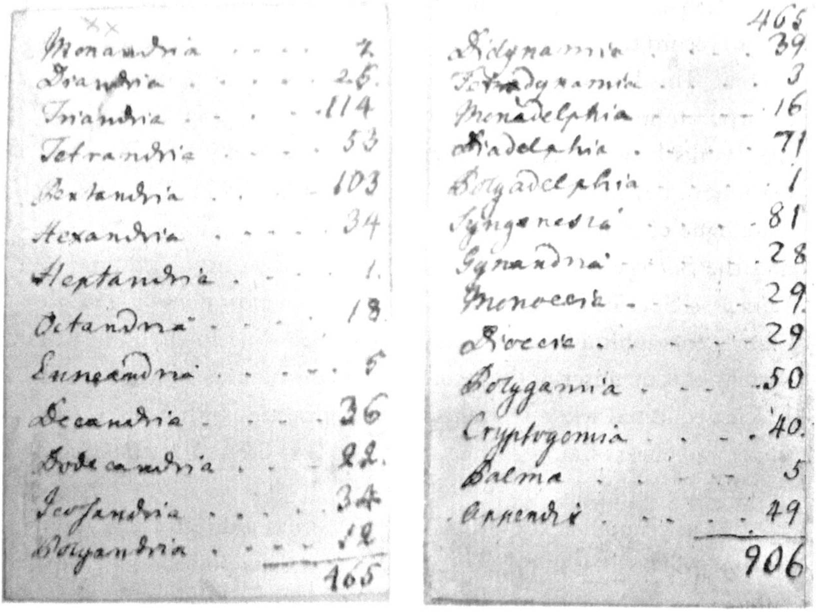

FIGURE 2.3. Daniel Solander's totals for new species collected from the eastern coast of New Holland, arranged according to each Linnaean class. These were recorded on a slip of card inserted into Solander's "Plantæ Novae Hollandia." By kind permission of the Trustees of the Natural History Museum, London.

and incorporating information into the Linnaean system and the collaborative hierarchy it enforced within Banks's team.[22] To overcome problems with the vast accumulation of descriptions and representative specimens, Solander took over the enumeration of species in New Holland. This was probably a result of Banks's desire to collect examples of as many new species as possible, declaring in his journal that this brought a "usual good success" once they reached Botany Bay, although the vast quantity of specimens meant they had to "find an excuse for staying on board to examine them a little."[23] Once they reached Possession Island, at the northmost point of New Holland, Solander enumerated the total number of new species under the twenty-four Linnaean classes. He then added these together to create an overall total of new species encountered in eastern New Holland. Banks's and Solander's accounting for species over several different mediums is a clear attempt to cope with the classification and cataloguing of an overwhelming number of plants that allowed them to trace information across several different paper formats.

The process of enumerating species linked Banks's specific geographical inventories to Solander's notebooks and manuscript slips, giving Banks access to this vast quantity of information. The close collaboration between Banks and Solander that had emerged since their time in the Society Islands drew the working practices of these individuals together, equalizing their levels of authority when recording the botanical material and instructing Banks's natural history staff. For example, although Banks took over the task of enumerating species after they reached New Guinea on September 1, 1770, these records were not kept as a separate manuscript. Rather, Banks's enumeration appears on a slip of paper that he inserted into Solander's "Index Plantarum Novæ Hollandia" notebook to combine this numerical account with Solander's systematic descriptions.[24] Banks described this collaborative relationship in an obituary of Solander in 1782: "We worked at the great table in the cabin with our draughtsman opposite. We directed his drawing, and made rapid descriptions of our natural history specimens while they were still fresh."[25] The structure Banks describes is laid out in figure 2.2 and as a result of their close collaboration, Banks decided to let Solander share in his planned publication of the botanical discoveries from the Pacific, declaring to Johan Alströmer that "Solander's name will appear jointly on the title page along with mine since everything was done jointly. Hardly a sentence was written while he lived to which he did not contribute."[26]

The distinct social hierarchy Banks established over his natural history staff enforced consistent working practices and standardized the means for collecting and interpreting information. Banks's use of his own finances to fund his staff ensured they were not contractually or financially associated with Cook, reducing the latter's authority over Banks's employees. Banks's approach to management, delegation of tasks, and keeping advanced records was familiar working practice to a Lincolnshire landowner. On a visit to Revesby Abbey, the country seat where Banks spent around three months each year, Arthur Young from the Board of Agriculture took note of Banks's meticulous filing practices and recordkeeping system.[27] Central to this was Banks's use of books, including his copy of Thomas Tusser's *Five Hundred Points of Good Husbandry* (1610), which Banks has had interleaved to accommodate copious notes.[28] Banks's parallel natural-historical hierarchy often conflicted with Cook's rigid naval regime and established a parallel natural-historical hierarchy. These conflicts became apparent when Banks attempted to prioritize natural history collecting above everything else. When attempting to fit out the *Resolution* for the second voyage to the Pacific, Banks requested for a superstructure to be added to the ship to accommodate his botanical equipment. This caused the ship to become top-heavy and Cook ordered for the removal of the superstructure, resulting in Banks's withdrawal from the second expedition.[29] Banks's and Cook's different social origins also caused conflicts. Banks, as a prominent landowner, was used to instructing and managing individuals he deemed to be of a lower social rank than himself. However, unlike his team of specimen collectors, Banks had a reduced influence over Cook and the rest of the crew who were employed by the Admiralty.

Banks viewed Solander as the most important individual aboard the *Endeavour*, describing him as "the Dr" in reference to his reputation as a teacher and scholar. Banks recognized that without Solander's familiarity with the Linnaean system and ability to manage paper to classify vast quantities of new species, much of the natural-historical work would have been very difficult to perform. This is reflected in his journal; Solander nearly died in a blizzard in Tierra del Fuego and when he collapsed due to the cold, Banks dragged Solander to the place where a fire had been lit: "The Dr on the contrary said he must sleep a little a little before he could go on and actually did a full quarter of an hour, at which time we had the welcome news of a fire being lit about a quarter of a mile ahead . . . With much difficulty I got the Dr to it."[30] The two African servants who had accompanied them, George Dorlton and Thomas Richmond, died of the extreme cold in the Tierra del Fuego on January 15, 1769. Cook remarked in his journal that Banks's failure to return to the ship that evening "gave me great uneasiness . . . However, about noon they returned in no very comfortable condition, and what was still worse 2 blacks, servants to Mr. Banks had perished in the Night with cold."[31] The fact that Banks saved Solander and left the two servants reflects the rigid social and economic value Banks placed on his employees. Banks was not only good friends with Solander but considered him to be indispensable for cataloguing, describing, and classifying the collection. In comparison, it was easy to replace servants and field assistants. Banks could even enlist Indigenous people to supplement these roles.

The final key figure responsible for classifying and recording species was Herman Spöring. Solander outlined Spöring's role in a letter he sent to Linnaeus from Rio de Janeiro on December 1, 1768: "A son of late Professor Spöring in Åbo is with me here, as a scribe; his name is Herman Diedrich, and he went to sea from Sweden in 1755 and has for the last 11 years stayed in London as a watch maker. The last 2 years I employed him to write for me."[32] Similar to many of Banks's later amanuenses who were employed to catalogue his metropolitan collection from the 1770s to the 1810s, Spöring had experience as a physician, naturalist (he was officially listed as an "assistant naturalist" in the ship's log), draughtsman, and scientific instrument maker.[33] Solander employed Spöring at the British Museum since 1766 because he had "some skill in natural history" and was of far greater use than Solander's earlier unnamed assistant, who could only "copy out his manuscript notes of the catalogue."[34] Spöring's experience of working with Solander and his knowledge of how to organize collections according to the Linnaean system defined his place within Banks's system. Spöring was a tier down from Banks and Solander in the hierarchic team, occupying a position on a level with the artists, and was employed for his wide range of transferable skills. For example, when the astronomer Charles Green's quadrant broke during an attempted observation of the transit of Venus in May 1769, Spöring was enlisted to assist with its repairs.[35]

Spöring's knowledge of scientific instruments allowed him to mediate between Banks's team and those officially employed by the Admiralty to undertake astronomical observations. Spöring was also a skilled technical draughtsman, skills that became apparent after the landscape artist Alexander Buchan's death in 1769. Many of Spöring's drawings have ruled lines and he often gives measurements for scale reflecting on his work as a scientific instrument and watchmaker in London. Spöring copied some of his illustrations and presented them to Cook, who as a navigator and cartographer, appreciated their mathematical accuracy.[36] These gifts were central for maintaining good relations with Cook, whose temper was frequently tested by the limited space in the great cabin of the *Endeavour* being consumed by specimens and the naturalists' consistent complaints about the short durations ashore. Despite his work on scientific instruments and drawings, Spöring's main task was to assist Solander as a secretary, as evidenced by his hand appearing in several indices in Solander's notebooks, manuscript transcriptions of botanical and zoological descriptions ordered according to specific geographical locations, and throughout Banks's interleaved copies of *Species Plantarum* (1762–1763) and *Systema Naturae* (1766).[37] These books were essential for achieving Banks's and Solander's aims of creating a global inventory of species, a process that broke down after Spöring's death in 1771. Banks recalled how "the death of our secretary prevented fair copies [of the descriptions] being made before we landed."[38]

The Resources of Natural History

Joseph Banks's rigid social hierarchy defined specific roles that interlocked with the different paper technologies used to manage information over the different scales of geography, time, and volumes of specimens. Many of these practices were transposed from situations Banks and Daniel Solander were familiar with in Europe. Banks emulated his approach to managing country estates. Solander utilized a set of familiar working practices that he had developed and tested at the British Museum when cataloguing reclassifying the collection amassed by Hans Sloane (1660–1753) according to the Linnaean system.[39] Contained in 336 bound volumes, Sloane's pre-Linnaean collection was not arranged according to a single system of classification. Solander developed an approach to identify, describe, and catalogue previously undescribed species he found on a daily basis while not disrupting the collection's historical physical arrangement.

In comparison to their previous work on dried botanical specimens, living plants were the main focus of study during the voyage. The very nature of the physical material being described, classified, and stored imposed a distinct chronological structure on the collecting process outlined in figure 2.2. From the very beginning these processes were aligned with Banks's pyramidal hierarchy and remained consistent until the later stages of the voyage. After constructing descriptions, Banks and Solander stored the specimens in piles of paper referred to as "books," compiled from stacks of old printed sheets that were folded over and loosely stitched along one side.[40] Banks described how the specimens "were entered directly into books in the form of the flora for each country we visited."[41] After observing and describing each specimen in its living state, Banks and Solander inserted it between the leaves of these books to dry and preserve as many physical features as possible. This approach to storing specimens dates from the late seventeenth century and was outlined in John Woodward's instructions

FIGURE 2.4. One of the pages from the last surviving "book" of *Endeavour* specimens, showing the title page of *Notes upon the Twelve Books of Paradise Lost* (1738). By kind permission of the Trustees of the Natural History Museum, London.

for collecting (1696), who suggested that "to preserve these samples of plants, put them each separately, betwixt the leaves of some large *Book*, or into a Quire of brown *Paper*, displaying and spreading them *Smooth* and even."[42] Similar practices were employed by Banks's contemporaries. For example, Alexander Anderson (1748–1811), curator of the St. Vincent Botanical Garden, sent endless requests for drying paper to William Forsyth (1737–1804), asking on August 1, 1786, "if you could send me some waste paper, such as old newspapers or old Books, to put specimens in I will be much obliged to you."[43] The rapid accumulation of specimens on James Cook's voyage meant Banks and Solander placed them into the books chronologically and resisted any attempts to order them systematically. Thus, a similar cataloguing system to that Solander developed to locate specific materials within the bound volumes of the British Museum's herbarium was needed to link specimens to field descriptions and add newly described species to the Linnaean framework.

The surviving examples of paper Banks and Solander used for the drying books come from a copy of *Notes upon the Twelve Books of Paradise Lost Collected from the Spectator* (1738). Each sheet is identical and contains the title page and preliminary sections of this work (figure 2.4).[44] This book has been titled in Solander's hand "Madeira III," showing it was the third bundle of plants collected from the Atlantic island of Madeira in 1768. The use of drying books to preserve physical specimens was far more practical for storing specimens aboard the ship than the separate sheets outlined by Carl Linnaeus in *Philosophia Botanica* designed to be "placed in a cupboard, which can be closed by two long folding doors, nicely corresponding to a vertical partition." Limited space and harsh environmental conditions aboard the *Endeavour* meant tall Linnaean herbarium cabinets designed to be 7.5 Paris feet high, could not be accommodated.[45] Folded over quires formed a practical solution for storing and preserving botanical specimens and were kept in Banks's strong chests to prevent damage from humidity, seawater, crewmen, and insects. Solander's paper technologies ensured these books were indexed and the species were findable during the voyage before their rearrangement to follow the specifications Linnaeus laid out in *Philosophia Botanica* in Banks's London collection.[46]

The paper used for Banks's and Solander's manuscripts, Parkinson's drawings, and the interleaved pages in their copies of *Species Plantarum* and *Systema Naturae* has the watermark of the Kent-based papermaker James Whatman the Younger (1741–1798), who produced the highest-quality paper available on British markets in the late 1760s. Whatman's firm had started to develop wove paper in the 1750s that was far smoother and more durable than laid paper—qualities Banks was looking for. Whatman's paper became so

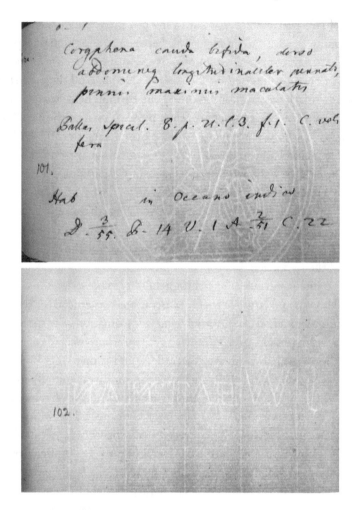

Figure 2.5. Typical watermarks of the papermaker James Whatman on two of Daniel Solander's zoological manuscript slips. Solander's notes outline a species of fish he and Joseph Banks caught in the Indian Ocean between Batavia and the Cape of Good Hope. By kind permission of the Trustees of the Natural History Museum, London.

popular among British buyers that it began to displace imports of paper from Continental Europe by the early 1770s.[47] The watermarks on this paper, a simple "JWhatman" and a seated figure of Britannia, can be found on the interleaved pages in Banks's copies of *Systema Naturae* and *Species Plantarum*, Solander's botanical and zoological manuscript slips, and his field notebooks (figure 2.5). Most of the paper Banks took aboard the *Endeavour* is laid paper containing chain lines of 1.25 centimeters apart and a smooth surface obtained through an ironing process.[48] The main link between the high-quality paper and that used for the drying books is the printer of *Notes upon the Twelve Books of Paradise Lost*. Between 1720 and 1760 the publishers of this work, J. and R. Tonson, hired the Birmingham-based printer John Baskerville to print many of their publications. By the late 1750s Baskerville, who was acquainted with Banks during the mid-1760s as a fellow member of the Lunar Society of Birmingham,[49] stocked several varieties of Whatman's paper, some of which Banks used on his trip to Newfoundland in 1766. Baskerville organized the production of the first book using Whatman's wove paper and often canceled a single leaf or whole gathering to omit errors, creating a large surplus of identical printed sheets similar to those used for the drying books aboard the *Endeavour*.[50]

The fine paper used for field notebooks, manuscript slips, interleaved books, and large sheets for illustrations was crucial for the rapid accumulation and classification of new information. The specimens in the drying books and Parkinson's illustrations were designed to be read alongside Solander's descriptions and support his Linnaean ranking of each species. Solander's descriptions were initially formulated in the field and recorded in his notebooks. Out of these, nine botanical and four zoological notebooks survive, all of which were bound in quarto format and compiled from sheets of Whatman's paper.[51] These were bound in thin card wrappers to create a light, transportable means for taking notes during excursions onto land and are representative of the standard high-quality notebooks available from London stationers. Descriptions were then transferred and refined onto other mediums of Whatman's paper, including Solander's slips that were cut from his quarto notebooks—the pages in these are double the size of the slips while many of the latter have gilded edges, signifying they were once part of a larger compilation.

Banks's and Solander's copies of Linnaeus's two-volume *Species Plantarum* and *Systema Naturae* have all been interleaved with large sheets of Whatman's paper.[52] After interleaving, the increased page count, size, and thickness added by sheets of Whatman's paper meant each title had to be bound in six quarto volumes. Interleaving was a common practice in the eighteenth century and could be commissioned from a myriad of London bookbinders. It was employed in dictionaries, encyclopedias, mercantile books of prices and imports, recipe books, and works of natural history to provide space for the addition of new information. Interleaving became a standard practice when adapting books for global voyages and was used by the employees of joint-stock companies, such as the East India Company, to collect and order information on a specific place or subject matter as they traveled.[53] The information Spöring added to the interleaved pages in *Species Plantarum* and *Systema Naturae* connects these books with related manuscripts and published works through annotated references, codes, numbers, and binomial names, carving out what Staffan Müller-Wille has referred to as a "paper space" for each species.[54]

Books on a Global Circumnavigation

The *Endeavour* library was compiled from a mixture of natural history books containing texts that date from the classical age to the 1760s. Many titles were listed by Carl Linnaeus in *Philosophia Botanica* and represent the works of authors he counted among the "Phytologists."[55] The rest of Banks's library was compiled from travel accounts by those who had embarked on intercontinental voyages. Working between printed books and manuscripts was a common practice employed by various travelers and naturalists. For example, the Linnaean apostle Pehr Kalm, who traveled through the interior of North America from 1748 to 1751, cited John Oldmixon's *The British Empire in North America* (1708) alongside various Linnaean works when discussing vermin in colonial New England and described how he recorded observations of a waterfall in a "pocket book."[56] Similar to Kalm, Joseph Banks and Daniel Solander developed a means for integrating their books and manuscripts as they traveled on the *Endeavour*. Banks later remarked that they were "well supplied with books on the natural history of the Indies."[57] Banks kept books and manuscripts in a bureau either in the cabin he shared with Solander or in the great cabin they shared with James Cook. For example, on the stormy night of January 6, 1769, when approaching Tierra del Fuego, Banks remarked: "My Bureau was overset and most of the books were about the Cabbin floor, so that the noise of the ship working, the books &c. running about, and the strokes our cotts or swinging beds gave against the tops and sides of the Cabbin we spent a very disagreeable night."[58]

Aside from the interleaved copies of *Species Plantarum* and *Systema Naturae*, the books Banks and Solander paid the most attention to date from the late seventeenth century to the 1760s. Many consist of what Daniela Bleichmar has described as "geographical" or "thematic inventories" and provided useful reference points for Banks and Solander to describe plants and animals.[59] These include Hans Sloane's *Natural History of Jamaica* (1707–1725), Leonard Plukenet's *Opera Omnia Botanica* (1720), and Engelbert Kaempfer's *Amoenitatum Exoticarum* (1712).[60] More contemporary books include the volumes of Georges-Louis Leclerc, Comte de Buffon's *Histoire Naturelle* published between 1749 and 1767 (1749–1804) and the first two volumes of Thomas Pennant's *British Zoology* (1768–1770).[61] It is important to note that for periodicals and multivolume works published over several years, the volumes that formed part of the *Endeavour* library were published before August 26, 1768, since it was impossible to obtain later works in the Pacific. However, these books were not, as Bleichmar and Annie Mariss have described, used and categorized as separate entities during expeditions.[62] Rather, books were integrated within Banks's and Solander's chain of manuscripts, images, interleaved books, and objects used for managing information.

The fundamental integration of books into broader practices of natural history is represented through the naturalists aboard the *Endeavour*'s consistent use of books to assess novelty and trace synonyms of species names across a variety of different formats.[63] Banks described how, when aboard the ship, he and Solander "completed our descriptions, and added synonyms using our library." In *Philosophia Botanica*, Linnaeus suggested "synonyms are variant names given to the same plant by phytologists," adding that naturalists should present synonyms in a column based on the age of the name, "ascending from the most modern to the most primitive."[64] Thus, when Solander devised a binomial he listed references to all of the earlier published names, starting with the most recent and ending with the oldest, while adding lists of any names sourced from Indigenous peoples.[65] This allowed Solander to assess the physical properties of species identified by previous authors, compare earlier descriptions with living specimens, and place his observation within its historical context.

Tracing synonyms through manuscripts, publications, and testimony from Indigenous informants was crucial for the construction of a new description and established the place of a species within the Linnaean hierarchy. All the books taken aboard the *Endeavour* contain extensive marginal annotations in Solander's and Banks's hands that ascribe each species described and figured a Linnaean binomial name. In the case of species Linnaeus had not previously described, Solander gave the new binomial recorded in his manuscripts. For species already published in Linnaean works, Solander added the specific page references. These notes intertwined printed books with Banks's and Solander's broader system for managing information and gave a chain of references that connects pre-Linnaean works to specific species, descriptions, and page numbers in *Species Plantarum, Systema Naturae*, and their manuscript descriptions. For example, in Banks's copy of Kaempfer's *Amoenitatum Exoticarum* (1712), Solander added Linnaean binomial names and page references to *Species Plantarum* next to every botanical description and illustration. These annotations allowed them to connect a series of names and descriptions for specific species throughout the full range of books and manuscripts in Banks's and Solander's *Endeavour* library, defining each Linnaean taxa within paper technologies, ranging from the physical object in the "book" of paper, through the manuscript indexes and descriptions, the interleaved Linnaean publications, and the annotated printed books.[66]

Despite the quantity and range of books in their library, Banks did occasionally wish he had more books aboard the *Endeavour*. These concerns came from matters of describing specific uses of plants by Indigenous peoples and are apparent from a long journal entry dated September 20, 1770, when Banks described the process local islanders in Savu, Indonesia, used to refine sugar. In his description, Banks mentioned that "I have been told that this very method was proposed in the Gentleman's Magazine Vol. p. many years ago but have not the book on board."[67] Although Banks's natural history library was relatively comprehensive, the example of refining sugar in Indonesia shows it was not as complete as he might have liked. Multivolume periodicals, such as the *Gentleman's Magazine* and *Philosophical Transactions* often graced the bookshelves of British country houses but were too extensive for Banks to accommodate on the ship. This predicament was not faced by other voyagers, such as the naturalists who participated in Spanish expeditions to South America, who received books and offprints of journal articles from Europe.[68] Banks had no means for accessing such materials after they departed from Rio de Janeiro. Thus, he left the section for the volume and page reference to the *Gentleman's Magazine* blank with the intention to perfect this account when he had access to the book.

The problems associated with a lack of reference books were commonplace on late eighteenth-century intercontinental natural history expeditions. For example, in 1790 the Scottish naturalist William Roxburgh (1751–1815) wrote to Banks from the Eastern India region of Samalkota, commenting on how his "great distance from every sort of help and with only a very limited library, renders your support doubly necessary."[69] Roxburgh's traveling library was sparser than Banks's as it was more difficult to carry books overland than aboard a ship even with the help of Indian servants. As a result of the lack of books, Roxburgh was unable to formally identify new species among the plant specimens he sent back to Europe. Roxburgh suggested that to overcome these problems the East India Company could "send out a collection of books for use of their servants in the medical line, which would enable & encourage many to become useful to themselves and their country."[70] Others were more specific in their requests for books. Writing to William Forsyth from New York, Alexander Anderson suggested, "I am at an infinite loss for want of Linnaeus nor can I get here for it would be a usefull & agreeable companion to me."[71] In comparison to Roxburgh and Anderson, who could rely on established trade routes, it was impossible for Banks to receive books once they rounded Cape Horn and passed into the uncharted waters of the Pacific. The *Endeavour* library became a floating natural history station. The information in the books aboard the ship, Banks's and Solander's botanical knowledge, and information obtained from Indigenous peoples remained the only means for describing and classifying the natural world.[72]

Previously Discovered Species

Many botanical and zoological species collected on this voyage had already been described by European naturalists. Improving earlier descriptions remained central to Joseph Banks's and Daniel Solander's mission, incorporating these into the Linnaean system and the development an understanding of their geographical distribution. Earlier publications in the *Endeavour* library were employed to assess whether a species was unknown to Europeans, incorporate it into the Linnaean system and list synonyms.

Perhaps the largest number of zoological species Banks and Solander collected and deemed new to the Linnaean system were the fish they caught from the ship.[73] Many aquatic zoological species were

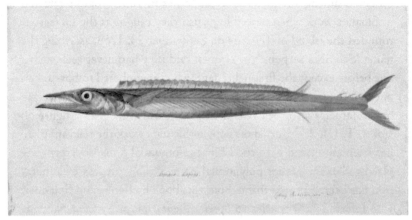

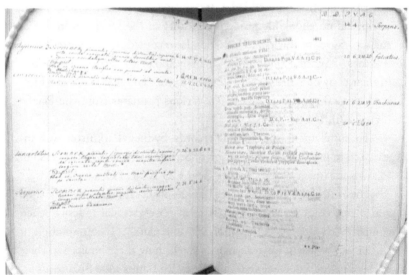

FIGURE 2.6. Top: Annotations on the copperplate image of the fish from Joseph Banks's copy of Hans Sloane's *Natural History* under which Solander ascribed the name *Scomber serpens*. Middle: Sydney Parkinson's drawing of *Scomber serpens* (*Gempylus serpens*), based on the specimen Banks caught on September 23, 1768. Foot: Banks's interleaved copy of *Systema Naturae* annotated by Herman Spöring. The annotation relating to the species *Scomber serpens* is on the bottom left. By kind permission of the Trustees of the Natural History Museum, London.

recorded in earlier publications by travelers and naturalists, even if they were not classified in Linnaeus's *Systema Naturae*. Examples can be found in Banks's and Solander's copy of Hans Sloane's *Natural History of Jamaica* (1707–1725), a book compiled from a systematic list and several hundred copperplate images of the flora and fauna Sloane observed during a trip to the West Indies in 1687.[74] Banks referred to Sloane's work when describing a fish they caught as the *Endeavour* rounded the island of Tenerife on September 23, 1768, ascribing the name "Scomber serpens; the seamen said they had never seen such a one before except the first lieutenant, who remembered to have taken one before just about these islands; S[i]r Hans Sloane in his passage out to Jamaica also took one of these fish which he gives a figure of, Vol. 1, T. 1, f. 2."[75] In Banks's copy of Sloane's work the relevant plate has been annotated with the Linnaean binomial *Scomber serpens*, replacing Sloane's earlier polynomial name.[76] This process of annotation has been repeated throughout this book by Banks and Solander, who referred each species to *Species Plantarum* or *Systema Naturae*. For example, Solander added a description of *Scomber serpens* to his zoological notebook alongside the locality of the specimen in one of the six "caggs" of spirits, information later copied into the interleaved volumes of *Systema Naturae* by Herman Spöring.[77] Other annotations in their copy of *Systema Naturae* refer to the "kongeroo" of New Holland, several species of birds, and various primates from the Pacific (figure 2.6).[78]

To supplement the textual records Sydney Parkinson was instructed to produce an illustration of *Scomber serpens*. On the verso of Parkinson's image, Banks noted, "Sept.r 23 1768 of Canary Islands."[79] The coloring of the image situates Banks's observation in a specific time and place, emphasizing a difference in observational practice when compared to Sloane's earlier monochrome plate.[80] The annotation of earlier works with Linnaean names shows how these books and their detailed copperplate engravings linked with Banks's and Solander's information-management system based on the order laid out in *Systema Naturae* (figure 2.6). In a letter sent from Madeira to Lord Morton, president of the Royal Society, Solander outlined these processes of recording marine life: "At sea we have been very fortunate in finding a great many Sea Productions, that I hope will be better cleared up by us, than have been by any one before, especially as Mr Banks's People have had an opportunity of drawing them when fresh and alive."[81]

Constructing descriptions while observing the living plants gave Banks and Solander opportunities to correct several Linnaean names and descriptions. Typical examples originate in the already well-explored Portuguese island of Madeira where the *Endeavour* made port between September 12 and 18, 1768. In these five days of collecting Banks and Solander gathered examples of 253 individual species, thirty-eight of which they identified as new to European natural history. Banks compiled the names in a systematic list arranged according to the 1762–1763 edition of *Species Plantarum* that he appended to the description of Madeira in his journal, a practice he developed when traveling in Newfoundland (1766) and England and Wales (1767–1768).[82] In Wales Banks compiled a systematic "catalogue of scarce Plants &c observd in the neighbourhood of Edwinsford august 1767" that, in a similar manner to Pennant's manuscripts explored in chapter 1, lists vernacular Welsh botanical terms.[83]

Banks described how he and Solander attempted to devote their six days on Madeira to the collection of botanical specimens. Despite this, their collecting was encroached upon by social events. One of the more notable was when the governor of Madeira interrupted their botanizing with an official state reception. Banks was not impressed by this interruption and the loss of a day collecting. This is apparent from an entry in which he described how he and Solander "contrived to revenge ourselves upon his excellency, by an electrical machine we had on board; upon his expressing a desire to see it we sent for it ashore, and shocked him full as much as he chose."[84] Banks's use of the electrical machine to shock this unfortunate Portuguese diplomat sheds light on the other scientific apparatus he had aboard. The "electrical machine" was made by the London instrument maker Jesse Ramsden (1735–1800) and produced static electricity through the rotation of a circular plate against leather pads. Banks's electrical machine had a dual purpose. Banks intended to test electricity on Indigenous populations, a practice he employed during his trip to Iceland in 1772.[85] However, the humidity damaged the leather pads on the electrical machine; Banks does not appear to have used it in Tahiti, and when he reached New Zealand it was severely damaged. The second purpose was to display the machine to the peoples he encountered and establish trading networks for similar instruments. For example, several of these electrical machines accompanied George Macartney on the first British embassy to China in 1793, an expedition heavily influenced by Banks, and formed a vital part of the products selected to open China to British trade.[86]

The first species in Banks's list of Macaronesian flora is *Canna indica*, next to which he cited the term *Linn* in reference to *Species Plantarum*.[87] On the interleaved page opposite the entry for *Canna indica*, Spöring has added notes that edit and expand the published diagnosis. This was something Banks and Solander had the authority to do after viewing the living specimen. In contrast to Linnaeus, whose descriptions of tropical species often relied on dried herbarium specimens, Banks and Solander had the opportunity to view the living plant in the field and record specific physical characters lost in the drying process to improve the Linnaean description.

A main reason for collecting specimens of species already published by Linnaeus was that the long-established practice of storing botanical specimens in "books" of waste paper was not always adequate for preserving all of the physical features of plants. As a result, Banks made considerable efforts to preserve specimens in books of paper. For example, in a journal entry written on May 3, 1770, five days after James Cook landed in Botany Bay, Banks described how:

> Our collection of plants had now grown so immensely large that it was necessary that some extraordinary care should be taken of them least they should spoil in the books. I therefore devoted this day to that business and carried all the drying paper, near 200 Quires of which the larger part was full [of plants], ashore and spreading them upon a sail in the sun kept them in this manner exposed the whole day, often turning them and sometimes turning the Quires in which were plants inside out. By this means they came on board at night in very good condition.[88]

The process of drying the quires reduced the chances of rot and insect damage induced by humid conditions that caused constant problems when storing books and specimens. When the *Endeavour* was approaching the equator just off the South American coast, Banks noted how "all kinds of leather became mouldy . . . this mould adhered to almost everything, all the books in my Library became mouldy so that they were oblig'd to be wiped to preserve them."[89] Humidity aboard ships remained a constant problem when transporting specimens and documents across the Atlantic and caused significant damage to specimen collections and shipboard libraries.[90]

FIGURE 2.7. Left: Sydney Parkinson's illustration of the *Alstromeria salsilla* (*Bomarea edulis*. Herbert). Courtesy of the Trustees of the Natural History Museum, London. Right: the preserved specimen collected by Joseph Banks and Daniel Solander from Brazil. Courtesy of the Trustees of the Natural History Museum, London.

Parkinson's illustrations supported Banks's and Solander's revision of Linnaean descriptions, names, and classificatory placements, emphasizing vital features lost during the drying process. A typical example is *Alstromeria salsilla*, a plant Linnaeus had already described in *Species Plantarum*, which Banks and Solander collected on December 7, 1768, when they secretly disembarked—against the commands of the Portuguese viceroy—on the island of Raza near Rio de Janeiro (figure 2.7).[91] *Alstromeria salsilla* was difficult to preserve owing to its fleshy stalks and flowers, meaning Parkinson made efforts to emphasize crucial classificatory features owing to the poor state of the specimen, elevating the importance of the image for supporting Solander's description.

Identifying, Recording, and Classifying New Species

The social hierarchy within Joseph Banks's team and its alignment with paper structures used to catalogue, classify, record, and collect information became essential in the process of describing species identified as new to European natural history. These practices adapted according to the different geographical scope of regions they explored, the number of species they encountered, and the quantity of information supplied by Indigenous people. All these factors placed considerable strain on approaches to managing information. As such, this section concentrates on the practices employed on the island of Tahiti, paying particular attention to the species Daniel Solander named *Convolvulus alatus*, which allows one to trace the processes used to define a new species.

The first manuscripts used to record new species were Solander's field notebooks, a series of small quarto volumes arranged according to each major geographical locality they visited, including Madeira, Brazil, Tierra del Fuego, Tahiti and the Society Islands, New Zealand, New Holland, Java, the Cape of Good Hope, and Saint Helena.[92] Solander used these notebooks to describe the physical characters of species as he encountered them, combining this with information he received after conversing with and observing Indigenous peoples. These manuscript notebooks have a similar physical format to Banks's journal and the bound commonplace books used by earlier naturalists. Solander's notebooks are full of loose slips of paper inscribed with rough pen and pencil jottings designed to be incorporated into the main text at a later date. These form vital translations between Indigenous and European botanical terminology. This information was integrated into the more polished descriptions on the bound pages before it was transferred into Solander's other manuscripts (figure 2.8). Phonetic transcriptions of Indigenous terms run throughout these notebooks and emphasise the level of contact the Europeans had with local people. Starting in Tahiti, Solander collected Indigenous terms throughout the Pacific, facing a lull at Botany Bay, where the local Aboriginal peoples avoided the Europeans, before picking up again at the Endeavour River. The most extensive examples are in the manuscript Solander titled "Plantæ Otaheitenses" and reflect the collaborative processes of gathering information during the *Endeavour*'s nearly three-month stay in Tahiti from early April to late June

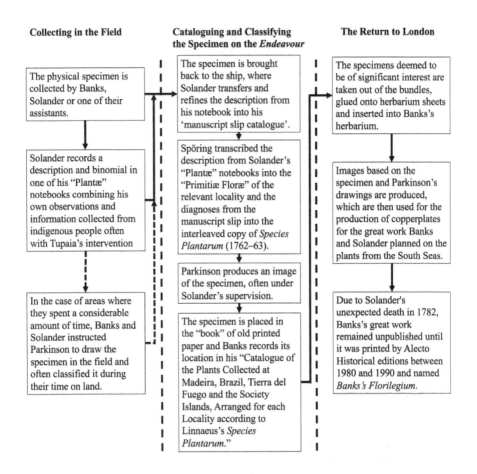

Collecting in the Field	Cataloguing and Classifying the Specimen on the *Endeavour*	The Return to London
The physical specimen is collected by Banks, Solander or one of their assistants.	The specimen is brought back to the ship, where Solander transfers and refines the description from his notebook into his 'manuscript slip catalogue'.	The specimens deemed to be of significant interest are taken out of the bundles, glued onto herbarium sheets and inserted into Banks's herbarium.
Solander records a description and binomial in one of his "Plantæ" notebooks combining his own observations and information collected from indigenous people often with Tupaia's intervention	Spöring transcribed the description from Solander's "Plantæ" notebooks into the "Primitiæ Floræ" of the relevant locality and the diagnoses from the manuscript slip into the interleaved copy of *Species Plantarum* (1762–63).	Images based on the specimen and Parkinson's drawings are produced, which are then used for the production of copperplates for the great work Banks and Solander planned on the plants from the South Seas.
	Parkinson produces an image of the specimen, often under Solander's supervision.	
In the case of areas where they spent a considerable amount of time, Banks and Solander instructed Parkinson to draw the specimen in the field and often classified it during their time on land.	The specimen is placed in the "book" of old printed paper and Banks records its location in his "Catalogue of the Plants Collected at Madeira, Brazil, Tierra del Fuego and the Society Islands, Arranged for each Locality according to Linnaeus's *Species Plantarum*."	Due to Solander's unexpected death in 1782, Banks's great work remained unpublished until it was printed by Alecto Historical editions between 1980 and 1990 and named *Banks's Florilegium.*

FIGURE 2.8. A chart displaying the processes used by Joseph Banks, Daniel Solander, Herman Spöring, Sydney Parkinson, and their field assistants to manage information.

1769.[93] This allowed Banks and Solander to examine multiple examples of each species and take advantage of a small geographical area to employ working practices resembling those used by parochial natural historians in Britain.

Convolvulus alatus is one of the plants Solander recorded in his "Plantæ Otaheitenses" notebook. Under the Linnaean name, Solander added the Indigenous term used to devise the Latin binomial, information he and Banks's team were able to source given their lengthy stay on the island (figure 2.9). This followed the instructions given by Linnaeus in his *Instructio Peregrinatoris* (*Instructions for Naturalists*

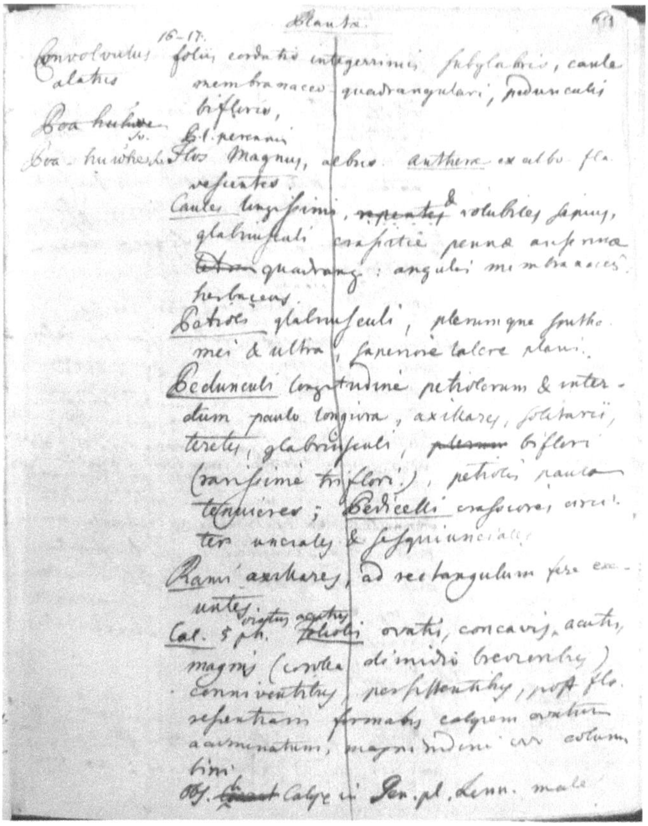

FIGURE 2.9. The page from Daniel Solander's "Plantæ Otaheitenses" notebook in which he describes *Convolvulus alatus* (now *Decalobanthus peltatus*). Courtesy of the Trustees of the Natural History Museum, London.

of Voyages of Exploration) (1759) who suggested botanical explorers should "note down the names given by the natives." The process of collecting transcriptions of Indigenous terms continued throughout the voyage. For example, on the peninsular near Te Wanganui-o-Hei referred to by Banks and others as Opoorage on the southern shores of what Cook later named Mercury Bay to commemorate his and Charles Green's observations of the transit of Mercury between November 4 and 14, 1769, the naturalists had consistent contact with the local Māori, who often collected specimens. These encounters were outlined decades later by Te Horetā (d. 1853), leader of the Ngāti Whanaunga, who recalled supplying the names for plants to the naturalists who disembarked from the *Endeavour*:

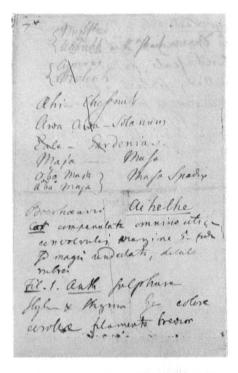

FIGURE 2.10. A typical slip of paper giving English and Latinate versions of different names for genera from Daniel Solander's "Plantæ Otaheitenses" notebook. Courtesy of the Trustees of the Natural History Museum, London.

They collected grasses from the cliffs, and kept knocking at stones from the beach, and we said, "Why are these acts done by these goblins?" We and the women gathered stones and grass of all sorts and gave to these goblins. Some of the stones they liked, and put them into their bags, the rest they threw away; and when we gave them the Grass and branches of trees they stood and talked to us, or they uttered the words of their language. Perhaps they were asking questions, and, as we did not know their language, we laughed, and these goblins also laughed so we were pleased.[94]

The time spent in Tahiti allowed the naturalists to build relationships with the local people, develop a significant command of the Tahitian language, and refine their approach to gathering information. Perhaps the most famous individual they encountered was Tupaia, a Raiatean priest who traveled with Banks and Solander aboard the

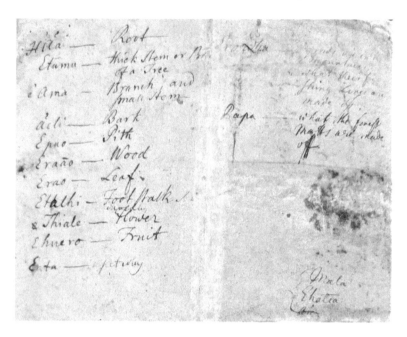

FIGURE 2.11. An example of Daniel Solander's rough field notes from the verso of the front cover of "Plantæ Otaheitenses." These translate Polynesian terms for different features of plants into English. Courtesy of the Trustees of the Natural History Museum, London.

Endeavour until his death in Batavia and acted as an essential mediator between Banks's team and the local people.[95] He often sat with Solander as they went through the specimens and manuscripts and accompanied the naturalists during excursions onto land. Tupaia's status in Tahitian society made sure he was soon incorporated into Banks's hierarchic team at a level akin to that of Solander so they could collaborate on describing species.

By including the Indigenous name for each species in his manuscripts, Solander was adhering to the guidelines Linnaeus laid down in *Critica Botanica* (1737): "I do not object to any nation retaining its own vernacular names for plants."[96] Linnaeus elaborated further in *Philosophia Botanica* (1751), commenting that "we adopt barbarous names if they were new-born, provided that we remake the words that have to be excluded, forming them from Greek or Latin."[97] Where the plants Solander collected in Tahiti were new genera and species, he followed Linnaeus's instructions and devised Latin binomial names and Linnaean descriptions by translating Indigenous terminology.

This practice is apparent from the jottings inside the front cover of Solander's notebooks and on the loose slips he inserted between the pages. These were used to translate Tahitian words that name a range of different physical features and specific names for plants into Latin and English (figures 2.10 and 2.11).[98] Solander's lists have a strong similarity to the vocabularies recorded by Banks and Parkinson, aligning the botanical notebooks with an interest in integrating Indigenous knowledge with European frameworks of natural history.[99] Many Indigenous words were used to formulate names for species and construct Linnaean diagnoses, showing how Solander enlisted the assistance of Indigenous peoples when structuring the entire description. Naming became a collaborative endeavor, with Solander making active efforts to both record information on new species and incorporate what Anne Salmond has referred to as "ontologies" of knowledge into botanical records and the Linnaean system.[100] These represent a two-sided conveyance of information, intertwining knowledge on new species with a whole variety of practices and uses, showing how Indigenous groups played a far more active role in the practice of natural history than has previously been recognized.[101]

The final pages of each "Plantæ" notebook have been used to compile systematic indices of all the species named and described in the earlier pages and chart their locations in the bundles of paper, the processes of recording these in a whole variety of different manuscripts, interleaved books, and as illustrations. The indices are in the hands of Solander, Banks, and Spöring and give both systematic and alphabetical keys to locate the more extensive descriptions in the notebook. This is because the species in the notebook are arranged according to the order in which Solander encountered them. Indices provide several orderings for the species described in the notebook and offer the means to locate the original description, specimens in the drying book, and illustration. The indices also serve as a checklist to mark the progress made by the other members of Banks's team, recording whether Parkinson had produced an illustration, the page containing the description in the "Plantæ" manuscript, the Latin binomial, the vernacular name, the number of the "book" of waste paper in which they stored the specimen, the number of specimens collected, and the specific locality in which it was collected (figure 2.12). The inclusion of Indigenous names emphasizes their centrality to the practice of natural history, serving as information that supported the status of

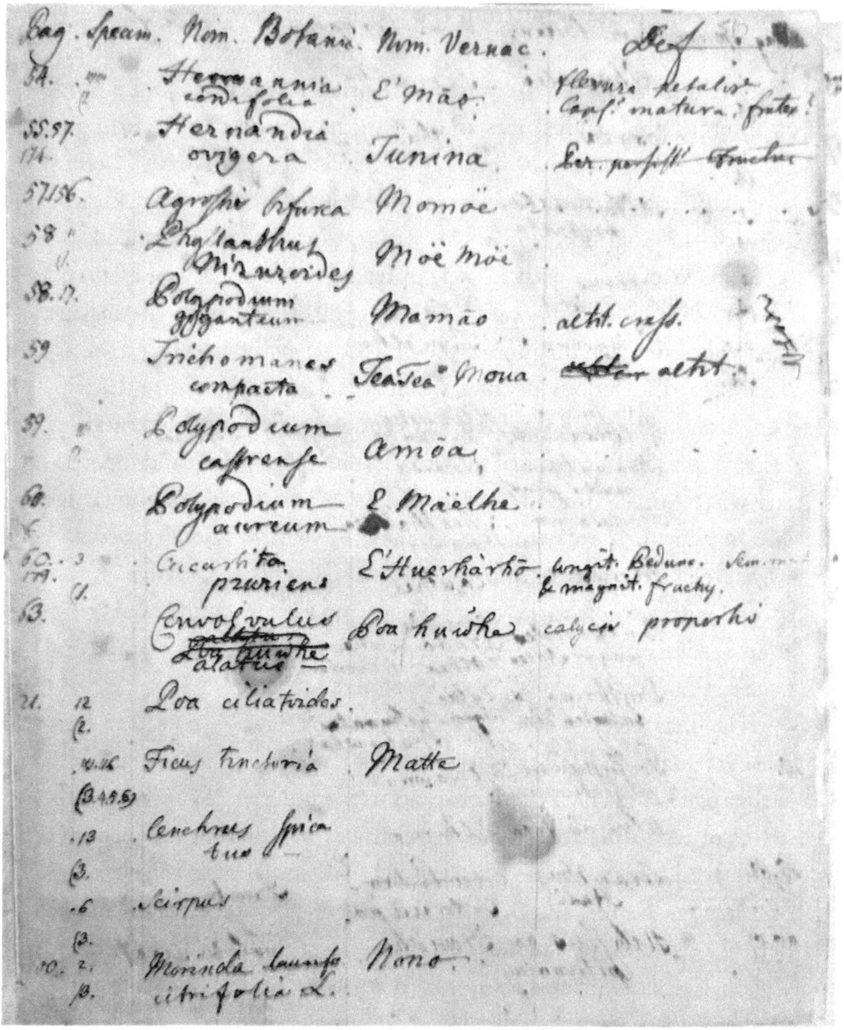

FIGURE 2.12. A page from Daniel Solander's and Herman Spöring's systematic index in "Plantæ Otaheitenses." The columns list the page number in the manuscript, the new binomial, a phonetic Tahitian name, and some definitive characters. Courtesy of the Trustees of the Natural History Museum, London.

the Latin binomial. These indices were the final part of this manuscript to be produced, probably when Banks, Solander, and Spöring transferred the specimens into the bundles of drying paper and began the process of transferring the information into more flexible paper formats.

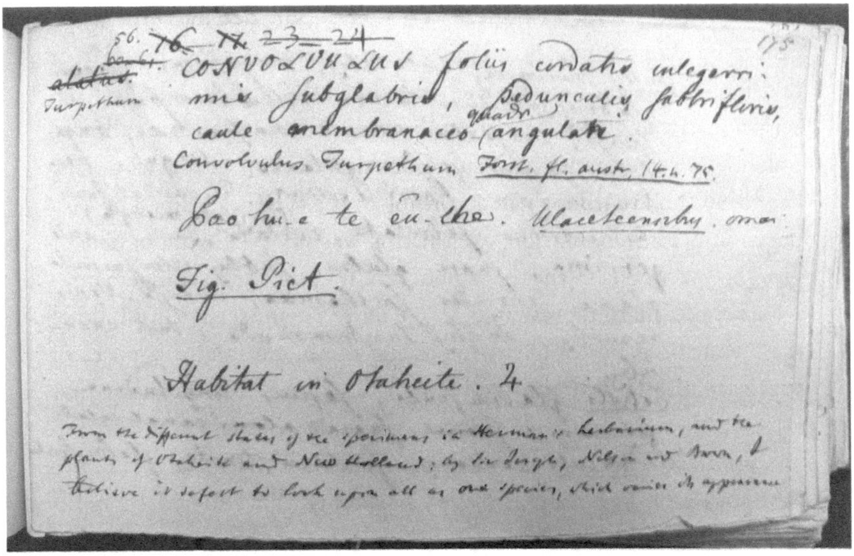

FIGURE 2.13. Daniel Solander's slip for *Convolvulus alatus*. The additions at the foot of the slip are in the hand of Jonas Dryander, Joseph Banks's curator from 1782 to 1810. Dryander compares this description with a specimen from the herbarium of Paul Hermann that Banks purchased in 1793. Courtesy of the Trustees of the Natural History Museum, London.

Transferring Information from Notebooks to Slips

A process represented throughout Daniel Solander's "Plantæ" manuscripts is the transfer of information into other repositories. This is represented by the vertical red line Solander crossed through every draft description, creating a clear visual indicator that his information had been transferred into the next manuscript in his record-keeping system: the "manuscript slip catalogue." A typical example is *Convolvulus alatus*, the description of which Solander copied onto a manuscript slip before inserting the specimen into a book of drying paper. Manuscript slips are commonly attributed to Carl Linnaeus, although Solander seems to have been using slips to manage information at the British Museum at least five years before Linnaeus realized their practicality. These pieces of paper, which have been bound into twenty-four volumes, were originally loose and encased in twenty-four Solander boxes, taking the form of hinged document cases allowing for the swift and continual addition and reordering of new slips relating to individual species. Each Solander box contained slips relating to a single Linnaean class. These were then subdivided into orders and genera by slips on which Solander wrote the names for these larger divisions before ending with each individual slip, which contains a description of a single species. The unbound nature of these slips allowed Solander to consistently add numerous descriptions of species during the voyage and reorder these according to the system outlined by Linnaeus in *Species Plantarum*, even if this was physically impossible to give the specimens in the drying books any systematic arrangement.

The information presented on Solander's manuscript slips follows the layout used by Linnaeus for listing the names and diagnoses of species in his publications (figure 2.13). On all the slips, including that for *Convolvulus alatus*, Solander gave a number that relates to the species numbers Linnaeus ascribed this genus in *Species Plantarum*. As Staffan Müller-Wille and Isabelle Charmantier have suggested, numbers served as a device to cope with the "information overload" many naturalists experienced in this period through linking species across different manuscripts, illustrations, and specimens.[102] Numerical references to *Species Plantarum* are followed by the binomial; the genus is capitalized on the right and the species is in lowercase on the left, similar to the typographical layout of Linnaeus's work. After the binomial Solander added a description or brief diagnosis, often a refined version of that recorded in his field notebook.

In addition to the systematic description, Solander transferred the Indigenous names for each species onto the manuscript slip. For *Convolvulus alatus*, Solander copied the name "Pao-hue tee e͡u-ihe" from "Plantæ Otaheitenses" under the diagnosis.[103] All of these vernacular names were sourced from communities encountered in the Pacific, a process facilitated by Tupaia, who obtained precise information on the Indigenous uses and annual life cycles for many of the plants. Solander's decision to retain the Indigenous name in the manuscript slips shows its importance for supporting the status of the binomials through securing a route to the Indigenous use of the plant.[104]

Over the following decades, these slips were updated by Solander and his successors. The first major addition made to the slip describing *Convolvulus alatus* came with the arrival of Omai (Mai) in London after James Cook's second voyage. Omai arrived on HMS *Adventure* in 1774 and was quickly introduced to Joseph Banks. Banks and Solander saw Omai as a valuable source of information on the Indigenous uses of plants from the Society Islands and revised significant parts of their manuscript descriptions according to the information he supplied. Evidence for this process can be found scattered throughout Solander's manuscript slips. In the case of *Convolvulus alatus*, Solander added another Indigenous name after that he had recorded in Tahiti before citing Omai as the source.[105] This exchange of information was immortalized in a famous portrait by William Parry that shows Omai standing on the left, Banks in the center, and Solander recording information in a manuscript.[106] This image is representative of the hierarchic social structure Banks maintained throughout the *Endeavour* voyage and when organizing his natural history collection in London—Parry has presented Banks as the main organizer and facilitator of the transfer of information from Omai to Solander, mirroring the relationship between Banks, Tupaia, and Solander when they traveled from Tahiti to Batavia.

The information obtained from Indigenous peoples on the economic potential of botanical species was integrated into Solander's manuscript slip catalogue. At the end of his descriptions of the morphology of *Convolvulus alatus*, Solander added a note: "Children suck out the sweet juice or water which generally is in closed within the Caylix."[107] Solander paid close attention to the Indigenous uses of plants to establish the potential economic virtues of new species, showing the connections between botanical exploration and the use of plants as economic resources. Sydney Parkinson made similar observations and extracted this information from Solander's notes. In the posthumous *Journal of a Voyage to the South Seas* (1773) Parkinson notes that the "native name" for *Convolvulus alatus* is "Taowdeehaow" and that "the stalks of this plant they give young children to suck."[108] The lack of reference to this species in Parkinson's surviving manuscripts suggests he extracted this information from Solander's work, leading to conflicts between Stanfield Parkinson and Banks about the publication of the *Journal of a Voyage to the South Seas* since it duplicated information, and thus undermined, the content of Banks's own planned publication on the plants they collected.

Similar names can be found throughout the manuscripts Solander produced after the *Endeavour* left the Society Islands. For example, when collecting in Tūranganui-a-Kiwa, which Cook renamed Poverty Bay, between October 9 and 10, 1769, where Tupaia proved his worth to Cook a few days earlier through avoiding a serious conflict with the local Māori,[109] Solander described the species *Piper myristicum* in his "Plantæ Australiæ" notebook.[110] Under this he recorded several Indigenous names for this species as the *Endeavour* circumnavigated New Zealand, relying on Tupaia, who quickly mastered the sound shifts between different Polynesian and Māori languages.[111] Examples include "Takauwa," which Solander recorded in Poverty Bay, and "Chawa Chawa," which Solander recorded in Mercury Bay. The Latin binomial *Piper myristicum* was devised directly from Solander's observations of its use by the Māori, who used it as an ointment. This, in addition to Solander's observations of its strong smell, were used to derive the term *myristicum* from the Greek *myristicos* in reference to its fragrance and use for making ointment. Thus, Solander's binomial and description was formulated as a direct result of the conversations Tupaia facilitated with local Māori and shows how the Latin names Solander ascribed in the field were designed to chalk out a path to the Indigenous uses of a species—vestiges of which were retained throughout the different layers of manuscripts.[112]

The detailed analysis of Indigenous terminology and uses of species followed the general mandate the Royal Society and Admiralty gave to those who traveled on global voyages. For example, in 1767 Samuel Wallis had been instructed to search for and identify "commodities usefull in Commerce."[113] Solander's observation of the use of *Convolvulus alatus* emphasizes a potential source of sugar for commercial markets.[114] Gathering this information was a collaborative process. For example, Parkinson, who took drawings of this species, also recorded in his journal that "the stalks of this plant they give young children to suck" alongside the Indigenous name and the Linnaean name that he copied from Solander's notebook.[115] This shows how Parkinson used information recorded in Solander's botanical manuscripts to construct his journal and illustrations, another collaborative document, information that crossed over with Banks's planned publication and a probable source for the dispute over the publication of Parkinson's account of the voyage in 1773.

The transfer of information on new species often took place back aboard the ship. This was undertaken by Herman Spöring who, once Solander had finished refining the descriptions on his manuscript slips, transcribed these into the interleaved copy of the 1762–1763 edition of *Species Plantarum* and the "Plantæ" manuscripts that refined and ordered the full descriptions according to the Linnaean

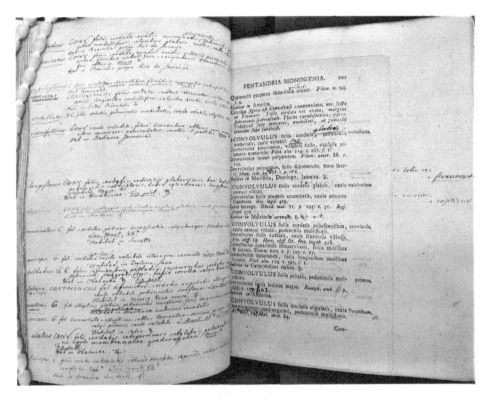

FIGURE 2.14. Herman Spöring's annotated name and diagnosis for *Convolvulus alatus* in the *Endeavour* copy of *Species Plantarum* (1762–1763) is the second entry from the end of the interleaved page. Courtesy of the Trustees of the Natural History Museum, London.

system. This book took a very different format to the various national floras annotated by Thomas Pennant and Gilbert White to define the variety of species found in specific parishes and counties. Rather, following the Linnaean programme, Banks and Solander intended for this interleaved book to encapsulate a global flora, constantly adding binomials and diagnoses of new species as they circumnavigated the globe. Spöring ensured that he inserted these descriptions into the Linnaean systematic framework. Diagnoses of new species were inserted opposite the printed entries concerning the relevant genus, and when the genus Solander described was entirely new, the large physical space given by the interleaved pages made it possible for him to add an entire section. Spöring transcribed the diagnoses for *Convolvulus alatus* from Solander's manuscript slip onto the

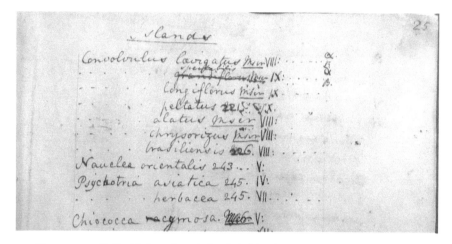

FIGURE 2.15. Folio 25 from Joseph Banks's "Catalogue," showing the different species of *Convolvulus* they collected in the Society Islands and their localities within the books of drying paper. Courtesy of the Trustees of the Natural History Museum, London.

page opposite the species Linnaeus had already classified under the genus *Convolvulus*.[116] A major omission was the phonetic transcriptions of the Indigenous names. In comparison to the more flexible manuscripts, this interleaved book was designed to produce a rigid Linnaean classification for species they encountered on the voyage (figure 2.14).[117]

The annotation of *Species Plantarum* was the final part of the process of assimilating new species into the Linnaean system when the specimens were still fresh, proving its capability for rapidly absorbing and converting information on a global variety of species. At the end of each description, Spöring has annotated "Mscr*," connecting this description with two main sets of manuscripts. The first are Solander's field or "Plantæ" notebooks and the manuscript slips Spöring used to copy the names and descriptions. The second relates to the physical position of the specimen in the books of paper recorded in Banks's manuscript "Catalogue of Plants." Banks's list connects the annotated entries in *Species Plantarum* to specimens in the drying books of waste paper (figure 2.15).[118] For example, in Banks's entry for *Convolvulus alatus*, a species linked to page 221 in *Species Plantarum*, shows that the labeled specimen was originally stored in book eight. The enforcement of a link between *Species Plantarum*, the manuscript slips, and the physical objects provides a fixed classification for new species, moving away from the vocabularies that had dominated the preceding manuscripts in this process. This method of cataloguing still functions for the one surviving bundle, or "book" of Macronesian flora compiled from uncut sheets of *Notes upon the Twelve Books of Paradise Lost.*

The descriptions in Solander's "Plantæ" notebooks, his manuscript slip catalogue, and Spöring's "Primitæ Floræ" manuscript and annotations in *Species Plantarum* are all followed by the phrase "Fig. Pict."[119] This indicates that Parkinson produced an illustration of the living plant to supplement the dried specimen and Solander's descriptions. Illustrations of the living plant reduced inaccuracies caused by basing descriptions and images on dried specimens. This resulted from the difficulties naturalists encountered when preserving fragile flowers in books of drying paper (plate 2). Parkinson's aim was to preserve a generalist account of these features, constructing images from multiple specimens and observations of the living plant to create a composite image that aligned with the descriptions.[120]

Books, Breadfruit, and the Visualization of Species

As we have seen, Joseph Banks, Daniel Solander, and their team of field assistants paid particular attention to the properties of species they deemed to have economic potential in their range of notebooks, paper slips, interleaved books, and manuscript copies. Nowhere is this better illustrated than in the case of breadfruit, a species Solander named *Sitodium altile* (*Artocarpus altilis*, Forst.). Banks often referred to breadfruit in his journal, observing the Tahitians use of this plant as a food source and its ability to produce fruit throughout the year.[121] These observations inspired Banks's later, and somewhat catastrophic, enterprises involving the movement of breadfruit from the Society Islands to the West Indies to provide a sustainable food supply for enslaved populations.[122] Although breadfruit was not "new" to European natural history, having been described and depicted by earlier naturalists such as Georg Eberhard Rumphius (1627–1702) in *Herbarium Amboinense* (1750) and in travel narratives by explorers such as William Dampier and George Anson, Carl Linnaeus did not describe or name this plant in *Species Plantarum*.[123] A result of their belief in the potential benefits this species could bring as a food source, and the fact that it was considered "new" to the Linnaean system, Banks and Solander made particular efforts to record it in different ways (figure 2.16).[124]

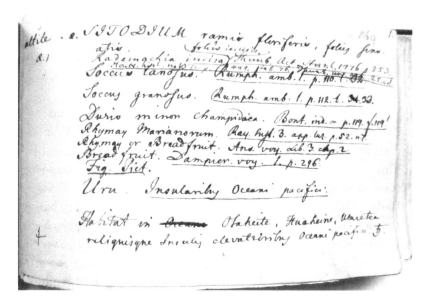

FIGURE 2.16A. Daniel Solander's slip describing *Sitodium altile*, in which he lists the previous published names. Courtesy of the Trustees of the Natural History Museum, London.

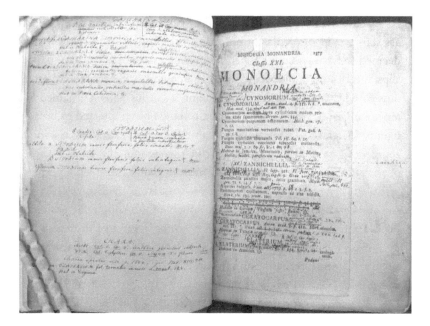

FIGURE 2.16B. The interleaved page in Joseph Banks's and Daniel Solander's copy of *Species Plantarum* showing Herman Spöring's additional annotation for *Sitodium altile* (breadfruit). Courtesy of the Trustees of the Natural History Museum, London.

FIGURE 2.17. Sydney Parkinson's two preliminary pencil sketches of the bread-fruit flowers (left) and fruit (right). These were used to produce the composite illustration in plate 3. Courtesy of the Trustees of the Natural History Museum, London.

Banks first mentioned breadfruit in his journal on April 18, 1769, and consistently counts it among "useful vegetables." This was a result of breadfruit being one of the main food sources in Tahiti; on April 20, 1769, Banks noted that "ever since we have been here [Tahiti] we have had more breadfruit every day than both the people and the hogs can eat."[125] Banks derived the name *breadfruit* from previous accounts. Dampier described "the Bread-fruit," which "grows on a large tree as big and high as our largest Apple-trees."[126] Banks had a copy of Dampier's *A New Voyage Round the World* (1697) in the *Endeavour* library that was cited by Solander in the list of synonyms on his manuscript slip alongside the works of Anson, Ray, and Rumphius to contextualize his new Linnaean name. In Banks's copy of Rumphius's work, Solander annotated the relevant descriptions and figures with

the binomial *Sitodium altile*, integrating these earlier plates and descriptions with his manuscript slip catalogue and wider information-
management system (figure 2.16).[127] These references have been placed
above the Polynesian term *Uru*, which Solander recorded in his notebook and manuscript slip, showing how he derived the new binomial
name from a mixture of earlier European publications, his personal
observation of the living plant, and information obtained from Indigenous peoples.[128]

In the interleaved copy of *Species Plantarum*, Solander's manuscript slip, and "Plantæ Otaheitenses," the annotation "Fig. Pict"
accompanies each description of the breadfruit. This indicates that
Sydney Parkinson was instructed to illustrate this species. During
their extended stay in Tahiti, Parkinson produced two preparatory
sketches of the living breadfruit tree that show the fruit at various
stages of development.[129] The main reasons for producing these illustrations were outlined by Parkinson in his journal, who commented
that "during our stay here [Tahiti], Mr. Banks and Dr. Solander were
very assiduous in collecting whatever they thought might contribute to
the advancement of Natural History; and, by their directions, I made
drawings of a great many curious trees, and other plants, fish, birds
and of such natural bodies as could not conveniently preserved entire,
to be brought home."[130] Parkinson's first image of the breadfruit displays the flowers and the second depicts the fruit in a mature state (figure 2.17). These images were produced after a detailed examination
of the tree itself and reflect Carl Linnaeus's guidance from *Philosophia
Botanica*, where it is suggested that "pictures should be drawn in the
natural size and position" and "in the case of the largest plants, when
the [actual] size cannot be depicted, it is best to offer a small branch."
Solander followed these rules to the letter, instructing Parkinson to
produce a range of field sketches. These all emphasize parts of the
breadfruit used to devise a Linnaean description, following Linnaeus's
guidance on the production of images: "The best pictures must show
all the parts of the plants, even the smallest part of the fruit body."[131]
This is reflected in Parkinson's illustration of the breadfruit, in which
he made a special effort to depict both the male and female flowers in
addition to the shape of the parts of fructification, essential features
for ascribing a Linnaean class, order, and genus.[132]

Parkinson was constructing his illustrations at the same time as
Solander was devising textual descriptions of this species; both were

observing the living plant and sharing information to ensure both the image and text supported one another and conformed to the rules Linnaeus laid out in *Philosophia Botanica*. The amalgamation of the two field sketches went toward producing a "typical" image of the breadfruit, providing a representation that could be used as a Linnaean standard for this species through depicting the parts of fructification at several main stages of development. Parkinson initially sketched this image in pencil before listing the colors at the foot of the sketch (plate 3a). He then produced the final color illustration (plate 3b). This was reduced in size when compared to Parkinson's preliminary field sketches showing that Banks and Solander prioritized the depiction of features essential for the classification of the species and more general physical characters over the precise dimensions of the living plant.[133]

The production of these images shows how Banks's hierarchic division of labor mirrors that specified by Linnaeus in *Philosophia Botanica*, who suggested that "a draughtsman, an engraver and a botanist are equally necessary to produce a praiseworthy figure."[134] Banks's employment of Solander (the botanist) and Parkinson (the draughtsman) ensured that the illustrations contained all the relevant features for classification. The production of a composite image for breadfruit, which strived to give an impression of three dimensionality and depict the living plant at all the stages in its development, was an essential part of the process of publication Banks envisaged for this species after his return to London in 1771.[135] A result of the importance of the breadfruit and its relevance to the official remit of the voyage Banks gave permission for this image to be published in John Hawkesworth's official three-volume account of the voyage (1773).[136] To produce the only botanical image in this publication Banks employed the engraver and draughtsman John Frederick Miller (1757–1796), who produced a pen and ink wash image based on Parkinson's completed illustration. This image was transferred onto a copperplate before it was printed in Hawkesworth's work (plate 3c and 3d). This completed the process outlined by Linnaeus in *Philosophia Botanica* and made Parkinson's illustration a standardized representation for the breadfruit that remained in use for the next century.

Returning from an Expedition

On March 13, 1771, the *Endeavour* rounded the Cape of Good Hope, and on July 12 the ship arrived in the English port of Deal. Since their stopover in Batavia, many of the crew had died from disease. Out of Joseph Banks's team, Tupaia died in Batavia while Sydney Parkinson and Herman Spöring died while crossing the Indian Ocean. As a result, the system for classifying, describing, and depicting species started to unravel on the homeward journey.

However, as Banks stated in an obituary of Daniel Solander that he sent to the Swedish naturalist Johan Alströmer (1742–1786), the majority of natural history work had already been commenced by the time they reached the islands of the South Atlantic: "Before we arrived home those [specimens] of Madeira, Brazil, Tierra del Fuego, the Islands of the South Pacific, and New Zealand were already in the presses. The descriptions of the little island of Saon, and the interesting Island of St Helena were finished."[137] Banks's view was reiterated by William Sheffield (c. 1732–1795), keeper of the Ashmolean Museum between 1772 and 1795, one of the first people to visit Banks and Solander after their return. In a letter to Gilbert White, Sheffield described "what is more extraordinary still, all the new genera and species contained in this vast collection are accurately described, and the descriptions fairly transcribed and fit to be put to the press."[138] Similarly, Thomas Martyn, professor of botany at the University of Cambridge, visited Banks and Solander in 1772 and described turning "over 3000 specimens of plants, 1000 of them new species; and coloured drawings of 700. all elegantly and accurately done upon the spot as were also very full descriptions."[139] The emphasis on the sophisticated organization and classification of the collection were a product of the processes employed over the course of this voyage, ranging from the collection of and illustration of specimens through to their detailed description and classification across a series of books and manuscripts. These integrated records created the basis for a book manuscript before the *Endeavour* reached Batavia—let alone the English coast.

The social hierarchy and standards of knowledge production Banks and Solander established over the course of James Cook's voyage remained in place for several years after they returned to London. The maintenance of a hierarchical team was essential for the continued organization of the collection and facilitating access to visitors attracted by the collection's fame inspired by events such as Cook's, Banks's, and Solander's audience with King George III and Queen

Charlotte in August 1771. Banks's home at New Burlington Street soon became a prominent meeting point for naturalists. The interconnectedness of the collections from the Pacific examined by Mai, Sheffield, Martyn, and naturalists such as Thomas Pennant, whose use of this collection was explored in chapter 1, presented numerous new species alongside sufficient information for those who visited to construct and publish descriptions. Naturalists' ability to undertake supervised examinations of this collection shortly after Banks's and Solander's return elevated their reputation within philosophical circles. In comparison to the materials Banks acquired from Newfoundland and Labrador in 1766, a region already known to Europeans for over a century, the novel nature of Banks's Pacific collections made them a central reference point for naturalists visiting London. This was a main reason for Banks's move to 32 Soho Square in 1778—a property with enough space to house the collection and facilitate access for Banks's trusted correspondents. Among these was Linnaeus the Younger, who used many of Solander's descriptions in his *Supplementum Plantarum* (1781). This included the genus *Banksia*, which Linnaeus named in Banks's honour.

A defined social hierarchy was central for ensuring the swift collection and recording of specimens and species over the course of the *Endeavour* voyage. Joseph Banks had learned the importance of this from the time of his father's death in 1761 when he worked alongside his mother and uncle to manage the country estates he inherited. As such, the hierarchies Banks was used to shaped the practice of natural history throughout this voyage. Figures such as Peter Briscoe and James Roberts, the servants from Banks's Lincolnshire estate at Revesby Abbey, served as a constant reminder of Banks's social position and returned to the estate in 1771. However, rather than remaining mere servants, their participation in the voyage ensured their promotion to managerial positions. By the time of his death in 1826, Roberts, who had been responsible for carrying Banks's natural history collecting equipment in the Pacific and Iceland (1772), was elevated to Banks's steward at his Lincolnshire estate. Defined hierarchical roles were essential for facilitating the swift collection, recording, and classification of species to bring a tentative structure to the flora and fauna of the Pacific and proved essential for bringing these diverse materials into the world of print.

Chapter 3

From Specimen to Print

Those walking down late eighteenth-century Fleet Street could not have failed to notice at number 63, on a bustling corner under the sign of Horace's Head, the shop of the natural history bookseller and publisher Benjamin White. Similar to the hundred or so other booksellers in the vicinity by 1790, White's shop was famed as a meeting point for "respectable" society and, from the 1750s, developed into a business that produced and sold publications "in the line of natural history, and other expensive books."[1] The "genteel" network White catered for demanded works that ranged from cheaper octavo and duodecimo volumes to expensive quartos and folios, books often privately printed for and funded by an elite group of subscribers. Copperplate images were displayed in White's shop window to entice genteel naturalists—who ranged from country vicars to gentry and wealthy aristocrats, into the dimly lit interior—a very different clientele to those who visited the vendors of Grub Street and John Newbery's shop famed for selling children's books.[2] The natural history books White sold were based on collections and paper technologies compiled during expeditions and proved valuable sources for traveling naturalists. Books were designed to be embedded within specimen collections to facilitate the organization, standardization, and communication of information. Conspicuous absences from White's stock included books produced by those who practiced the "genteel hobby" of owning their own press, publications at the pinnacle of genteel networks and never designed to be shelved by a Fleet Street bookseller.[3]

But how are we to understand what was in White's shop—and what was not in it? White's reputation brought him several genteel authors, whose views shaped the physical construction, financial management, and range of natural history books the firm published. This chapter examines the contrasting strategies used by Joseph Banks, Thomas Pennant, and others such as Gilbert White to produce natural history publications. The differences between these individuals' interests in different branches of natural history and the various audiences and print runs they had in mind shaped their approaches to constructing these works. Natural histories were based on physical collections and carried all the social connotations associated with these. This division was increased by Banks's rapid social elevation after his return from James Cook's voyage in 1771, the expansion of his country estates, and subsequent income by the 1780s. In comparison, Pennant's wealth and social advancement did not increase at the same rate. This important temporal shift, combined with Banks's bad experience of commercial publishing in the early 1770s, caused Banks's and Pennant's views on the production and dissemination of natural history publications to drift apart.

The publishing enterprises of Banks, Pennant, and their contemporaries coincided with a period of great expansion in the book trade and publishing industry. This originated in what Adrian Johns describes as a "bookshops boom" in the 1740s following a lapse in copyright acts and reduction of state censorship. Decreased regulation resulted in a proliferation of accessible and varied publications, such as novels, plays, almanacs, Bibles, encyclopedias, and dictionaries, ranging from expensive imperial folios to small volumes sold for a few pennies. The development of the book trade and production of prints reflects a growth in commercial markets John Brewer relates to the overall expansion of "modern commerce and refinement."[4] Richard Sher has since connected the increased variety of books to "a burgeoning culture of material consumption and commercialisation."[5] In comparison to publications designed for public consumption, on which the majority of scholarship has concentrated, natural histories proved to be far more complicated and expensive to produce and circulate. Many late eighteenth-century natural histories encapsulate the emergence of what William St Clair has described as "on commission" publishing in the context of the early nineteenth century, a process by which authors paid a publisher for the production of their works and allowing a publisher to take a 10 percent commission on sales.[6] Many natural history books were financially unviable to publishers such as Benjamin White, primarily because of the need for numerous expensive images and the levels of control authors and compilers maintained throughout the production process.

Natural history books remained complex productions throughout this period and involved many naturalists and the teams they managed using collections of objects, images, and associated manuscripts to compile information set in letterpress and reproduced as copperplate images. The complexity of these books creates a very different set of working practices when compared to the more commonly studied works of poetry, philosophy, and politics of the late eighteenth century and even some text-based works of natural history and natural philosophy. The analysis shows how the practices of producing natural history books differ from standard approaches to this period that aim to show wealthy patrons replaced by publishers in the book production process.[7] In comparison, the production of natural history books was regarded as a genteel process and relied on independently wealthy authors who could fund and organize the production of their own work.

Authors and compilers maintained stringent control over the content and finances of natural history books, integrating these with the practice of compiling a physical collection, while publishers remained peripheral in the production process. Publishers were only called upon when it came to distribute natural history books to polite society, gaining specific publishing houses genteel reputations. This process allowed landowning elites to define themselves as patrons and practitioners of natural history who compiled and distributed the products of wealth and leisure, distancing themselves and the process of knowledge production from the literary hacks and commercial publishers of Grub Street.[8] Authors' authority over the finances, materials, and practical production processes ensured many natural history books do not conform to general commercial publishing models set out since the 1980s that elevate the role of publishers and booksellers in the production process, ensuring the profitability of the books they produced and the author's financial success. Although there has been some more recent criticism of these claims by Geoffrey Turnovsky and others, these accounts tend to concentrate on the interests of eighteenth-century French authors in maximizing the dissemination of their works.[9] This and the next chapter reframe more recent criticism to explore the processes of producing natural history books in Great Britain, examining the profound impact interests of individual authors had on these practices, reflecting the combined concerns of natural historians and a landowning elite.

The views of the wealthy individuals who produced and financed natural history books informed the physical construction and content of these publications. Naturalists maintained rigid authority over their books and shaped these through the artisans they employed; the different systems of classification used to describe, depict, and order species; their approaches to financing books; and the availability of resources. Rather than producing books for personal financial gain, naturalists published to publicize a collection, elaborate its theological meaning, and display social status.[10] Illustrations remained at the pinnacle of these processes of translating a collection into print. The financial costs associated with the illustration of specimens and transfer of these onto copperplates depended on specific specimens and were regarded as more important than letterpress for emphasizing the physical features required to place a species within a system of classification. In the preface to William Aiton's *Delineations of Exotick Plants* (1796), Banks stated that "each figure is intended to answer itself every question a Botanist can wish to ask." Although Martin Rudwick has suggested images of objects "stood in for the real thing," most attempts to show how they were combined with a specific system of classification and physical collection have examined the period before the major taxonomic reforms of the 1730s.[11] Expensive images, which often formed the majority of the page count in these books, were designed to present a general representation for the identification of a species, although these were aligned with the ordering and physical appearance of objects in a collection, making private collections major reference points for natural historians.

Thomas Pennant and *British Zoology*

In December 1777 the author of *Westminster Magazine* described *British Zoology* as a "lasting monument to Mr Pennant's capacity, both as a Naturalist and as a polite writer."[12] By the late 1770s, Thomas Pennant had an established reputation as a prolific and informative natural history and travel writer. Between 1766 and 1780 he published a total of fourteen volumes, including seven multivolume works on natural history. These books contained a mixture of text and copperplate illustrations designed as visual accompaniments to the letterpress. Many images in Pennant's works were produced by Moses

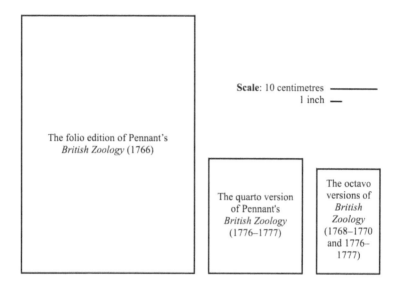

Scale: 10 centimetres ———
1 inch —

The folio edition of Pennant's
British Zoology (1766)

The quarto version
of Pennant's
British Zoology
(1776–1777)

The octavo
versions of
*British
Zoology*
(1768–1770
and 1776–
1777)

FIGURE 3.1. Size comparisons between the three formats of Thomas Pennant's *British Zoology* published in 1766 (left), the 1776–1777 quarto printing (center), and the 1768–1770 and 1766–1777 octavo (right).

Griffith and Peter Paillou and engraved by Peter Mazell. *British Zoology* was perhaps the most important and extensive of these publications, reaching four main editions produced in a range of different formats between 1766 and 1812. The first edition was published in 1766 and printed by J. and J. Marsh as an imperial folio; the second edition was printed in octavo format in London and Chester and published by Benjamin White, whose firm worked with Pennant and his son to distribute all subsequent editions; the third edition was primarily printed in octavo (although a limited number of copies were printed in quarto); and the last edition of 1812 was printed in octavo.[13] The different editions of *British Zoology* represent the full range of marketable natural history books available in the period (figure 3.1).

To subsidize expensive natural history publications, Pennant used the profits from his more widely distributed works, including his *Tour in Scotland and Voyage to the Hebrides, 1772* (1774). In comparison to his travel books, the copperplate images in *British Zoology* were often based on specific specimens from Pennant's collection, a feature apparent once these objects and images are compared. The number of images increased with every edition of *British Zoology*, reflecting additions to Pennant's collection after successive tours across Britain, revisions to previous observations, and the investment of the small profits from earlier editions into the production of more images. One of the most extensive editions of *British Zoology* was compiled from four volumes and published between 1776 and 1777 by Benjamin White. The printing of the letterpress was outsourced to William Eyres of Warrington (1734–1809), as it was easier for Pennant to supervise production with Warrington being far closer to Downing than London. One thousand five hundred copies of this edition were produced, the majority printed in octavo, although five hundred copies with wide margins were printed in quarto.

The main subjects covered in the initial volumes of *British Zoology* (1776–1777) were quadrupeds and land birds in the first volume and water birds in the second volume, both arranged according to the system devised by John Ray. Ray based his classification of birds and quadrupeds on their habitation of land or water before grouping them according to their general physical resemblances. Quadrupeds were arranged in relation to the differences between paws, cloven hoofs, and the correlation between these physical characters and the habitation of specific areas. Pennant's adherence to Ray's system for birds is apparent from their general ordering, starting with passerine birds with sharp beaks and claws designed for clinging onto the mountain ledges, tracing these through to ducks and geese with webbed feet and bills used for filter feeding in wetlands. The third volume of *British Zoology*, on reptiles and fish, and the fourth volume—on Crustacea, Mollusca, and Testacea—were arranged according to the system of classification and nomenclature laid out by Carl Linnaeus in the 1766 edition of *Systema Naturae*.

The use of different systems across the volumes of *British Zoology* emerged from the alternate approaches used for observing animals in the field and the arrangement of Pennant's collection at Downing Hall. For example, nearly all fish, crustaceans, and molluscs inhabit rivers, ponds, or the sea, and it was often difficult to assess their social attributes. Pennant used a system that placed a singular reliance on physical characters, describing how "I have borrowed from him [Linnaeus] the *Latin* trivial names" relying on the 1766 edition of *Systema Naturae* to append binomials to the descriptions of species.[14] In comparison, the system devised by Ray outlined in the introductory section of this book placed a far greater reliance on different environmental conditions preferred by different species, correlating these

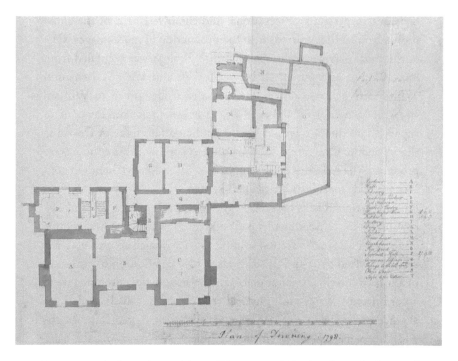

FIGURE 3.2. A floor plan and key of Downing Hall, Flintshire, the home of Thomas Pennant (1726–1798), produced around the time of Pennant's death in 1798. The library, hall and parlour at the front of the house marked A, B and C housed the books and zoological collections while the ethnographic objects, including those collected on Cook's voyages, were kept in the "smoking parlour" next to the great staircase marked D. By permission of North East Wales Archives.

alongside different physical attributes and their methods of surviving in specific conditions. This was emphasized in Pennant's collection. Prior to its donation to the British Museum (Natural History) in 1913, Pennant's taxidermy bird collection was displayed on top of the bookcases in Downing Hall. One of the last observers of the collection before the contents of Downing Hall were cleared and the property sold remarked that "the birds are in glass cases, which since the death of Thomas Pennant in 1798 have ornamented the tops of the bookcases in the Library and the print-room."[15] Ornithological specimens were arranged in dioramas to mimic the original natural surroundings where Pennant observed the species and had them depicted in his publications.[16] In comparison, the other branches of zoology, most notably shells and crustacea, were kept inside wooden specimen cabinets since they were arranged according to the Linnaean system with its concentration on physical characters and limited reliance on external environments. The aforementioned arrangement of the birds and books emphasizes the close connections between these objects, with Pennant using the connected collection of manuscripts discussed in chapter 1, his own recollections, correspondence, and information in printed books to construct descriptions for his publications. When Pennant's library was sold in 1913 the *London Morning Post* noted that "the books formed what may be called his 'working library,' and among them are choice examples of the works of nearly all the old writers on natural history. Most of the volumes are freely annotated by Pennant, who appears to have made himself master of their contents in the fullest sense."[17] The importance of this collection is outlined in a plan of Downing Hall taken at the time of Pennant's death in 1798 that depicts the spaces devoted to the library, parlor, hall, and print room that held Pennant's collection as some of the largest and most significant in the property (figure 3.2).

An image from *British Zoology* based on a specimen from Pennant's collection is that of the little woodpecker. This species was first depicted in the folio edition published in 1766 and a condensed illustration was revised for the later editions. Pennant noted the scarcity of the little woodpecker in his description, "it has all the characters and actions of the greater kind, but is not so often met with."[18] Designed to depict a scarce species, this and many of the images published in the 1776 edition were drawn and occasionally engraved by Moses Griffith, who observed birds alongside Pennant in the field while integrating aspects from Pennant's taxidermy collection.[19] Griffith always attempted to present a standardized depiction of a species, although this often shared features with Pennant's specimen, as becomes apparent when the image and specimen of the little woodpecker are compared (plate 4).

Griffith made a significant effort to emphasize specific features used to define the placement of the little woodpecker in Ray's classificatory system. These relate to Pennant's description of a "strait strong angular bill" and the "climbing feet," that have been attached to an almost vertical branch by the taxidermist, a feature replicated in the image.[20] The close relationship between Pennant's specimen and Griffith's illustration aligns the publication with a physical collection, bringing order to a group of objects arranged to reflect their habitation of the north Welsh woodlands. Associated images and descriptions increased the value of Pennant's specimen collection for naturalists, connecting these different entities to create a reference point for the classification of this species.[21]

Griffith and Peter Mazell added grid lines to these drawings to assist in the process of transferring the image onto a copperplate for publication in the 1776–1777 edition of *British Zoology*. Griffith charged Pennant £1 for each illustration and the final engraved copperplate cost £1/1, giving an initial outlay of £2/1 per image. To put this in perspective, the illustrations and plates for *British Zoology* provided Griffith with sufficient income to support himself, as a skilled artisan, and his family for several years. By 1780, £1 could afford the equivalent of six days skilled labor.[22] An essential feature of Ray's system for the classification of birds was the division between land and water birds. This is emphasized in the image and description of the middle and little woodpeckers; Pennant was careful to show their habitation of woodland, describing the specific localities where he and his correspondents had observed and caught specimens in Britain.[23] In comparison, the images in the second volume of *British Zoology* all depict water birds. These are associated with an idealized water source, such as the heron depicted on the edge of a pond, or the red-breasted goosander shown swimming on water. This arrangement reflects the specimens from Pennant's collection. The surviving specimen of the red-breasted goosander is in sitting position and there are traces of blue paint on its sides, suggesting it was presented in a diorama to representative of Griffith's illustration for *British Zoology*.[24]

The relationship between the physical arrangement of Pennant's taxidermy collection and *British Zoology*, exemplified by the order of the objects, reflects aspects of Ray's taxonomic scheme. These align the printed book with the physical specimens to promote Pennant's collection and Ray's system as the best means for arranging this branch of natural history. Ray's use of preferred living conditions to classify species was translated into the plates used for *British Zoology* and allowed Pennant to have the taxidermy arranged in naturalistic poses. This presents a sharp contrast to ordering systems outlined by Linnaeus, causing some naturalists to reject the approach to depicting species in an idealized setting to emphasize the broader divisions of Ray's system. For example, when commenting on Peter Brown's *New Illustrations of Zoology* (1776), a book to which Pennant lent his patronage, Joseph Banks described the ornithological images as "horrid, bad, and drawing by a fellow /as he terms it:/ that knows nothing of natural history."[25] In comparison to more artificial approaches, Ray's system allowed Pennant to show that although nature could be ordered to some extent, aspects of the more animated branches, such as ornithology, always escaped the ordering activity of humankind. This reflects the views of many critics of the Linnaean system, perhaps the most prominent being Georges-Louis Leclerc, Comte de Buffon, who Pennant met when traveling through France in 1765.[26]

One of the most significant additions to the 1776–1777 edition of *British Zoology* was the new volume on molluscs and crustaceans. Published in 1777, this volume contains ninety-three plates depicting specimens from Pennant's collection. Pennant attributed the late publication to "the shameful delay of the Warrington Printer." The plates are grouped together at the end of the volume and are ordered according to the Linnaean system, an arrangement that mirrors the composition of Pennant's physical shell collection. Pennant's shells were kept in small card or tin boxes laid out in the wide shallow drawers of his two tall wooden specimen cabinets.[27] These were then ordered according to the series of specific morphological characters outlined by Linnaeus in the 1766 edition of *Systema Naturae*. Linnaean names aligned Pennant's collection and publication with an orderly system that could be distributed across polite society in the form of a book. The Linnaean relationship between the physical specimens, textual counterpart and images were made clear in Pennant's preface: "I HAVE paid implicit respect to the Swedish NATURALIST, in my classing of VERMES and SHELLS. I have on another occasion, given my sentiments to that wonderful man, (after RAY) the greatest illuminator of the study of nature."[28] Most of the plates published in the final volume of *British Zoology* were engraved by Peter Mazell, who charged Pennant £1/6s for the copper and engraving. The combined cost of the ninety-six copperplates was £120/1s and included the plates used for the vignette title page and appendix.[29] The *Monthly Review* described these as being "elegantly engraved" forming "a capital ornament to a branch of natural history which contains by far the least interesting objects, when compared with the fore going classes." The separation of most species onto individual copperplates shows a move toward the presentation of a practical guide that enforces a Linnaean classification onto the shells of Britain.[30]

Ninety-three of the plates in the fourth volume of *British Zoology* form a very different physical block of information when compared to the letterpress pages. These plates, printed on more durable, thicker paper than the letterpress, have been grouped together at the back of the volume and are perhaps the most striking feature of this publication. This arrangement was deliberate, as seen by Pennant's note at the end of the index: "The Binders are requested to place all the Plates at the End," representing a significant structural change when compared to the first three volumes where the images have been interleaved with the related letterpress descriptions. Reasons for this choice are twofold. Initially, William Eyres's delay in producing the letterpress resulted in the text and images of the book being produced at very different times and in different places. Secondly, the separation of the series of images to the text promote a very different kind of reading associated with Linnaean natural history when compared to the earlier volumes, allowing one to visualize the changing physical characters of species as the Linnaean framework progressed. This volume was produced in 500 quarto copies, printed on large paper at a cost £1/4 each, and 1,000 octavo copies that cost Pennant fourteen shillings per volume, a significant reduction in price when compared to the 1766 edition. Copies were then presented as gifts or sold and distributed by White who marketed the octavo copies at £2/8 and the quarto copies at £4/4, a standard markup for a late eighteenth-century bookseller.[31]

White's main role was to distribute these books; matters such as sourcing materials, artisans, and the financial management of the production process were left to Pennant, who kept meticulous accounts for his publications throughout the 1770s (figure 3.3). The total expenditure on all the copies of the final volume of the 1776–1777 edition of *British Zoology* was £289/16, a cost that included Pennant's "expenses of collecting" and the money he gave Benjamin White for distributing copies to booksellers.[32] The ninety-six copperplates were the largest financial outlay for this volume, the next being the cost of the drawings, which totaled £40, followed by the £37/13 Pennant paid to the paper supplier John Curtis for a mixture of thin paper for printing letterpress and more durable high-quality paper for printing the copperplates. These were printed by the specialist intaglio printer William Hixon (1735–1802), who was simultaneously appointed by the Royal Society of London and the Society of Antiquaries of

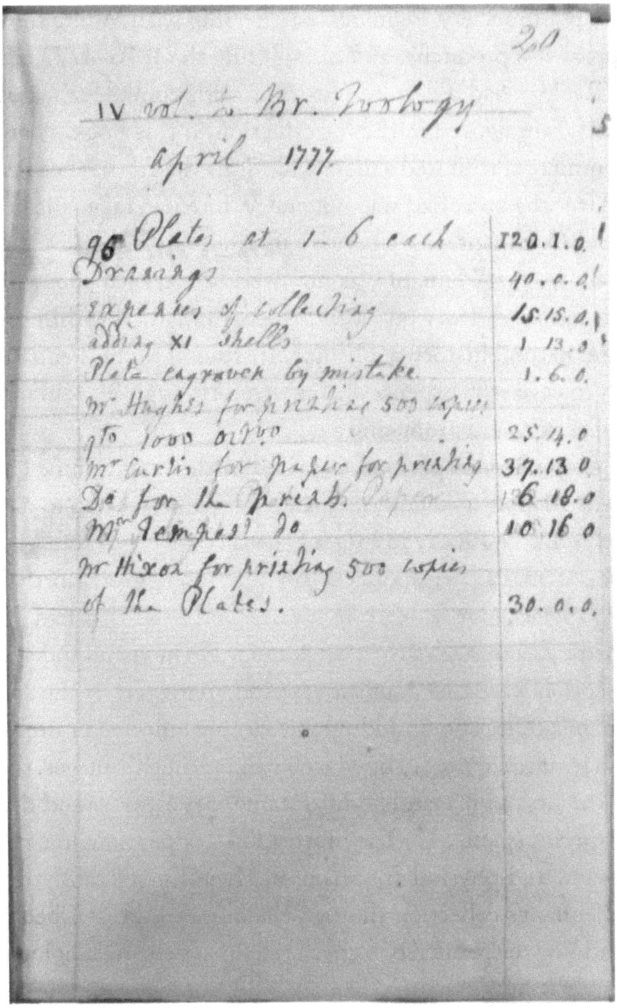

20

IV vol. Br. Zoology
April 1777

96 Plates at 1..6 each	120.1.0
Drawings	40.0.0
expences of collecting	15.15.0
adding XI shells	1.13.0
Plate engraved by mistake	1.6.0
Mr Hughes for printing 500 copies 4to 1000 8vo	25.4.0
Mr Curtis for paper for printing	37.13.0
Do for the prints	16.18.0
Mr Tempest Do	10.16.0
Mr Hixon for printing 500 copies of the Plates.	30.0.0

FIGURE 3.3. Thomas Pennant's accounts for the fourth volume of *British Zoology* (1777). Reproduced by kind permission of the Warwickshire County Records Office.

London to print the copperplates for their journals. Pennant's publication of several illustrated articles in the *Philosophical Transactions* from the 1750s to the 1770s, on matters such as corals (1756), "Pinguins" (1768), and "tortoises" (1771), put him into contact with Hixon as authors had to finance and arrange the production of engravings to accompany their articles.[33]

After production, most copies of *British Zoology* (1776–1777) were given to Benjamin White, whose function was to distribute Pennant's books to polite society. Pennant's partnership with White shows how his approach to producing and distributing the 1776–1777 edition of *British Zoology* had a certain crossover with commerce, as outlined in the diagram figured in the introduction to this book. A crossover with commerce facilitated a more wide-scale distribution of Pennant's work. After the book had sold out and White had taken his booksellers' margin, Pennant was left with a total profit of £40/4.[34] This small return shows that Pennant was not overly concerned about profiting from his natural history publications remaining well within the scope of genteel natural history publishing. Instead of profit, Pennant needed sales to cover the expenses he incurred through his tours, compiling a collection, and producing a book.

An example of a specimen depicted in the final volume of *British Zoology* is *Mytilus umbilicatus*, a name Pennant ascribed to a shell he received from the Reverend Hugh Davies (1739–1821) of Beaumaris, Anglesey, a main correspondent on the zoology of North Wales. Pennant suggested it was "a rare species, and new" that had not been published by Linnaeus in *Systema Naturae*.[35] The name and systematic description were based on the physical characters of the specimen in Pennant's collection, including the "deeply inflected or umbilicated hinge," a feature supported by Mazell's and Griffith's images, thus producing the first Linnaean binomial name, description, and depiction of this species (plate 5).[36] The illustration, copperplate image, name, description, and physical specimen of *Mytilus umbilicatus* are linked across Pennant's collection through the numbers he ascribed to each object, adding the printed book to the chain of paper technologies used to govern his physical natural history collection. For example, Pennant inked the number 76 onto his specimen of *Mytilus umbilicatus*, a number associated with this species across his manuscript, printed names, images, and descriptions (plate 5). Unlike images of shells produced for earlier natural history publications, such as those by Martin Lister, who added a shaded plain to situate the time and place in which the specimen was depicted,[37] external shadows have been omitted from the images published in the fourth volume of *British Zoology*. The only shading that appears is in the contours of the specimens themselves to give a sense of three-dimensionality. This places a singular emphasis on the object and follows guidelines laid down by Linnaeus, who stipulated that there should be no distraction from the main subjects of the description if it accompanied by supplementary images.

The images of shells are very different from those in the earlier volumes of *British Zoology* that were based on the setting and physical arrangement of the taxidermy in Pennant's collection. This was due to the vital differences between Ray's and Linnaeus' classificatory systems. Ray's system accounted for the physical attributes of birds and quadrupeds alongside their natural habitat, social customs, and the sounds they made. In comparison, the Linnaean system relied on correlations between physical characters. Pennant seems to have chosen this approach given that shells remained a less "animated" zoological class when compared to birds and quadrupeds. However, Pennant and Griffith seem to have made little effort to produce any kind of "typical" or "average" image of *Mytilus umbilicatus* to define its physical characters.[38] Rather, the image presented in the copperplate is a two-dimensional reproduction of Pennant's specimen, reproducing specific features, such as the staining on the inside of the shell, making Pennant's specimen a kind of "classification type" and elevating the importance and usefulness of Pennant's collection for natural historians. The differences in the arrangement of the drawings of specimens belonging to the different zoological classes published in *British Zoology* reflects Pennant's methods of keeping these specimens in cabinets and glass cases. As a result, each volume encapsulated aspects of the physical arrangement and specific systems of classification Pennant used to arrange his private museum at Downing Hall, making this collection a definitive model for the zoology of Britain.[39]

Thomas Pennant's Luxury Publications

The emergence of commercial publishing markets during the second half of the eighteenth century not only increased the production of smaller quarto and octavo works, but fostered demand for expensive luxury publications. This is represented by the 1766 edition of *British Zoology*, a large imperial folio volume and Thomas Pennant's first significant publishing venture. Released in parts from 1762, the final bound copies were produced in 1766. Pennant remarked on how he was "ill advised to publish [the 1766 edition] on large paper; had it originally been in 4 to. the school would have considerably benefitted from it."[40] This is a reference to the Welsh charity school at Clerkenwell Green, to which Pennant had promised to donate any profits from this work. These were derived from the fees of 114 subscribers, among whom were naturalists, aristocrats, and gentry from across

Europe and North America. These individuals viewed financial donations to an educational establishment alongside patronizing a work of natural history as fulfilling a major gentlemanly ideal; indeed, Pennant remained anonymous throughout the production process acting through his agent, Richard Morris, so as not to erode his reputation through association with a commercial venture. Pennant later donated the dividends from more successful publications to charitable causes.[41]

The vast costs associated with printing 132 folio copperplate images created consistent problems for Pennant and Morris. The enterprise bankrupted Morris and the whole process proved a long, arduous financial catastrophe mirroring the production of other large copperplate natural history books funded by subscription such as Robert Morison's *Plantarum Historiae Universalis Oxoniensis* (1680). Each copy of *British Zoology* cost £3/10 to produce (before binding), and by 1768 the remaining copies were priced at "eleven guineas, half bound," although they were still "sold for the benefit of the *Welch* charity-school." Additional plates added a premium to the process charged by booksellers. For example, Benjamin White marketed the new edition of Patrick Browne's *Civil and Natural History of Jamaica* (1789), a book that contained forty-nine large copperplate engravings, for £2/12/6 in 1793.[42] Pennant's 132 copperplate images each cost roughly £12 to produce. The smallest image depicts the hedgehog on a plate measuring 18.54 × 13 centimeters and the largest depicts the great speckled diver, measuring 31 × 51.3 centimeters.[43]

George Edwards (1694–1773), a leading ornithologist, maintained an extensive correspondence with Pennant in the decade before his death in 1773. Pennant sent Edwards "curious birds that you [Pennant] chuse to have done [painted]." To produce these images, Edwards drew "on 2 half sheets of imperial your 3[d] Birds and one new bird I mentioned to you in my last and I believe you may Pronounce them the last I shal ever draw."[44] A result of Edwards's retirement, he only produced five of the images published in the 1766 edition of *British Zoology*.[45] Pennant relied on the London-based artist Peter Paillou (c. 1720–1790) for the other 128 illustrations, who produced twenty-eight images when he stayed with Pennant at Bychton Hall July–September 1762. Paillou's visit cost Pennant £28/16 and gave him an opportunity to observe and take preliminary sketches of Pennant's zoological collection.[46] These were completed in Paillou's

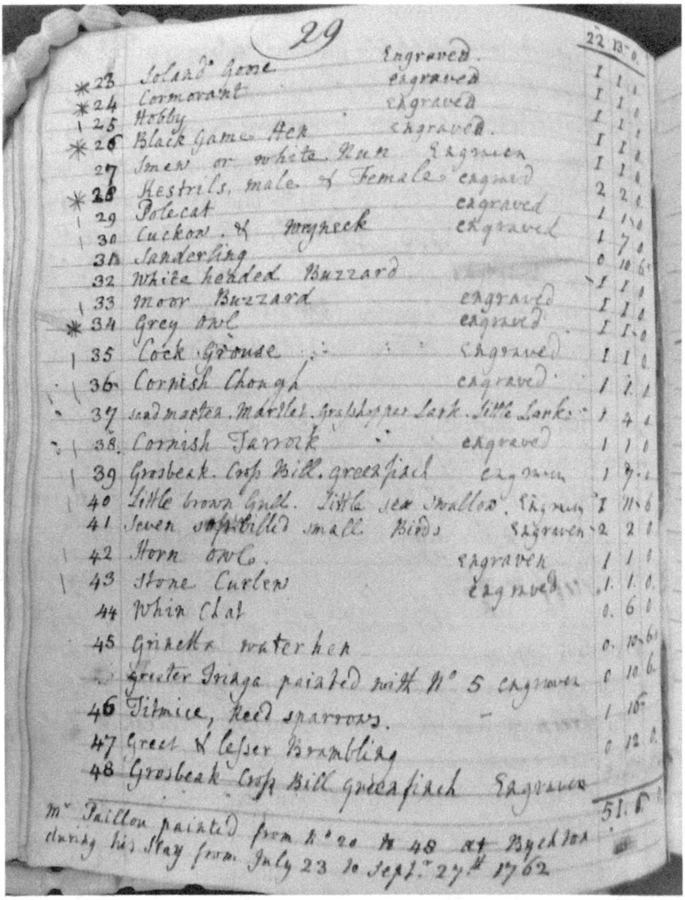
29

22 13 0

*23 Soland Goose — Engraved — 1 1 0
*24 Cormorant — engraved — 1 1 0
| 25 Hobby — engraved — 1 1 0
*26 Black Game Hen — engraved — 1 1 0
27 Smew or white Nun — Engraved — 1 1 0
*28 Kestrils, male & Female — engraved — 2 2 0
| 29 Polecat — engraved — 1 1 0
| 30 Cuckow & Wryneck — engraved — 1 7 0
31 Sanderling — 0 10 6
32 White headed Buzzard — 1 1 0
| 33 Moor Buzzard — engraved — 1 1 0
*34 Grey Owl — engraved — 1 1 0
| 35 Cock Grouse — engraved — 1 1 0
| 36 Cornish Chough — engraved — 1 1 0
37 Sand martin. Martlet. Grashopper Lark. little Larks — 1 4 0
| 38 Cornish Jarrock — engraved — 1 1 0
| 39 Grosbeak. Cross Bill. greenfinch — engraved — 1 7 0
| 40 Little brown Gull. Little sea swallow. — Engraved — 1 11 6
41 Seven soft billed small Birds — Engraved — 2 2 0
| 42 Horn Owl. — engraved — 1 1 0
| 43 Stone Curlew — engraved — 1 1 0
44 Whin Chat — 0 6 0
45 Grinetta water hen — 0 10 6
greater Ivaga painted with No 5 engraved — 0 10 6
46 Titmice, Reed sparrows. — 1 16
47 Great & lesser Brambling — 0 12 0
48 Grosbeak Cross Bill greenfinch — Engraved

51. 5

Mr Paillou painted from No 20 to 48 at Bychton during his Stay from July 23 to Sept. 27th 1762

FIGURE 3.4. Thomas Pennant's "List of Drawings" showing the amounts he paid Peter Paillou for images produced at Bychton between July 23 and September 27, 1762. By permission of Llyfrgell Genedlaethol Cymru/The National Library of Wales.

London studio and served as essential validators for the precise details described in the text of *British Zoology*.[47] As in later editions, these images aligned Pennant's book and collection alongside the system of John Ray, emphasizing a mixture of species habitats and physical characters. For example, in the case of the cuckoo, Pennant described how "both [the] male and female are faithfully given in the plates," images that depict the male and female cuckoos sitting on stylized branches to indicate their habitation of land.[48] The female cuckoo in the 1766 edition is in an almost identical pose to that presented in the 1776 edition, indicating that the same specimen was used to produce these different images. Although certain elements of these might have been copied, the difference in size and the addition of the wryneck to the plate in the 1776 edition suggests the artists relied on Pennant's specimens for both images. Each image cost between twelve shillings and £2/2 to produce before they were engraved, although most cost the standard price of £1/1 (figure 3.4).[49] The cost of the copper and Mazell's engraving added approximately £10 to each image.

Pennant intended for the expenses incurred during the six years it took to produce this publication to be offset by the 114 subscribers. However, Pennant's expenditure did not just include processes related book production. In his list of "disbursements," Pennant added bills associated with his natural history collecting. Examples include "an otter" that cost two shillings and sixpence on May 20, 1763; the "carriage of the Tin case to London" in October 1761 for one shilling and sixpence; a bill dating from June 24, 1761, for payment to a "Mr. Williams for coloring 5 Doves pd. Him before for 2 Dozen specimens"; and the expenses incurred from every letter sent in relation to this publication.[50] This list of disbursements shows that Pennant considered the activities of collecting in the field, assembling a collection, and constructing a publication to be part of the singular process of gentlemanly natural history that were impossible to separate even when it came to matters of financial expenditure. The expenses accrued through this publication cost Pennant and his agent, Richard Morris, roughly £1,600. Although it was subsidized by the subscribers, this price exceeded all potential profits made by this ambitious first publication.

The economic problems caused by *British Zoology* ensured that Pennant was cautious about his use of images in the first two volumes of his next large folio publication, a series of volumes with the collective title *Outlines of the Globe*. Based on India, the first volumes of *Outlines of the Globe* were published shortly before Pennant's death in 1798. Two subsequent volumes were produced between 1799 and 1800 and managed by Pennant's son and heir to the Downing estate, David Pennant (1763–1841). The first two volumes, titled *The View of Hindoostan*, contained only twenty-four copperplate images that depict the topography and natural history of India and was marketed at £2/12/6 by John White. This was a substantial reduction when compared to *British Zoology* (1766) as it was not so complicated to produce; rather than factoring in his own expenses for expeditions and collecting, Pennant sourced the images and descriptions from his

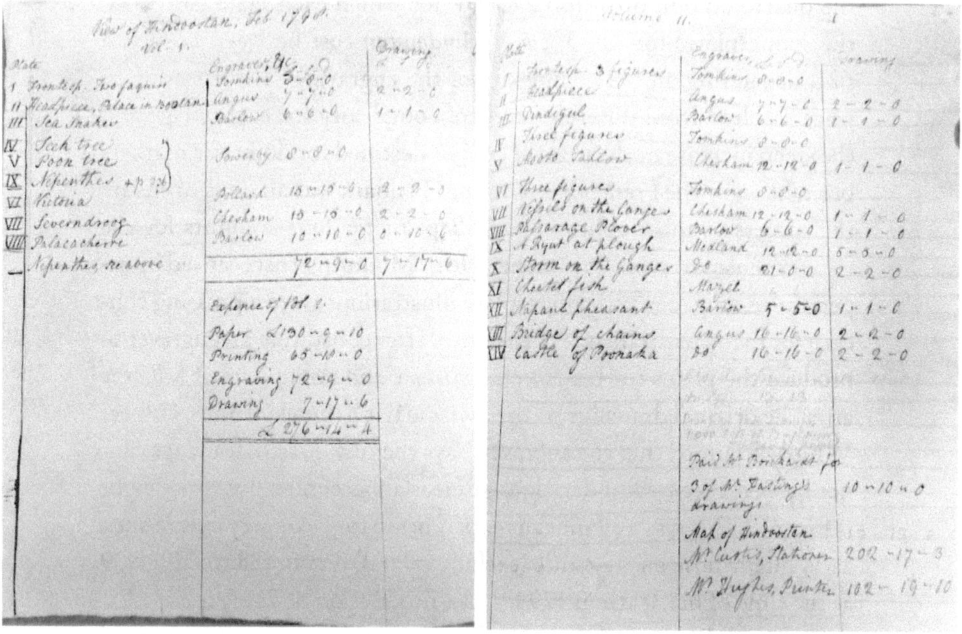

FIGURE 3.5. Thomas Pennant's accounts for *A View of Hindoostan* (1798). By permission of Llyfrgell Genedlaethol Cymru/The National Library of Wales.

correspondents and earlier publications. This was a similar approach to that used for *Indian Zoology* (1790), in which Pennant reproduced a series of images sent by the governor of Dutch Ceylon, John Gideon Loten, who compiled a large collection of natural history illustrations during his time in India. The plates for *Indian Zoology* proved so expensive that Pennant shared the cost with Joseph Banks, Johann Reinhold Forster, and Loten, and only twelve were engraved as "the undertaking appeared so arduous that the design was given over."[51]

When constructing *The View of Hindoostan*, Pennant either had an esteemed natural history or topographical artist produce each original drawing or he purchased these from dealers. Examples include the "3 of Mr Hastings's drawings" Pennant purchased from a "Mr Borckardt" for £10/10/0. These images originated in the collection of the notorious East India Company administrator Warren Hastings (1732–1818), whose impeachment trial lasted for seven years before his final acquittal in 1795. Illustrations were then engraved onto copperplates that measured between 19.05 × 24.13 centimeters and 20.32 × 25.4 centimeters. In comparison to Pennant's other works, such as the quarto edition of *British Zoology*, for which each plate cost £1/1, the copperplates for *The View of Hindoostan* cost between £6/6 and £21, depending on the complexity of the engraving. The variation in price resulted from different levels of complexity for each image and the steep increase in the price of copper destined for domestic markets brought on by the French Revolution.[52] Pennant was anxious to ensure the success of this publication and kept meticulous accounts for the total expenses to avoid any financial loss. Mounting costs caused by expenditure on paper and copperplate illustrations were a pressing concern for many naturalists. For instance, after consulting an engraver to produce the plates for his *Natural History and Antiquities of Selborne* after the original drawings by the Swiss artist Samuel Grimm, Gilbert White exclaimed that the engraver "says that my quarto drawings cannot be well executed under eight guineas a piece: now five times eight is forty!"[53] This was a significant expenditure for a country curate such as White, whose entire estate (excluding land) amounted to £29/16/9 at the time of his death in 1793.[54] The production of *The View of Hindoostan* cost Pennant £276/14/4 for the first volume and the second cost £495/18/1, a price increase that resulted from a higher page count and almost double the number of copperplates (figure 3.5).[55]

A high percentage of the total expenditure on Pennant's *View of Hindoostan* was devoted to paper. This was expensive, not least because the book was printed in folio, each sheet being folded once to form four letterpress pages. Pennant purchased printing paper for his publications from John Curtis (d. 1801), who charged £130/9/10 for the paper used to print the first volume of *The View of Hindoostan* and £202/17/3 for the second, enough money to employ a skilled artisan for three and a half years. The smooth wove paper used in this book became commonplace in the 1790s.[56] Expenses incurred on paper is evidenced by a letter Curtis sent to Luke Hansard (1752–1828), who was responsible for printing Pennant's *View of Hindoostan* and all subsequent volumes of *Outlines of the Globe*, on July 6, 1799: "At your request I write on a sheet of fine wove printing Royal which recommend for your work about to be put to Press; it is the same sort as the Hindöstan Work was Printed on. I have at present Fifty Reams by me, & if you approve This Paper I will reserve it for your use only & get as much more made at the Mills as may be required. The lowest price Per Ream will now be 40/6—a material use upon all Papers having just taken place."[57] Hansard forwarded this letter and specimen of printing paper to David Pennant to confirm the order for Pennant's *View of India Extra Gangem, China and Japan* (1800). Curtis's comment on "a material use upon all Papers" refers to the consistent rise in duties on paper of the finest quality to three pence for every pound in weight by 1803.[58] An increase in duty was part of the British government's attempts to finance the wars on the Continent, a vast source of expenditure until Napoleon Bonaparte's defeat in 1815.

The wide margins of *The View of Hindoostan* show Pennant's interests in grangerizing this work, producing a luxury publication similar to his other folio books. The genteel recipients of limited edition books with wide margins were Pennant's correspondents and fellow extra illustrators, such as Horace Walpole, Lord Bute, and Richard Bull.[59] The printed versions of *Outlines of the Globe* could then be treated by Pennant and his fellow grangerizers in a similar manner to Pennant's original manuscript volumes of this work, "on which uncommon expense has been bestowed in ornament and illuminations."[60] Gilbert White used the same papermaker as Pennant for his *Natural History and Antiquities of Selborne* (1789) and there are similar wide margins around the text. Similarities between these books are a result of Benjamin White's knowledge of Pennant's suppliers and shows that Gilbert White intended for *Natural History and Antiquities of Selborne* to take a similar physical format and appeal to the same network as the quarto and folio volumes Pennant published in the 1780s and 1790s.

High-quality wove paper was ideal for printing fine letterpress and the large, expensive copperplates used by Pennant in *The View of Hindoostan*.[61] These plates were produced by eight of the most notable engravers of late eighteenth-century London, including Peter Mazell, Francis Chesham (1749–1806), Thomas Medland (c. 1765–1833), William Angus (1752–1821), and James Sowerby (1757–1822). Each engraver was selected for a specific skillset. For example, Chesham, who specialized in landscape views, engraved the plates showing views of "Mooto Tahlow" and "Boats on the Gangees"; Angus, who specialized in architecture, engraved plates showing the "Bridge of chains" and several Indian palaces; and Sowerby, one of the most famous botanical draughtsmen and engravers of the period, who had previously undertaken work for William Curtis and James Edward Smith, produced the plates and illustrations for the "Teek Tree," "Poon Tree," and "Nepenthes."[62]

FIGURE 3.6. James Sowerby's illustration of the poon tree from Thomas Pennant's *A View of Hindoostan* (1798), including the original proof (center) and final engraving (right). By permission of Llyfrgell Genedlaethol Cymru/The National Library of Wales.

The natural history plates in *The View of Hindoostan* were less expensive than those depicting landscapes, a result of the complexity associated with the engraving of the latter. Views of topographical features tended to fill a designated frame on the copperplate and often required the engraving of a larger proportion of the total surface area to create perspective. This process was far more complex and time consuming than the engraving of a single natural history specimen on an otherwise blank plate. A typical example is the plate Francis Chesham engraved titled "Vessels on the Gangees." Although this was somewhat smaller in total surface area when compared to the natural history plates, it cost Pennant £12/12. This amounted to approximately three months wages for a skilled artisan at the time.[63] It seems this was a reason why Gilbert White received a high quote for the images intended for publication in his *Natural History and Antiquities of Selborne*. Many images, such as the large frontispiece White commissioned from Grimm, consisted of complicated views and took a long time to engrave. In comparison, the natural history illustrations and engravings of specimens in Pennant's *View of Hindoostan* concentrate on a single object that was often displayed as an isolated feature in the centre of the plate. For instance, Chesham charged Pennant £15/15 for his engraving of "Sevendroog" in the Savandurga Hills, Bangalore, whereas Sowerby only charged £8/8 for each botanical engraving (figure 3.6).[64]

Sowerby's work as an artist and engraver had first come to Pennant's attention in 1776 when he was preparing John Lightfoot's *Flora Scotica*.[65] This gained Sowerby the reputation of an accomplished botanical artist and engraver who had a clear understanding of naturalists' desires when they wished to produce botanical illustrations and plates according to the Linnaean system. The display of a singular object in the natural history illustrations was a result of Pennant's desire to follow Linnaean conventions when depicting these plants. As with the depiction of shells in *British Zoology*, the Linnaean system of classification relied on the physical characters of the different specimens; for plants, a major emphasis was on depicting the flowering parts, especially the stamens and pistils, enabling those who followed the Linnaean system to ascribe a class, order, genus, and species. All external characters, such as the preferred habitat of a species and the color of the specimen, were omitted to remove any distraction from the specific physical characters. Examples include the elevation of the detail of the flowers in Sowerby's gray-wash image of the poon tree, emphasizing specific features used to define this species from other members of the same genus. Pennant reserved botanical images in *The View of Hindoostan* for species he believed to be of economic importance. After the engraving process proofs were taken to ensure the plate was of a sufficient quality, after which a text engraver added the Linnaean binomial name. Pennant's arrangement of the original illustration, proof engraving, and a more advanced proof secures his ownership over this particular image. As Anthony Griffiths has suggested, the second stage proof before the addition of any lettering or numbering was of great interest to collectors, and Pennant's alignment of the original image, proof, and final print in his copy of *The View of Hindoostan* gave his collection an increased status while solidifying his intellectual authority over these materials.[66] Pennant's process represents a similar chain of production to that used by Banks when preparing various botanical publications from the 1770s to the 1810s, elevating the status of a private collection in social and intellectual circles.

The Great Failure? Joseph Banks's Plants from the South Seas

Shortly after returning from the *Endeavour* voyage, Joseph Banks and James Cook submitted their journals to the editor John Hawkesworth, who had been employed by the British Admiralty to construct an official narrative. However, Hawkesworth's *Account of the Voyages*, published by Strahan and Cadell in 1773, embarrassed Banks and the other voyagers by interspersing the narrative with hints of the combined sexual and botanical delights they had encountered in Tahiti and the other Pacific islands.[67] Banks soon became the object of satirical attacks in pamphlets such as *An Epistle from Obera, Queen of Otaheite, to Joseph Banks, Esq.* (1774), *An Epistle from Mr. Banks Voyager, Monster Hunter, and Amoroso, to Obera, Queen of Otaheite* (1773), and cartoons such as "The Fly Catching Macaroni" and "The Simpling Macaroni." Banks quickly became a public example of an explorer who had cast aside all European morals and values to indulge in the utopian society of Tahiti.[68] The catastrophe of Hawkesworth's account was extended after naturalists, such as Georges-Louis Leclerc, Comte de Buffon, relied on and incorporated inaccurate ethnological descriptions into his natural history publications, while others, such as Erasmus Darwin, likened the reproduction of plants to a Tahitian marriage ceremony to challenge established European social protocols.[69]

The damaging impact of Hawkesworth's *Account of the Voyages* on Banks's social standing and reputation led him to maintain stringent control over all aspects of the process of publishing and distributing the results of the natural history expeditions in which he was involved for the next fifty years. Banks exercised caution when publishing images from his collection, such as that of the breadfruit which appeared in Hawkesworth's controversial *Account of the Voyages*. This created conflicts with Banks's contemporaries, who viewed the rapid expansion of the print trade as an opportunity to circulate material throughout polite society. Perhaps the most notable of these disputes was with Thomas Pennant, who was preparing *Arctic Zoology* (1784–1785) for the press in the early 1780s, a work renamed from *American Zoology* after the defeat of the British forces during the American Revolutionary War and the independence of the United States in 1783. Pennant quoted from several descriptions Banks had accumulated during journeys to Newfoundland and Iceland (see chapter 1) as well as using images Banks had commissioned from John Frederick Miller (1759–1796), an artist who had traveled on the Icelandic expedition in 1772.

After Banks's expedition returned to London, Miller retained some of the illustrations for private exhibition and sale. Pennant purchased several of these and, after asking Banks for permission to publish, received an angry letter dated May 4, 1783: "I must inform you, without entering into the Question of the propriety of buying things circumstanc'd as these were, that I have always consider'd them as stolen from me in a most unhandsome as well as illegal manner, & have held myself ready to Prosecute Miller, if he should publish them in any shape whatever: so circumstanc'd, I trust that you will excuse me for refusing my consent to their publication."[70] Banks regarded the sale of these images by Miller, the main target of his threat of prosecution, as a direct attack on his personal property. However, Banks retained considerable ill feeling toward Pennant which, rather than originating from Pennant's wish to publish these materials, originated from the way Pennant intended to publish. Pennant viewed it as his task to make the information collected by Banks and other correspondents accessible through books with moderate print runs designed to be distributed to members of polite society in North America and Western Europe.[71] In his response to Banks, Pennant stated: "My assiduity &

labour for so great a length of time makes me consider myself as a public man, meriting the assistance of my friends. I therefore hope, had those drawings been with you, I should not have been denied my request. My Preface has the sanction of good judges, therefore must not be supressed." Banks's reluctance to submit his natural history findings before the "public" and "good judges" was a direct result of the fiasco brought about by the popular press after the publication of Hawkesworth's *Account of the Voyages*. In his response to Pennant, Banks expressed his distaste for public opinion: "I do not /*consider*/ / think/ the Public /*as*/ likely to patronise any symptoms of pugnacity which I /*might*/ /may/ exhibit unless I have ample reason given me for submitting / *myself* / to their Tribunal."[72] As a result, Banks and Pennant developed very different ideas on publishing that created a deep disagreement about the production and distribution of natural knowledge.

The very particular views Banks developed about publishing the results of his expeditions are best represented by one of the most notorious and ambitious publications of the late eighteenth century. This was compiled from the botanical information Banks, Daniel Solander, and their team of assistants had collected on Cook's voyage between 1768 and 1771 and renamed *Banks' Florilegium* in the late twentieth century. Throughout the production process Banks and Solander instructed a team of secretaries, artists, and engravers who produced 743 copperplates based on the illustrations of living plants Sydney Parkinson produced in the Pacific. This publication has received considerable historical attention since the early twentieth century. Although it has often been assumed that Banks intended to fund this book through a mixture of subscriptions and his own income with an intention to sell copies through a commercial publisher to recuperate any financial loss, there is no evidence that Banks ever intended for this work to enter the public sphere. This was a result of the fiasco that resulted from the publication of Hawkesworth's *Account of the Voyages*, leading Banks to fund the entire production process out of his personal income. Banks intended to create a private publication with a limited print run, the quality and content of which he could control at every stage in the production process. Banks could then distribute a restricted number of copies that were never intended for sale, but designed for a closely knit genteel network. This attitude became well-known in philosophical circles. For example, when writing to Georg Forster, the Swedish naturalist Anders Sparrman—who had also traveled on Cook's second voyage—remarked that he would "prefer to send somebody less avaricious of collections and the merit of them than he [Banks] is."[73]

Comparable to Pennant's *British Zoology* (1766), the large copperplates Banks commissioned all measure 46.4 × 31.9 centimeters and were designed to be ordered according to the Linnaean system before being bound in fourteen folio volumes. The textual component was to be derived from the revised descriptions Herman Spöring transcribed into the "Primitæ Floræ" manuscripts and the interleaved copy of *Species Plantarum* (1762–1763). Although this is one of the most widely acknowledged and researched publications of this period, little effort has been made to understand its precise construction and Banks's motivations for funding such a project.[74] The vast size and number of images ensured that the production process was both expensive and complicated, since the final product was designed to unite manuscripts, interleaved printed books, physical specimens, and images under the Linnaean framework. This process incorporated the skills of numerous individuals whose activities Banks governed according to a similar hierarchical structure to that he established aboard the ship.

Banks based the physical structure of this work on publications by his contemporaries and earlier Linnaean books that contained numerous copperplate images. Examples include Carl Linnaeus's *Hortus Cliffortianus* (1737), Johann Miller's *Illustratio Systematis Sexualis Linnæ* (1777), and William Curtis's *Flora Londinensis* (1775–1798), a work published in six fascicles designed to be bound in two volumes.[75] Banks was familiar with these books and had some acquaintance with the means for sourcing materials and the production processes used by his two contemporaries. *Flora Londinensis* made a financial loss throughout its production; Curtis subsidized this work, which was privately printed, with the profits made by his *Botanical Magazine*. Benjamin White distributed the parts to 318 subscribers.[76] In comparison to Curtis, who was known for generating a significant income from his botanical publications, Banks distanced himself from all aspects of the commercial publishing industry throughout the production process of all the books he managed. Unlike Pennant and Curtis, Banks did not seek subscribers to finance the images of plants from the South Seas. This was a result of Banks's intention to create an impression of himself as a patron and authority of natural history, considering published outputs to be a symbol of independence and gentlemanly leisure. Thus, he funded the entire production of this work from the income generated through country estates in Lincolnshire and the Midlands and never intended, and had no need, to recuperate this expenditure through sales and subscriptions.

Rather than relying on the commercial publishing industry to produce the images of plants from the South Seas, Banks intended to print these plates at his own expense either by using a rolling press installed in the engravers rooms under the library in Soho Square or through privately commissioning a small-scale printer. For example, it was a common practice for eighteenth-century engravers to publish large prints on a small scale. Examples include those Banks hired to engrave the South Seas plants, such as Charles White, who produced high-quality prints from his premises in Chelsea.[77] Banks designed a similar, although perhaps somewhat more ambitious, setup to Horace Walpole's printing program. Walpole used his private printing house based at his home at Strawberry Hill near Twickenham to define himself as a gentlemanly connoisseur, and in 1774 had his long-standing printer, Thomas Kirgate (1734/5–1810), print *A Description of the Villa of Mr Horace Walpole* on his private printing press. This book had a print run of only one hundred copies. Walpole intended for these to be distributed among a few very close friends during his lifetime and drew up a list of eighty people to whom he wished to bequeath a copy after his death.[78] However, Walpole's publications were compiled entirely from letterpress and thus far cheaper to produce than the 743 copperplates Banks was having prepared. The only plate Walpole commissioned for this work was the vignette title page, which was probably outsourced to a specialist intaglio printer.

Two privately printed publications, which contain numerous copperplate images comparable to those Banks planned for the South Seas plants, were Sir William Hamilton's *Campi Phlegrae* (1776), on the volcanos of the Two Sicilies, and John Stuart, 3rd Earl of Bute's *Botanical Tables* (1785). Hamilton, the British ambassador to the Neapolitan court, believed large illustrations were more important than letterpress for conveying information on the spectacular volcanic eruptions he witnessed. By funding and overseeing the production of fifty-seven images, Hamilton produced a luxury publication that outstripped the quality of anything published in periodicals or by commercial booksellers. However, by 1776 the costs of this work came dangerously close to the threshold of Hamilton's diplomatic allowance, meaning that he permitted limited sales to recuperate expenditure.[79]

Bute's *Botanical Tables* probably resonated more closely with Banks's intention for the South Seas plants as Bute never intended to lose control of his readers by risking sale on commercial markets. Bute funded this nine-volume work, containing 654 copperplate illustrations, entirely at his own expense and only twelve copies were produced. These were destined for specific individuals who would recognize the work as a symbol of personal association and intimacy, appreciate its physical content, and Bute's continued patronage of "useful knowledge."[80] Banks took over Bute's role as the main overseer of the Royal Botanic Gardens, Kew in 1772 and it seems that he sought to emulate Bute's style of publication to integrate himself within aristocratic circles.[81] However, Banks was more ambitious than Hamilton and Bute. Although the plates for the South Seas plants were a similar size to those in Hamilton's *Campi Phlegrae*, Banks commissioned 743 plates, 686 more than Hamilton's fifty-seven, and Bute's *Botanical Tables* were printed in quarto, a work that required smaller, cheaper plates when compared to Banks's grand folio.

The Process of Private Production

To understand Joseph Banks's approach to the private production of a work on his South Seas collections, it is necessary to examine his means for producing less ambitious publications. For example, in 1781 Banks organized the production of William Houstoun's *Reliquiæ Houstounianæ* (1781). Houstoun (1695–1733) was a Scottish physician and botanist who, under the patronage of Hans Sloane, had been employed by the province of Georgia to collect plants in Mexico and the West Indies. In 1730 Houstoun sent Sloane; Philip Miller (1691–1771), head gardener at the Chelsea Physic Garden; and John Martyn, professor of botany at Cambridge, a large assortment of botanical specimens. After Houstoun's death in Kingston, Jamaica, in 1733, Miller received many of his specimens, manuscripts, illustrations, and a selection of copperplates Houstoun had engraved in the West Indies. Banks purchased these manuscripts and plates after the sale of Miller's library on April 12, 1774 for £2/2.[82] Houstoun's copperplates relate to specimens that had been incorporated into Miller's herbarium, the original drawings and notebook of manuscript descriptions Banks also purchased (figure 3.7).[83] However, the production of this publication was less complex than that required for the plants from the South Seas. For example, Houstoun had already undertaken most of the work up to the point of engraving the copperplates. The only task required of Banks was to employ a letter engraver to add the most up to date Linnaean binomials, have the plates printed, and produce the letterpress title page and descriptions that were based on Houstoun's notebooks. Although there was an unusually long wait between the engraving of Houstoun's plates and their printing, they represent a smaller-scale version of a process Banks hoped to emulate with the South Seas plants. *Reliquiæ Houstounianæ* allowed Banks to test the process of transferring original illustrations into print and the skills of the various artisans he employed to produce the letterpress and intaglio impressions.

It seems that Banks had the equipment to print intaglio plates in the basement of Soho Square. For example, on August 20, 1791, Banks implied that his engraver, Daniel Mackenzie (fl. 1770–1800), who had been employed at Soho Square since the 1770s, had the means to quickly print out and distribute numerous copies of copperplate images from the plates in his collection.[84] As a result, Banks ordered his staff to privately print the intaglio plates that were kept on bookcases in the room under the library and herbarium at Soho Square and appended these to the letterpress components of books such as *Reliquiæ Houstounianæ* (1781). In the inventory of Banks's library taken in 1820, the British Museum's librarian recorded the presence of "104 copies of Letterpress" for *Reliquiæ Houstounianæ*.[85] The absence of the same number of copies of intaglio prints show that Banks placed a bulk order for the letterpress with a commercial printer and then had the plates printed in the basement of Soho Square whenever a new copy was required. The difference between outsourcing the letterpress to a commercial printer and Banks's strict control over the copperplates reflects his views that images were the most important component of a natural history publication that he did not want to risk exposing to the public sphere. For example, in 1789 he commended James Bruce, the Scottish naturalist and explorer of Ethiopic Africa, for publishing images without descriptions, as botanists "could learn from them without any assistance."[86]

By the early 1790s, Banks's approach to privately printing intaglio

FIGURE 37. Left: William Houstoun's original drawing of the species *Aster aurantius* (L.). Center: the copperplate that shows the plant *Aster aurantius*, engraved by Houstoun in the West Indies. Right: an impression taken from the copperplate on the rolling press under the library and herbarium at 32 Soho Square. Courtesy of the Trustees of the Natural History Museum, London.

plates while outsourcing the letterpress to a commercial printer had become a standard practice for all the publications with which he was involved. For example, Banks privately published *Icones Selectæ Plantarum, Quas in Japonica Collegit et Delineavit Engelbertus Kaempfer Ex Archetypis In Museo Britannico* in 1791, and directed the engraving, printing of the copperplates, and publication of William Roxburgh's *Plants of the Coast of Coromandel* between 1795 and 1819, Francis Masson's *Stapeliæ Novæ* (1796) on South African succulents, and William Aiton's *Delineations of Exotick Plants Cultivated in the Royal Gardens at Kew* (1796). The letterpress components for all these books were printed by William Bulmer (1757–1830), whose relationship with Banks went back to the 1770s when Bulmer printed Hawkesworth's account of Cook's voyages. In 1792 Banks appointed Bulmer as the printer for the Royal Society's *Philosophical Transactions* due to his development of various high-quality workshop practices and standards of production.[87] Banks managed the production of this journal in a similar way to his copperplate books; the letterpress was produced by Bulmer whereas the images were outsourced to an intaglio printer or commissioned by independent authors.[88] Banks's preference for Bulmer's work is evidenced by his letter to the former printer of the *Philosophical Transactions*, John Nichols: "Mr Bulmer has the credit for bringing into extensive use what is technically called *Fine Printing* . . . It consists of new Types, excellent ink, improved Printing Presses, a sufficient time allowed to the Pressman for extraordinary attention, and last, not least, an inclination in the Employer to pay a considerably advanced price." Banks boasted that Bulmer's printing was "the best in the Country." In addition to employing Bulmer as printer, the paper size of the *Transactions* was increased, an approach soon followed by many learned societies such as the Linnean Society and the Royal Society of Edinburgh.[89]

Banks's preference for Bulmer's "Fine Printing" shows that he never compromised on quality, a reason why he regarded projects such as the publication of the plants from the *Endeavour* voyage to be economically unviable before they even began. Bulmer did not just print the individual letterpress components for Banks's botanical books. By the early 1790s, Bulmer printed parts of every single publication with which Banks had some sort of involvement. From 1796 to 1800 Banks commissioned Bulmer to print Jonas Dryander's catalogue of his library and publications associated with the British Museum, Royal Society, and individual pamphlets for meetings. Given this close relationship with Banks, it seems probable that Bulmer was selected as the main printer for the letterpress component of the plants from the South Seas that could be appended to a limited private print run of the plates produced on the rolling press at Soho Square.[90]

The Problem of Color

Expensive copperplate images, similar to those Joseph Banks and Daniel Solander planned for the South Seas plants, became more common from the mid-eighteenth century. Examples include the 446 plates published in Albertus Seba's *Locupletissimi Rerum Naturalium Thesauri* (1734–1765) based on preserved specimens in Seba's natural history collection. The images in Seba's work were printed in black ink before being painted over in color. The artist was left to give the animals and plants lifelike qualities, resulting in many of the specimens being placed in somewhat stylized poses with an overly vibrant coloration. A consortium of different publishers and a long list of subscribers was needed to finance Seba's work, a similar strategy to that used by George Edwards for his *Natural History of Birds* (1743–1771) and Thomas Pennant for the 1766 edition of *British Zoology*. Each subscriber had to part with 160 guilders, although, primarily due to Seba's death in 1736, it took until 1765 to publish the final volume.[91] In comparison to these earlier authors, Banks had sufficient private income to bankroll such a work. As a result, Banks was able to adhere to strict Linnaean conventions throughout the production process, omitting features from the printed illustrations, such as color, that would normally be added after images had been printed to increase a publication's commercial prospects.

When on the *Endeavour* voyage, Banks and Solander instructed Parkinson to ensure the images emphasized the relevant physical properties outlined in their Linnaean descriptions. After their return, Banks and Solander continued working in line with Linnaean conventions to produce their book, extending each layer they had calved out for an individual species in the paper technologies developed on the voyage into print. Banks made his intention to adhere to Linnaean conventions clear in a letter he sent to Carl Linnaeus the Younger (1741–1783) on December 5, 1778: "The plants of my intended publication will be arranged according to his [Linnaeus's] Strictest rules. Such as are of Genera described by him will have his names. The new ones, which I think will almost outnumber them, will be named Either in honour of distinguished Botanists, or, according to the Rules in Philosophia Botanica, by names derivd from Greek."[92] This commitment led Banks to emulate the conventions outlined by Linnaeus in *Critica Botanica* (1737) and *Philosophia Botanica* (1751) when it came to image production. Many of the specific conventions of botanical illustration emphasized in these books, such as stressing particular physical features and showing the flowers and fruits at different stages of development, had already been adhered to by Sydney Parkinson in the original illustrations and all that remained was to translate these into print. Banks emulated the stylistic conventions displayed in Linnaeus's *Hortus Cliffortianus* (1737), the only botanical publication Linnaeus produced that contains numerous copperplate illustrations, these being financed by the Amsterdam banker George Clifford. Many of the illustrations reproduced in *Hortus Cliffortianus* had been based on the living specimens from Clifford's garden and presented an example for Banks to use when reproducing Parkinson's illustrations.[93] These emphasize a very different style of image when compared to those produced for Seba's *Locupletissimi Rerum Naturalium Thesauri* and Hans Sloane's *Natural History of Jamaica* (1707–1725).

Perhaps the main Linnaean convention Banks adhered to when preparing the images of the South Seas plants was his intention to print these in monochrome. Although Parkinson's illustrations and the finished versions made up after the voyage were produced in color, the two contemporary sets of prints taken from the plates engraved for Banks's Pacific collection were printed in black. Three of Banks's publications, *Reliquiæ Houstounianæ* (1781), *Icones Selectæ Plantarum* (1791), and Georg Forster's *Icones Plantarum* (c. 1800), the plates for which were all produced in the basement of Soho Square, received a similar treatment.[94] Printing in black follows the rules laid down by Linnaeus in *Critica Botanica* and reaffirmed in *Philosophia Botanica*, in which he commented that "*colour* within the same species is remarkably sportive, and so is of no value as a distinguishing character."[95] Linnaeus believed that color varied too much between different examples of the same species and was too dependent on external factors such as soil quality and the availability of light. This created confusion and ran the risk of unnecessarily multiplying species. In addition, problems experienced when it came to producing standardized pigments ran the risk of a distinct variation in the coloration applied to plates that depict the same species. As Kärin Nickelsen has suggested, plates were often distributed in an uncolored state to make them affordable and allow purchasers to have their copies colored at a later date according to their own specifications. Banks's intention to print these illustrations in black facilitated the distribution of standardized Linnaean depictions while allowing those who received copies to reorganize the plates according to whatever taxonomic scheme they deemed appropriate.

Banks and Solander had been attempting to standardize Parkinson's botanical illustrations according to Linnaean conventions since their time in the Pacific. Characters such as the shape of the flowers and fruits and the position of the leaves were essential for placing each species within the confines of the Linnaean system, these being specific features emphasized in Solander's original descriptions and duplicated across full range of paper technologies. The desire to reduce external influences was further emphasized by the blank background designed to curtail aspects of the natural and cultural environment that became intertwined with visions of the Pacific as a utopian and egalitarian society from the 1770s. This was similar to the approach used in William Aiton's *Delineations of Exotick Plants*, in which Banks described how the images were arranged to emphasize "the internal structure of the flower, respecting the shape and the comparative size of its component parts."[96] This combined aspects of the specific specimen in Banks's collection with Linnaean iconographic practices. It seems probable that Banks employed the engravers to produce proofs from the rolling press in Soho Square to check the quality of their work and that he intended for William Bulmer to print the associated letterpress. Banks's views on the coloration of copperplates were supported by botanists in the early nineteenth century, a number of whom believed that color detracted from the image of a plant.[97] Examples include William Wilson, who suggested that "accurate dissections of, & essential characters of genera and species would I am convinced do more towards the propagation of the science, than all the coloured figures that ever were or will be published."[98]

The densely laid nets of parallel and crosshatched lines used to engrave Parkinson's illustrations were designed to give an impression of three-dimensionality through a broad spectrum of different blacks and grays. This shows the unsuitability of the engravings for color printing that resulted in the loss of several important tonal effects and perspective when these plates were printed and published as *Banks' Florilegium* in the 1980s. The inclusion of shadows to induce perspective was a standard practice of eighteenth-century engraving used to emphasize a particular quality and quantity of light and situate an object in a specific time and locality.[99] When it came to botany, shadow on the flowers, fruits, and leaves was used to emphasize the specific physical characters necessary for classifying a species according to the Linnaean system. The physical makeup of the engravings combined with the fact that Banks planned to adhere to Linnaeus's "strictest rules," shows that the original intention was for this publication to be printed in black leaving a white background, following the example exhibited by Georg Dionysius Ehret in *Hortus Cliffortianus* (1737). Printing in black avoided any potential variations in color, controlling the quality of the images and setting a standard for conveying information.

Preparing to Publish the South Seas Plants

The "arrangement" Joseph Banks specified to Carl Linnaeus the Younger took two forms. The first is the physical layout of each figure in accordance to the Linnaean system. The second relates to the ordering and binding of these images to reflect the arrangement outlined by Linnaeus in *Species Plantarum*. To gain the greatest understanding of the planned production, the following concentrates on one specific species Banks and Solander collected from Botany Bay between April 29 and May 5, 1770. This is *Banksia integrifolia* L., a species from the genus Linnaeus the Younger named after Banks. This is one of the four species of this genus Banks and his team collected from this locality, plants that inspired Cook's famous remark on how "The great quantity of new plants etc. Mr. Banks and Dr. Sollander collected in this place occasioned my giving it the name of Botany Bay."[100] Banks and Daniel Solander recognized this plant as a new species and ascribed the name *Leucadendrum integrifolium*, classifying it among the Proteaceae (an order well represented at the Cape of Good Hope). Herman Spöring then copied these descriptions into the interleaved copy of *Species Plantarum* (1762).[101] This name remained unchanged in Banks's and Solander's manuscripts until 1782, when Linnaeus the Younger, who had observed specimens, images, and descriptions of this species at Soho Square in 1781, published it under the name *Banksia* in *Supplementum Plantarum* (1781).[102] Linnaeus the Younger would have viewed the images of *Banksia integrifolia* in the variety of different states set out in plate 6 to rename this spices, taking into account the hierarchy of physical attributes outlined in each. In some cases, Parkinson's sketch and the printed version reveal more of the plant's physical characters, including the shape of each small flower and its position in the larger composite flower that were deemed essential for classifying and renaming this species in the early 1780s. In a letter to Carl Peter Thunberg, Banks described the younger Linnaeus's visit to Soho Square: "Mr Linné is here and I see a good deal of him he certainly has a good deal of Botanick Knowledge." Linnaeus often copied information from Solander's manuscript slips onto his own slips and used this to compile publications.[103]

As we have seen, many of Banks's staff died on the return voyage. Sydney Parkinson and Herman Spöring were just two of over seventy (out of ninety-four) seamen who died aboard the *Endeavour* between Batavia and London. Thus, to transfer the rigid social hierarchy Banks had applied to different tasks associated with the practice of natural history on the voyage to his London library, replacements had to be found. These included the German naturalist, surgeon, and secretary Sigismund Bacstrom (1750–1805), who filled Spöring's vacant position.[104] Bacstrom was employed by Banks from 1772 to 1775 and possessed skills in natural history and instrument making, along with the ability to capture and preserve birds "in the most lively natural manner . . . and also in so durable manner as to last for years."[105] These practical skills and the ability to perform secretarial work made Bacstrom an ideal candidate to accompany Banks's party on his trip to Iceland in 1772 and to assist Solander with the publication of the plants from the South Seas.

Another individual Banks hired to assist Solander was Jonas Dryander (1748–1810), who arrived in London in 1777 after Bacstrom had left Banks's service. Dryander's main task was to respond to nomenclatural changes that occurred since the mid-1770s. An example is Dryander's replacement of the name *Leucadendrum* with the term *Banksia* across the full range of paper technologies after the publication of this genus in Linnaeus's *Supplementum Plantarum* (1781). The case of *Banksia integrifolia* presents a prime example; in the margin next to the description in Solander's original field notebook,

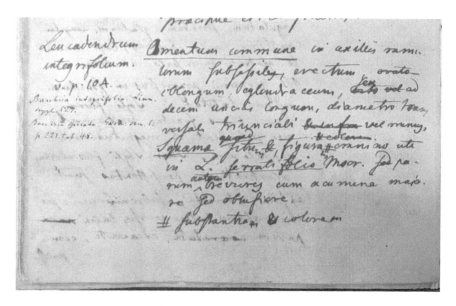

FIGURE 3.8. Jonas Dryander's annotations replacing *Leucadendrum integrifolium* with *Banksia integrifolia* in Daniel Solander's field notebook. By kind permission of the Trustees of the Natural History Museum, London.

Dryander added the name "*Banksia integrifolia*. Linn" and a citation to *Supplementum Plantarum*. Dryander repeated this on the representative specimens by replacing the labels (figures 3.8 and 3.9).[106]

In 1772 Linnaeus wrote to John Ellis at the Chelsea Physic Garden, commenting that Solander was "now putting his collection in order, having first arranged and numbered his plants, in parcels, according to the places where they were gathered and then written upon each specimen its native country and appropriate number."[107] The linking of the numbered specimens, descriptions, and images, a process that had been commenced aboard the ship, was essential for connecting these different physical aspects of Banks's *Endeavour* collection when removing the specimens from the "books" of waste paper and affixing them onto single herbarium sheets. These could be ordered according to the method proposed by Linnaeus in *Philosophia Botanica* and aligned with the annotated descriptions in the interleaved copy of *Species Plantarum* (1762–1763) and Parkinson's images that had been ordered and bound to reflect the ordering of species within the interleaved book.

Many of these systematic changes were commenced after Banks's and Solander's return from Iceland in 1772 and coincided with the most intense period of preparing these images for the press. In a letter

FIGURE 3.9. Jonas Dryander's new label on a specimen of *Banksia integrifolia* that refers to *Supplementum Plantarum*. By kind permission of the Trustees of the Natural History Museum, London.

to the artist and engraver Peter Perez Burdett, who had previously attempted to gain employment under Banks as one of his engravers, Benjamin Franklin observed how Banks "employs 10 engravers for the Plates."[108] To fulfil his role as secretary, Bacstrom kept a long manuscript titled "Catalogue of Drawings of Plants from Cook's First Voyage," in which he recorded the production processes for the completion

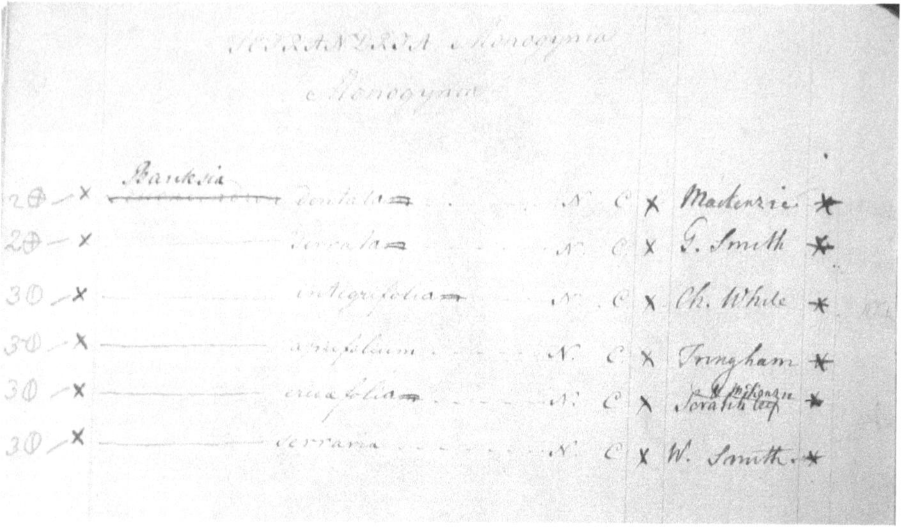

FIGURE 3.10. The entries concerning the genus *Banksia* in Sigismund Bacstrom's "Catalogue of Drawings of Plants of Cook's 1st Voyage, 1768–1771." The corrections are in Jonas Dryander's hand. By kind permission of the Trustees of the Natural History Museum, London.

of Parkinson's unfinished images, the engraving of the copperplates while Dryander's hand can be seen updating the nomenclature in line with Linnaeus's *Supplementum Plantarum* (figure 3.10).[109]

In comparison to Pennant, who paid careful attention to the total price of each copperplate he published, Bacstrom did not record the costs, suggesting that Banks never intended for this work to be sold and was unconcerned about the expenditure on each individual component. Rather, this manuscript is purely systematic, extending and connecting the network of paper technologies developed during the *Endeavour* voyage to the printed images and broader book production process. However, Banks paid Daniel Mackenzie, who also produced many of the engravings of the plants from the South Seas, a total of four guineas per plate to engrave Aiton's *Delineations of Exotick Plants* (1796). Owing to Mackenzie's death around 1800 Banks had to employ Franz Bauer (1758–1840) to finish the engravings. Unlike Mackenzie, Bauer was not exclusively working for Banks, meaning it was more difficult to monopolize his output. It seems that Mackenzie and the other engravers received a similar standard payment for each plate engraved for the South Seas plants.[110] Therefore, Banks would have spent approximately £2,975 on the labor of engraving 743

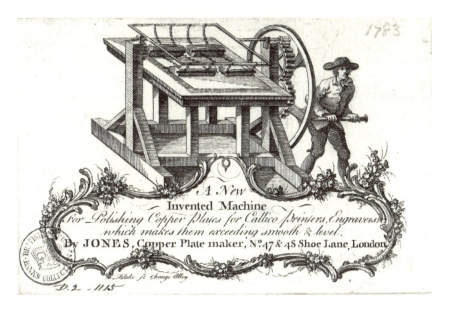

FIGURE 3.11. Richard Jones's trade card from an album compiled by Sarah Sophia Banks. © The Trustees of the British Museum.

copperplates. This price does not account for the copper sheets, costs of private production, ink, paper, and the services of the artists who produced completed versions of Parkinson's illustrations.

Bacstrom's approach to recording the production process becomes apparent in the case of *Banksia integrifolia*, an illustration completed during the 1770s. Parkinson was unable to complete the original field sketch due to pressures on his time and resources brought about by the vast quantity of material Banks and Solander collected on the eastern coast of New Holland (figure 3.10). The symbol × in Bacstrom's "Catalogue" shows that a complete colored illustration was produced by Frederick Polydore Nodder (fl. 1770–1800), who based this image on the original line drawings and instructions left by Parkinson. Nodder was one of the most renowned natural history artists of the period, contributing to lavishly illustrated publications such as George Shaw's *The Naturalist's Miscellany* (1789–1813), on the engraved title page to which Nodder noted that he was "Botanic Painter to Her Majesty."[111] The initials "N. C." in Bacstrom's manuscript indicate the locality Banks and Solander found this specimen, in this case, "Nova Cambria" or New South Wales. Other examples were labeled as "*T.d.F.*" for Tierra del Fuego, "*N.Z.*" for New Zealand and "*Mad.*" for

Madeira, divisions originally outlined in Solander's field notebooks. The symbol "✳" indicates that a copperplate based on the completed image was produced by the engraver listed in the final column.[112]

Engravers accounted for eighteen of the staff employed to work on the South Seas plants, among whom were two of the most accomplished in their field: Daniel Mackenzie, who produced the copperplate for *Banksia dentata*, and Gabriel Smith (1724–1783), who produced the plate for *Banksia serrata*. Six engravers were employed to produce plates for the genus *Banksia* alone, a process that took several weeks. The copper used for these plates was obtained from Richard Jones (fl. 1772–1788), a supplier based in Shoe Lane, London, whose stamp appears on the verso of all the plates in Banks's collection.[113] Jones was an innovative copperplate supplier; his business card from 1786 depicts "A New Invented Machine for Polishing Copperplates for Callico Printers, Engravers which makes them exceeding smooth & level" (figure 3.11). The use of new machines to produce copperplates almost certainly attracted Banks to Jones's firm. Banks paid close attention to Jones's business, which merged with that of William Pontifex following Jones's death in 1788 to become Jones and Pontifex. This is apparent from the trade cards collected by Banks's sister, Sarah Sophia Banks (1744–1818), from 1780 until her death.[114]

Jones and Pontifex's trade cards cast light on Banks's relationship with London tradespeople through the numerous letters and notes they added to the verso of the cards. For example, in 1796 Pontifex addressed a letter to Banks, noting that "the utmost care and attention shall be given to any Orders you please to honour me with," going on to discuss the newly invented copper brewing kettles depicted on the recto. Given Banks's continued interest in machines and industrial innovation, Jones's company proved to be far more appealing than other coppersmiths. Although the copper for printing plates was probably beaten with hammers until the desired smoothness was obtained, it seems that Jones's machine produced a set of uniform copperplates, a reason why many of the plates Banks purchased are similar in size, thickness, and weight. These prepared copperplates could then be sold to private individuals, engravers, and publishers.[115]

Joseph Banks and the Plants of Japan

By the 1780s, Joseph Banks's interests had shifted from the discoveries he made in the South Seas onto the increasingly important British

trade interests in the Far East, most notably in the economic benefits associated with the opening of a trade route to China and Japan.
Interests in Asia contributed to Banks's loss of interest in the publication of the plants from the South Seas, an event that coincided
with spiralling costs, an economic depression caused by the American
Revolutionary War, and Daniel Solander's death in 1782. Writing to
the Dutch natural philosopher Martijn van Marum in 1792, Banks
attributed the delayed production to the diversion of his attention to
outfitting scientific voyages to the East Indies and the establishment
of Kew Gardens.[116] Banks's interests in the Far East were partially inspired by Carl Peter Thunberg, a student of Carl Linnaeus, who visited
Banks and Solander on his way back from Japan in 1778. On his visit
to London Thunberg examined the Japanese plant collections held by
the British Museum and exchanged East Asian specimens with Banks
for duplicates collected on the *Endeavour* voyage. Thunberg described
how the collection of the British Museum was built "on a very large
and extensive scale" and declared that it was "Kæmpfer's Manuscripts
and Collection of Herbs together with the Drawings and Designs,
were the articles, which gave me the greatest pleasure to see here."[117]

The collection of the seventeenth-century Westphalian physician
and employee of the Dutch East India Company, Engelbert Kaempfer
(1651–1716), who had traveled to Japan in the 1690s, inspired Banks's
engagement in the botany of the Far East. By the 1780s and 1790s,
these interests began to correlate with the agendas of the British government, drawing Banks's interests in the economic potential of East
Asian botany, particularly that of Japan and China, to the fore and
contributing to the abandonment of his plan to publish the species he
had collected in the South Seas. Japan had been closed to Europeans
since the Sakoku Edict of 1635, although one of the few European
companies still permitted to trade was the Dutch East India Company, Vereenigde Oostindische Compagnie (VOC), whose operations
were confined to the enclave of Dejima in Nagasaki. Kaempfer was
one of the few Europeans to have carried out systematic surveys of
the natural history of Japan when he visited Dejima between 1690
and 1692.[118] Much of Kaempfer's collection was collected illegally,
aided by his appointed translator Imamura Gen'emon, who copied
numerous secret maps and gathered specimens during their annual
procession to Edu (Tokyo), a journey made to pay homage to the Japanese court. Kaempfer used his Japanese collection to compile his work

Amoenitatum Exoticarum (1712) after his return to Germany. After Kaempfer's death Hans Sloane purchased his manuscripts and specimens in 1723.

The main items that drew Banks's interest from Kaempfer's collection was a folio of 217 illustrations of Japanese plants and two volumes of herbarium specimens. These came to Solander's attention when he was employed to reclassify the natural history collections of the British Museum in 1763. Many of Kaempfer's illustrations relate to specimens in the herbarium and were drawn from the living plants by local artists in Japan.[119] Given Banks's growing interest in East Asian botany from the mid-1780s, these images were of great importance as they presented the first detailed inventory of Japanese plants. The increasingly isolationist policy of the Japanese government led to Banks's publication of Kaempfer's drawings as *Icones Selectæ Plantarum* (1791). For this, Banks maintained much of the infrastructure he had established to produce the plants from James Cook's first voyage, hiring Daniel Mackenzie, the main engraver for his previous project, and ordering the copperplates from Richard Jones and William Pontifex. However, Banks had learned from the spiralling costs associated with the images of plants from the South Seas. In comparison to the *Florilegium*, the plates Banks had engraved for Kaempfer's drawings were double-sided, halving the initial expenditure on the copper. Saving money on copper was essential given the expediential increase in the price of this commodity alongside the onset of the French Revolution.[120] These engravings were produced on a range of large and small copperplates. The large plates were the same size and weight as those used to engrave the plants from the Pacific and the smaller plates were half this size.

The engravings designed to represent Kaempfer's illustrations were produced in mirror image of the original drawings, ensuring that, once printed, they produced a precise reproduction of the original (plate 7). This shows a distinct historical sensitivity toward Kaempfer's manuscript. Banks had not seen living examples of the original specimens in the field and given that Kaempfer's specimens in the British Museum's herbarium were fragmentary—Solander had described them as being "made up from spare specimens"—Banks wished for each published plate to remain true to the drawing from Kaempfer's collection.[121] Thus, unlike the images Banks commissioned in the South Pacific, Kaempfer's illustrations were not adapted

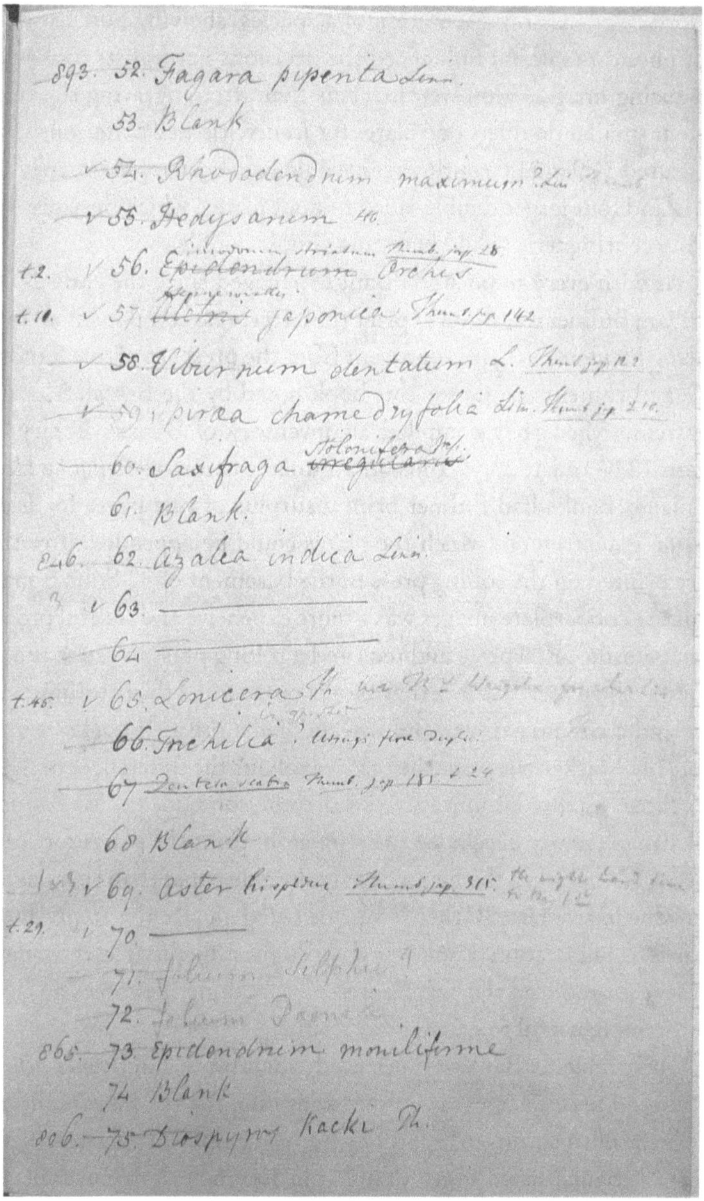

FIGURE 3.12. A page from Joseph Banks's handwritten list of Japanese plants annotated and corrected by Jonas Dryander. The first column refers to the table number in *Icones Selectæ Plantarum*, the second numbers each drawing in Engelbert Kaempfer's collection, and the third contains Banks's binomial name. *Epidendrum orchis* (*Limodorum striatum*) is depicted in plate 2. By kind permission of the Trustees of the Natural History Museum, London.

to form a generic representation of a species, showing how the available physical material influenced the decisions naturalists took when producing images. However, it seems that after engraving the twenty-four smaller double-sided plates for *Icones Selectæ Plantarum*, which measured 31.9 × 21.6 centimeters, would have cost approximately £12 each, and four larger double-sided foldout plates, which measure 46.4 × 31.9 centimeters, would have cost almost double this.

As with every publication Banks managed since the early 1770s, William Bulmer was hired to print the letterpress component of *Icones Selectæ Plantarum*. This is apparent from the presence of one hundred title and contents pages for this book noted by the British Museum librarians when they compiled an inventory of Banks's library between 1820 and 1823.[122] Thus, in a similar manner to *Reliquiæ Houstounianæ*, Banks had Bulmer print a surplus of title pages for *Icones Selectæ Plantarum*, to which the plates could be appended after they were printed on the rolling press in the basement of 32 Soho Square. Printing copperplate images was a more expensive and skillful process than printing letterpress and required a rolling press to crush the paper into the plate. The fact that Banks outsourced the printing of the title and contents pages to Bulmer and had a private intaglio printer, probably Mackenzie, print the plates explains the differences between the paper used to produce these different components. Banks used Whatman's newly developed wove paper to print the plates for *Icones Selectæ Plantarum*, a thicker and more durable paper than that used to print the letterpress. Rather than this being an attempt to produce a luxurious publication, Banks used the highest-quality paper available because it produced the best impressions, making the copies he chose to distribute useful to natural historians.

The problems Banks experienced when producing *Icones Selectæ Plantarum* become apparent after examining the list he compiled to select the illustrations to reproduce from Kaempfer's collection (figure 3.12). [123] Banks listed every drawing in Kaempfer's manuscript and assigned each plant depicted the most up to date Linnaean binomial name according to the 1779 edition of *Systema Plantarum* (the successor in Banks's collection to the 1762–1763 edition of *Species Plantarum*), edited by Johann Jacob Reichard (1743–1782), and Thunberg's *Flora Japonica* (1784). When there was no known Linnaean name listed for the species represented in the drawing, Banks derived his own binomial from the Kaempfer's pre-Linnaean manuscripts and

Greek descriptive terms, probably with the assistance of his curators, Dryander and Solander, who had examined this material alongside Thunberg in 1778. Dryander's involvement in this project is apparent from the annotations he added to Banks's list, going through and updating the nomenclature used for each species in addition to crossing through species they did not wish to reproduce in *Icones Selectæ Plantarum*. Dryander then added the figure numbers for the reproduced image next to the relevant binomial.

Over the course of this project, Banks ordered Mackenzie to produce a total of fifty-nine engravings on thirty double-sided copperplates. The verso of the last copperplate for *Icones Selectæ Plantarum* was used to engrave an image based on the illustrations Georg Forster (1754–1794) produced on Cook's second voyage to the Pacific. Banks reproduced only 59 out of Kaempfer's 217 drawings, possibly a result of the vast increase in the cost of paper and copper in the early 1790s, in addition to Thunberg's publication of thirty-nine images of Japanese plants in *Flora Japonica*, some of which represented species depicted by Kaempfer. However, Banks's main reason for limiting this project was his desire to reproduce images that would be useful to other botanists. Therefore, Banks and Dryander elected to publish images of plants that had been described but not previously figured or had not received attention from Linnaeus and his successors while showing Banks's authority over the British Museum's collections.[124] The addition of this book to existent publications on Japanese and East Asian botany served to update a large body of pre-Linnaean material, uniting the books of Kaempfer, Thunberg, and Banks to form an interlinked, transportable repository that proved useful to naturalists travelling on British expeditions to East Asia.

The approaches genteel naturalists such as Joseph Banks and Thomas Pennant took to produce their publications gives a range of different avenues of production that can be traced through the chart figured in the main introduction. Individuals took a variety of different routes to produce natural history books, although all worked hard to ensure these remained well within the scope of approved genteel practices. The crucial qualification for these was not to appear overtly commercial, privately funding the production process while portraying oneself as a connoisseur for quality and skill. Banks's and Pennant's differing methods reflect their divergent levels of personal wealth, standing in genteel society, and motivations for distributing knowledge on the natural world. Pennant's catastrophic financial loss when producing the 1766 edition of *British Zoology* encouraged him to initiate a certain "crossover with commerce," moving through the practices outlined in the smallest oval in the introductory chart (figure I.1). These were essential for producing a surplus of copies for the open market, a crucial purpose of which was to allow Pennant to regain his expenditure on the production process. In comparison, Banks's much larger landholdings and income ensured that, once the necessary materials had been purchased, it was possible for him to self-fund the entire production process and not worry about financial losses. Different genteel approaches reflect on Banks's and Pennant's social statuses from the 1770s when, in comparison to Pennant, Banks's elevation to a baronetcy and presidency of the Royal Society engaged him with aristocratic interests in private publishing shared by figures such as Lord Bute and Horace Walpole. Banks's international fame meant that publishing would only have a minimal impact on his advanced network. Pennant, in comparison, remained in North Wales, gradually growing his landholdings and natural history collection.

When compared to many late eighteenth-century authors of printed books, genteel naturalists maintained stringent control over the content, physical makeup, and materials that went into producing a publication. This sets naturalists aside from those who produced "on commission" books through paying a publisher a set amount to manage and produce a book.[125] For example, Pennant did not contract with Benjamin White to produce books such as *British Zoology*. Rather, he saw White's firm as a window through which he could distribute books to members of genteel society beyond his established network of correspondents. In comparison to many on-commission publishers, White did not have any involvement in editing Pennant's text, sourcing printing paper, copper for the plates, and the various artisans responsible for producing illustrations, engraving plates in addition to the intaglio, and letterpress printing. Rather, these, and the finances for the books, were the sole concern of Pennant, who managed all the production processes. Both Banks's and Pennant's sourcing of the materials for their books and close supervision of the artisans they employed allowed them to structure the physical makeup of the book according to their own designs, following the contours of their physical natural history collections to ensure the plates and corresponding letterpress were arranged to support various classificatory systems.

Images manifest these genteel processes, emphasizing naturalists' private wealth that bought them the leisure time to instruct the various artisans on the best means for depicting species from their collections and reproduce these in print. Rather than producing images in workshops or studios, the artisans worked under close supervision in Banks's and Pennant's private homes. These commissions were of huge importance to figures such as Daniel Mackenzie, who lived and worked in the engravers' room under the library of 32 Soho Square from the 1770s until his death in the late 1790s. Mackenzie spent his entire career producing Banks's botanical plates and his work is scarce outside of Banks's collection. Similarly, Moses Griffith lived with his family in a cottage on Pennant's estate for decades. This shows the level of control figures such as Banks and Pennant exerted over the production process, employing artisans who relied on these projects for their long-term income. Through privately funding artisans, naturalists could ensure the images they produced conformed to the strict taxonomic schemes they followed in their collections and the letterpress components of their books, meaning there was no incentive to adapt these for commercial markets. It also meant the artists and engravers had a long-term engagement with the physical material and the philosophical concepts naturalists wished to illustrate.

Naturalists' possession of images in their various states of production solidified their ownership over their intellectual content while increasing the prestige of their private collections. This was often done through the volume of these materials Banks and Pennant owned, many of which constituted the first visual depictions of newly described species. For example, by 1800 Banks had 879 individual printing plates capable of printing over one thousand images stored in the presses under the library at 32 Soho Square. Similarly, in 1793 Pennant published a table of all the plates he had commissioned describing "the services I did to the professors of the art of engraving, by the multitude of plates performed by them for my several works."[126] Pennant gave an enumeration of 802 plates, a list that did not include the printed portrait based on his own likeness by Thomas Gainsborough, and the books he published after 1793—including in his *History of the Parishes of Whiteford and Holywell* (1796) and *The View of Hindoostan* (1798). This shows how Banks's and Pennant's private production of natural history books went well beyond the scope of many private aristocratic publishing houses, producing hundreds of copperplate images at vast expense. The distribution of these books played an important role in securing their positions in genteel society and networks of natural history, their approaches to which are explored in the next chapter.

PLATE 1. Top: George Stubbs's *Cheetah and Stag with Two Indians*, commissioned by the governor general of Madras, George Pigot, in 1765. © Manchester Art Gallery / Bridgeman Images. Bottom left: the image of the "cheetah" copied from the above painting by Thomas Pennant's artist Peter Paillou. Courtesy of the Linda Hall Library of Science, Engineering & Technology. Bottom right: the published version of this image from Pennant's *Synopsis of Quadrupeds* (1771). Private collection.

PLATE 2. Left: the specimen of *Convolvulus alatus* (*Merremia peltate*, now *Decalobanthus peltatus*) Joseph Banks and Daniel Solander collected in Tahiti. By kind permission of the Trustees of the Natural History Museum, London. Right: Sydney Parkinson's image of *Convolvulus alatus* (*Merremia peltata*). Courtesy of the Trustees of the Natural History Museum, London.

PLATE 3. Top left: Sydney Parkinson's preliminary pencil sketch of the breadfruit that Solander ascribed the binomial *Sitodium alitle* (now *Artocarpus altilis*). Top right: Parkinson's final colored illustration of the breadfruit. Bottom left: John Frederick Miller's pen and ink wash image of the breadfruit based on Parkinson's color illustration. By kind permission of the Trustees of the Natural History Museum, London. Bottom right: the copperplate impression from John Hawkesworth's *Account of the Voyages* (1773). Courtesy of the Linda Hall Library of Science, Engineering & Technology.

PLATE 4. Left: Thomas Pennant's specimen of the little woodpecker. This was removed from its original lifelike arrangement in a glass case during the early twentieth century. © Jonathan Jackson, Natural History Museum, London. Center: the original drawing of Pennant's little woodpecker by Moses Griffith. The grid lines were used to transfer this image onto a copperplate. Courtesy of the Linda Hall Library of Science, Engineering & Technology. Right: the copperplate image based on the specimen and original drawing published in Thomas Pennant's *British Zoology*, vol. 1 (1776). Private collection.

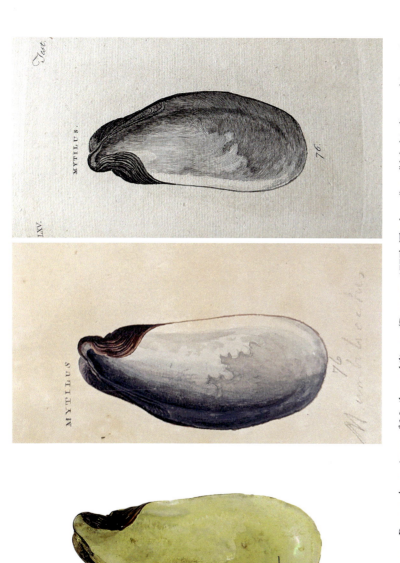

PLATE 5. Left: Thomas Pennant's specimen of *Mytilus umbilicatus* (Pennant, 1777). The later "type" label indicates this specimen was used to formulate the first widely accepted published description of this species. By kind permission of the Trustees of the Natural History Museum, London. Center: the original drawing of this specimen, by Moses Griffith, tipped into Pennant's copy of *British Zoology* (1777). By kind permission of Arader Galleries, New York. Right: The final published image, engraved by Peter Mazell from Pennant's *British Zoology*, vol. 4 (1777). Private collection.

PLATE 6. Images of *Banksia integrifolia* showing the transition from the original field sketch to the monochrome impression. Top left: Sydney Parkinson's original field sketch, produced in Botany Bay, 1770. Top right: Frederick Polydore Nodder's finished drawing. Bottom left: the copperplate, in this instance engraved by Charles White (chrome plated in the 1980s). Bottom right: the proof impression of this plate printed on a rolling press in the engravers room of Soho Square. Courtesy of the Trustees of the Natural History Museum, London.

PLATE 7. Left: Engelbert Kaempfer's original drawing, produced in Japan c. 1690. © The Trustees of the British Library, London. Center: the copperplate engraved by Daniel Mackenzie later printed in *Icones Selectæ Plantarum*. Courtesy of the Trustees of the Natural History Museum, London. Right: plate 2 from Joseph Banks's copy of Kaempfer's *Icones Selectæ Plantarum* (1791), which depicts the species *Limodorum striatum*. Courtesy of the Trustees of the Natural History Museum, London.

PLATE 8. Left: Carl Linnaeus's copy of the 132 plates from Thomas Pennant's *British Zoology* (1766). The small, annotated binomials at the foot of the image are in Linnaeus's hand and relate each species to 1766 edition of *Systema Naturae*. By permission of the Linnean Society of London. Right: a colored version of the same plate from a copy sold through a bookseller. Biodiversity Heritage Library/Harvard University, Museum of Comparative Zoology, Ernst Mayr Library.

PLATE 9. Left: the copperplate used to print the image issued by Joseph Banks in c. 1776 that depicts the *Cinchona officialis* according to the Linnaean system. Courtesy of the Trustees of the Natural History Museum, London. Right: a printed impression of *Cinchona officinalis* from this copperplate. © The Board of Trustees of the Royal Botanic Gardens, Kew.

PLATE 10. Thomas Pennant's annotated copy of *British Zoology*, vol. 1 (1776) open on the pages concerning buntings to show the specimen sent by Hugh Davies and interleaved notes. By kind permission of the Trustees of the Natural History Museum, London.

Chapter 4

From Print
to Distribution

I admire your Prudent conduct in the County as much as I Object
to the absurdity of David Pennant his father Thom had written
about birds and beasts till he realy beleived he was of some Political
importance. I hope however the Lesson he has receivd will be of
use to him & to his son. I wonder that the lineage of Cadwallader
does not feel a Boiling of Blood that such like new men should
attempt to write themselves into that Antiquity of Family which
Welshmen know how to value when it is Genuine.

—Joseph Banks to John Lloyd of Wigfair, September 5, 1796

Different methods of producing natural history books motivated authors' and compilers' approaches to distribution. Controlling distribution was central for maintaining a position in genteel society, and
attitudes toward it varied alongside personal income, social standing,
approaches to building a network of correspondents, and elevating
one's reputation in polite circles. Contrasting views are spelled out by
Joseph Banks in a letter to the Welsh antiquarian John Lloyd of Wigfair disapproving of Thomas Pennant's wide distribution of publications to increase his reputation as an authority on natural history and
establish his position as a notable figure in Welsh history. Pennant's
approach to disseminating knowledge was very different to that exercised by Banks, who avoided commercial markets throughout the
processes of production and distribution, exercising a program similar
to that of aristocrats and gentlemen poets who maintained control
over the production of a limited number of copies.[1] Many naturalists

integrated their approaches to distributing publications with their own interpretation of genteel natural history. Rather than entering a commercial market to be sold to any individual who could afford to purchase a copy, the recipients of natural history books were chosen by authors for specific purposes.

It is important to understand the selective distribution of natural history publications in the context of eighteenth-century ideas of improvement relating to internal British affairs and ever more important broader global contexts. These concerns were influenced by specific Anglophone interpretations of cameralist physiocratic thought that shaped individual interests in improvement and their associated collaborative networks.[2] Examples explored here—including Thomas Pennant, Joseph Banks, Gilbert White, and William Jackson Hooker—had a specific range of individuals in mind for their publications, distributing these to interact with select groups in Great Britain, Continental Europe, and the wider world. The need to mediate between different geographical scales—and to link individuals across these—influenced the systems of classification naturalists used. Classificatory systems united specific networks with agendas ranging from increasing agricultural production in Great Britain to those who sought to expand global empires. For example, Pennant intended for *British Zoology* to "suggest so many hints towards enlarging and improving our manufactures and agriculture," ideas that applied to elites who ranged from country vicars to the wealthiest landowners. For this book to appeal to a wide range of individuals, Pennant used the system developed by John Ray, believing this approach was familiar to "every country gentleman" because of the attention Ray had paid "to his own country in particular."[3] The reliance of Ray's system on a range of physical and external characters, such as the sounds animals made, feeding habits, and the usefulness of a particular species to industry and agriculture, combined the genteel act of reading Pennant's book with notions of improving the productivity of country estates and private land.[4]

Domestic improvement was of particular importance to figures such as Banks, who owned over nine thousand acres in Lincolnshire by 1790. After his journey to the Pacific, Banks began to combine vested interests in national improvement with a global network that became entrenched in wider British imperial expansion. The distribution of books was central to Banks's imperial aims and wish to bond a "learned empire," a diverse intercontinental network that inspired his consistent use and defence of the Linnaean system. As Banks and Daniel Solander had proved on James Cook's voyage, the reliance of the Linnaean system on select physical characters created a global standard for the classification and communication of botanical information. Books and associated systems of classification bonded networks over designated geographical areas to fit the motivations of their authors or compilers, creating specific frameworks for the simultaneous collection, communication, and redistribution of natural knowledge.

Gifting books was central for bonding diverse networks of naturalists, landowners, politicians, and others. A practice that had emerged in aristocratic courts during the sixteenth and seventeenth centuries, the distribution of books as gifts took a new shape in the late eighteenth century. This was a result of the relative growth in commercial publishing houses that amassed several authors—including Tobias Smollett, Samuel Johnson, Alexander Pope, among others—substantial personal fortunes and provided incentives for other literary writers who did not make such large returns.[5] However, the genteel natural history authors discussed in this chapter gifted publications to emphasize their financial independence from patrons and the state while interacting with and patronising a communication network. Rather, the patrons to whom many dedicated books were those who shared their collections and gave permission for the publication of descriptions and images. As a result, the gifting of books was often viewed as an effort to unite personal autonomy and social integration; books could be seen as products of independent thought even if they were gifted and dedicated to a patron. The linking of a network through gifts was central for defining a particular intellectual identity, influencing a recipient's opinion of the author's credibility, social rank, and the content of their work.[6] For example, recipients were obliged to treat books received as gifts from authors with a degree of civility whereas those obtained on the open market could be criticized in personal communications, reviews, and new books. British naturalists sought to defend their work from unfavorable reviews brought about by public sale, either through regulating the number of copies that found their way onto the open market or distributing the entire print run as gifts. Presenting publications allowed natural history writers to distance themselves from those who wrote for a profit, believing they had the greater task of improving knowledge of the natural world in learned circles rather than lining their own pockets.

Readers of Thomas Pennant's Natural Histories

Throughout his career, Thomas Pennant followed the late eighteenth-century ideal of publication as a means of gaining intellectual credit in genteel circles.[7] This was a direct result of the interest many landed individuals had in Pennant's work, believing its content could be used to improve their own property. Pennant's books became well-known, not just among naturalists, but among royalty, the country gentry, the emerging mercantile elite, and the clergy. The dissemination of Pennant's works was complemented by his desire to establish a program of national improvement through increasing access to zoological knowledge, ensuring that his work was disseminated throughout the "reading public" of Britain, Continental Europe, and English-speaking North America. In Pennant's seminal work, *British Zoology*, an active effort was made to ensure that "the names [of animals] shall be given in several *European* languages," providing columns of translations that contained ancient British (Welsh), French, German, and Italian names for species. Pennant also authorized the production a German translation.[8] However, Pennant was always careful to employ printers who had genteel reputations, describing how "I am happy in a gentleman-like worthy printer" when discussing the conduct of the New Bond Street publisher Robert Faulder, who took over the production of several of Pennant's works in the late 1780s following a dispute with Benjamin White.[9] By ensuring the genteel reputations of his publishers, Pennant was able to distribute his works to a specific sector of elite society and distance the physical books and their readership from the literary hacks and cheap publications of Grub Street.

Pennant saw publishing as his task to communicate his findings and those of his correspondents to benefit and generally improve British society. This was not only Pennant's decision. By the early 1770s, many correspondents expected him to publish and distribute books to a specific sector of society, a task Pennant entrusted to genteel publishers such as White and Faulder. For example, the clergyman and print collector James Granger (1723–1776), wrote to Pennant from the London home of the Earl of Bute, in 1772, commenting on his forthcoming trip to Antwerp, adding that "you, Sir [Pennant], would not travel with such [an] expedition. If you had, you must give the public very meagre Books. On the contrary, your exact observation, nice Judgement and diligent application have converted matter overlooked by other Travellers."[10] Similarly, the Lüneburg translator John Timaeus wrote to Pennant in 1793, commenting that he wished "to pay my respects to Mr. Pennant by whose valuable writings I have been greatly benefitted," offering to translate Pennant's *Literary Life* into German.[11] Pennant, and many of his correspondents, believed the distribution of natural-historical knowledge throughout polite society would ensure that "they would find their ideas sensibly enlarged, till they comprehended the whole domestic œconomy, and the wise order of providence," bringing about political and economic harmony.[12] By promoting the utility of his books to landowners across Britain, Pennant made his work applicable to the improvement projects initiated by a range of aristocrats, yeomen farmers, clerics, and other freeholders, bonding a broad group of individuals.

Pennant's other motivation for publishing on a wider, albeit controlled, scale was that any surplus income from sales could be offset against expenses incurred by research expeditions, natural history collecting and the books he presented as gifts. This ensured that he did not face financial problems brought about by these interests, such as those experienced by his friend and correspondent Emanuel Mendes da Costa, who was sent to debtor's prison in 1767. This was a consequence of Pennant not having such large revenues from his estates when compared to many other naturalists, commenting in 1798 that "good as my fortune is I live in these times with difficulty," adding that the war and government duties had seriously reduced his income.[13]

Pennant's use of publications to refund his travel expenses is apparent from the meticulous accounts kept for *A Tour in Scotland and Voyage to the Hebrides, 1772* (1774). This journey cost £296, an amount Pennant included in the list of expenses associated with the production of the final published work, on which he also recorded his payments of £50/18/6 to the letterpress printer; £115/7/6 to the intaglio printers for producing 1,500 copies; the costs of each original illustration and copperplate, such as the £8/8 Pennant paid Peter Mazell for copying the famous image of Fingal's Cave on the island of Staffa (now known as the Isle of Staffa) from Banks's collection (the total expenditure for all forty-two plates amounted to £225/1/0); alongside the costs associated with the postage of volumes to booksellers throughout Britain. The disbursements for Pennant's *Tour in*

Scotland and Voyage to the Hebrides totaled £1,256/14. The contemporary purchasing power of this amount equates to the sum required to employ a skilled artisan, such as Moses Griffith, employed for over thirty years. As noted in chapter 1, Pennant's lack of interest in profiting from his work is very different from other contemporary Scottish tourists, such as Samuel Johnson, who reduced the costs associated with his trip to ensure the profitability of his published account that he relied on for his income. By contrast, Pennant's ambition to break even is reflected in the large natural history team that accompanied him around Scotland and the Hebrides and the numerous high-quality copperplate illustrations that appeared in the book.[14]

To recuperate expenditure from his natural history collecting and the production of *A Tour in Scotland and Voyage to the Hebrides*, Pennant intended for numerous copies to be sold on the open market. Although he only ever intended to break even, replacing the capital he had invested in this project, Pennant did accumulate a small surplus through sales, inscribing "clear gains 41. 18. 0." at the foot of his accounts. This small additional income after all production and collecting costs was not something Pennant could or would rely on to fund his living expenses.[15] Rather, it was a sign that utilizing commercial markets stimulated the distribution of his books across genteel society with no major personal expenditure, providing a suitable amount to reinvest in the production of new editions, expeditions, and specimens.

Pennant had a very different authorial outlook when compared to other writers of the period, many of whom relied on the profits from sales and contracts of employment from booksellers to edit or produce publications. For example, Oliver Goldsmith received eight hundred guineas for his eight-volume *History of the Earth and Animated Nature* (1774), a book commissioned by the publisher Francis Newbery (1743–1818) and produced purely to be sold at a profit. Similarly, John Hawkesworth (c. 1715–1773) was paid £6,000 to edit Cook's and Banks's journals of the *Endeavour* voyage for *An Account of the Voyages* (1773) by the London publishers Strahan and Cadell.[16] However, Pennant set his publishing program aside from those of figures such as Goldsmith and Hawkesworth, who genteel writers regarded as little more than literary and scientific hacks, only utilizing commercial markets to distribute his work and cover his expenses. Surplus income was invested into new editions, expeditions, and collections.

Pennant was assisted by Benjamin White at a time when publishers had largely eclipsed the role of wealthy patrons for financing and distributing books. Unlike many books in the period, Pennant controlled the finances for this work and White's only task was to distribute copies to a genteel network. This relationship is apparent from White's sole ownership of the copyright for Pennant's works. When a book was not financed by its author, it was common for the copyrights of expensive illustrated books to be shared among consortiums of publishers, often known as "congers," who all agreed to fund a certain percentage of a publication. This was the case for the new edition of Pennant's *British Zoology* (1812), which required investments from twelve publishers.[17] White's ability to distribute books combined with Pennant's independent wealth resulted in the swift publication and relative financial success of his *Tour in Scotland and Voyage to the Hebrides*. This was in stark contrast to attempts by other authors to publish works that contained numerous illustrations. For example, a main factor that contributed to the bankruptcy in 1765 of the cartographical engraver and publisher Thomas Jefferys was the expenses incurred from the simultaneous production of large copperplate engravings depicting a detailed field survey of England.[18] In comparison, Pennant's private income, knowledge, and collaboration with White resulted in the successful distribution of his books to a genteel audience.

Subscribing to Thomas Pennant's *British Zoology*

Thomas Pennant's approach to natural history publishing was less commercial than that used for his topography and travel books. This is apparent from his response to James Edward Smith, president of the Linnean Society, in 1798, after members of the society had accused Pennant of profiting from his publications: "The claim they make on me they have no sort of right to make: & to comply with it, schooled as I am as an author would in such times as these be no small inconvenience. I will lay the plain truth before you. I never yet sold the copy of any work for more than my expenses, my prerequisite was the retaining of 20 copies of every Edition to give away; at present it happens that I have exhausted my supply except in the case of Hindoostan."[19] Pennant used the commercial publishing industry to secure a gift economy of natural history books, establishing his reputation as a genteel author. Many of his later books were entirely self-funded while utilizing aspects of the growing publishing industry to assist with production and distribution. Others were the product of collaborations. Pennant distributed the costs of *Indian Zoology* (1769), describing how "the expense of the plates was divided between Mr. Banks, now Sir Joseph Banks, *Baronet*, John Gideon Loten, Esq; a governor of *Ceylon*; and myself." Even sharing the expenses proved too difficult for *Indian Zoology*, with Pennant describing how only twelve plates were published before "the design was given over."[20] However, the costs associated with the production and distribution of Pennant's larger, more expensive copperplate books went beyond his total income from estates in North Wales, so that he needed to establish a readership who was willing to pay for books before their distribution. This was done by recruiting subscribers who were willing to pay a set amount in several instalments to fund production. A result of the narrow profit margins for such an enterprise, it was only possible for Pennant to reserve a small percentage of copies as gifts to solidify a network of useful contacts.

Pennant's large folio edition of *British Zoology* (1766) is a typical example of a work funded through subscriptions, describing how he "supported the far greater part of the expense. I lost considerably by it, notwithstanding several gentlemen contributed."[21] These "gentlemen" constituted 114 subscribers, including both men and women, who each paid three subscriptions of £2/2/0 per copy. Copies were ordered between 1761 and 1766 and reflect the time it took to engrave all 132 plates. Fees gathered through subscriptions were added to Pennant's already large expenditure on this project and an initial donation made by the Honourable Society of Cymmrodorion, who had formed a private partnership with Pennant to publish this book (figure 4.1). A result of having to resort to the ungenteel task of asking for money to assist with publication, Pennant managed the finances for *British Zoology* anonymously. Pennant maintained anonymity by working through his agent, Richard Morris (1703–1779), and the secretaries of the Cymmrodorion Society, to request the funds from subscribers. For example, in 1771 Pennant wrote to his agent in Scotland, George Paton, on the subject of a late payment for *British Zoology*, commenting that "not sending & paying for the fine copy of Br. Zoo. Is both a loss & trouble to Mr. Morris."[22] By distancing himself from financial matters, Pennant was able to portray himself to the subscribers as a patron, rather than a naturalist who needed patronage.

The list of fee-paying subscribers gives a clear impression of the

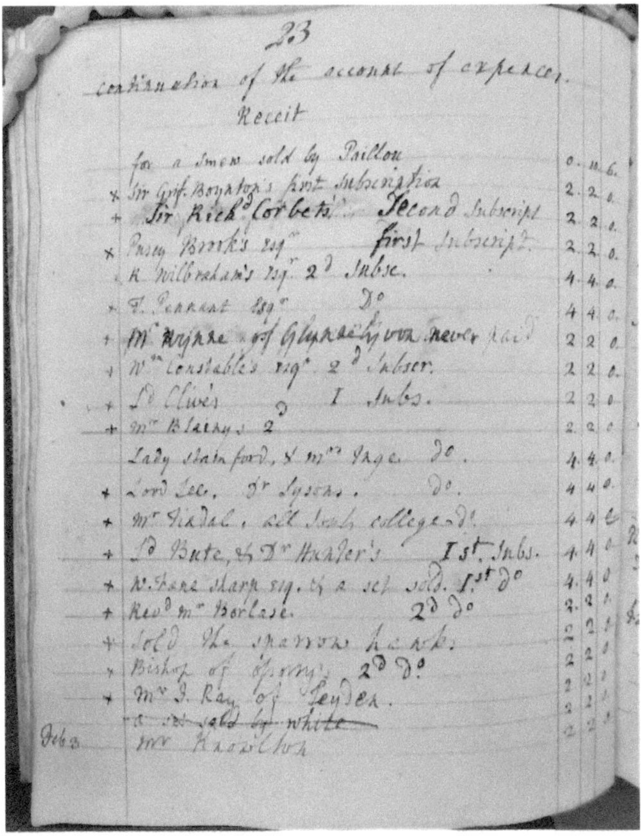

FIGURE 4.1. A section from Thomas Pennant's list of subscribers who paid for the 1766 edition of *British Zoology* between September 1762 and January 1763. By permission of Llyfrgell Genedlaethol Cymru/The National Library of Wales.

intended audience for *British Zoology*. These included a range of institutions, such as All Souls College, Oxford; wealthy aristocrats; gentry; rural vicars; prominent natural historians; and metropolitan elites (figure 4.2). Many came from the highest levels of British and European society, such as Robert Clive (1746–1774), famed for establishing the British East India Company's hold over Bengal; Robert Petre (1742–1801), 9th Baron Petre; John Stuart, the Earl of Bute; and Georges-Louis Leclerc, Comte de Buffon. Figures such as Bute, Clive, and Petre's vast landholdings and their desire to improve these to increase and secure their incomes was a main inspiration for their interests in the practical management of nature. In 1764 Bute purchased the four-thousand-acre estate of Luton Hoo in Bedfordshire, his wife

FIGURE 4.2. The distribution of the British subscribers to Thomas Pennant's *British Zoology* (1766). For the key, see the appendix.

inherited lands in West Yorkshire that brought an annual income of £17,000, and in 1768 Bute inherited over twenty thousand acres in South Wales. Others included gentry such as Pennant's neighbor, Lord Roger Mostyn (1734–1796); wealthy clerics such as Richard Newcome (1701–1769), Bishop of St Asaph; and a range of natural historians. Among these were William Hudson, under librarian at the British Museum; John Hill, author of *The Vegetable System* (1759), a book privately published with Bute's patronage; Benjamin Stilling-fleet, author of *Miscellaneous Tracts Relating to Natural History, Husbandry and Physick* (1762), which translated several Linnaean works;[23]

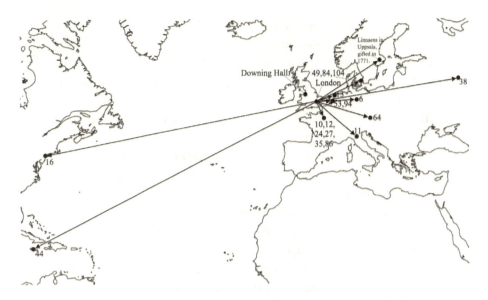

FIGURE 4.3. The international distribution of Thomas Pennant's *British Zoology* (1766). For the key, see the appendix.

Peter Collinson, who led a vast North American natural history network; and Ashton Lever, owner of the Leverian collection. These individuals were some of the main supporters of Linnaean botany during the 1760s, although many continued to follow Ray's system for the more "animated" branches of nature; an approach to classification that related to the improvements many landowning subscribers initiated on their country estates. Among the "animated" branches of natural history were quadrupeds and birds, the main subjects of the 1766 edition of *British Zoology*.

In addition to securing a network across Britain, which concentrated on England and Wales as Pennant had not yet commenced his journeys through Scotland, the folio edition of *British Zoology* found its way into the hands of naturalists across Continental Europe and the Americas (figure 4.3). Those based in Europe reflect a chain of correspondents Pennant developed during his "grand tour" or "Journey to the Continent" in 1765. In comparison to many British genteel grand tourists, who contemporary observers described as "swarms" that travelled through France and Switzerland to Renaissance cities and the ruins of Imperial Rome in the Italian States, Pennant's journey covered a broader geographical area and had a specific purpose. He did, however, travel a well-trodden route taken by many notable naturalists, visiting collections to secure connections with prominent naturalists and enlightenment figures. In France he made personal acquaintances with Buffon, with whom he had "the most agreeable and rational conversation," resulting in the latter's subscription to *British Zoology*. Pennant also met Voltaire, whom he described as "that wicked wit."[24] In Bern he met Albrecht von Haller, who "showed the utmost alacrity to promote my pursuits," and in Nuremburg Pennant met the physician, botanist and collector Christoph Jacob Trew (1695–1769), a "venerable patron of natural history" who subscribed to *British Zoology*.[25]

After this Pennant traveled down the Rhine River to Cologne before reaching the Dutch Republic, making acquaintances with Peter Simon Pallas in The Hague, sending him copies of "all the works I had published" after returning to Britain. Pallas and Pennant maintained a correspondence until the latter's death in 1798. In Leiden Pennant met Laurens Theodoore Gronovius, another correspondent and subscriber to *British Zoology*.[26] These individuals' well-established networks ensured news of Pennant's publication was distributed throughout Europe; it was probably Pallas, from 1767 the naturalist to the Russian court of Catherine the Great in St. Petersburg, who suggested Paul Grigoryevich Demidov (1738–1821) subscribe to *British Zoology* prior to the opening of his mineralogical museum in Moscow. Similarly, copies found their way to Ferdinando and Laura Bassi (1711–1778) in Bologna with whom Pennant had maintained a correspondence since the 1750s.[27] This journey, subsequent subscriptions to *British Zoology*, and lengthy correspondences it established secured Pennant's reputation as an authority of natural history in Continental European networks.

Two copies of *British Zoology* were sent across the Atlantic to subscribers and correspondents in the Americas. One recipient was Ashton Blackburne in Queens County, New York. Pennant had been introduced to Blackburne through his sister, Anna Blackburne, who built a large collection of American bird specimens at Orford Hall, near Warrington, out of material her brother sent from New York. Blackburne also maintained correspondence with Carl Linnaeus, Pallas, and Johann Reinhold Forster. Writing to Pennant in 1778, Anna Blackburne made note of "Dr. Pallas proposal I shall be very glad to exchange some north american ~~Birds~~, Lizards, toads, frogs, insects & some other articles from thence w:ch the Doct:r wants, if you think he will make a proper return of Siberian Beasts & Birds."[28] Blackburne's comments show how Pennant acted as a broker between her North American correspondents and his contacts in Russia and Continental Europe, securing a transatlantic natural history network. Her role was so important to Pennant that, in *A Tour of Scotland and Voyage to the Hebrides*, he described: "Mrs. Blackburne his daughter extends her researches still farther, and adds to her empire another kingdom: not content with the botanic, she causes *North America* to be explored and its animals, and has formed a *Museum* from the other side of the Atlantic."[29] Although Anna Blackburne's "empire" of natural history was one of the most important in Britain, her subscription to *British Zoology* was placed under the name of her father, the noted botanist John Blackburne (1694–1786). Male relatives ordering books on the behalf of women seems to have been a standard practice when subscribing to publications, although Pennant's does contain four exceptions, including a "Mrs Inge," probably Anne Inge, of Thorpe Hall, Staffordshire, and "Lady Waters," both of whom had a source of income separate to that of their male relatives. These present a few exceptions in a world where, although it is clear that women did contribute to natural history programs in both domestic and institutional settings, the overall products were deemed to be masculine.[30] For example, although Anna Blackburne owned and annotated numerous books from the library of Orford Hall, examples being copies of Pennant's *British Zoology* discussed in chapter 1 and a copy of John Berkenhout's *Outlines of the Natural History of Great Britain and Ireland* (1769–1772), all of these are marked with her brother and father's bookplates. Anna Blackburne did not sign the books she used in spite of her annotations being the most prolific.[31] Similarly, despite the extensive contributions Pennant's wife, daughters, and sisters made to his natural history collecting, illustrating, and publishing program, the books Pennant circulated were gendered masculine.[32]

Although most subscribers to *British Zoology* had distinct interests in natural history and the improvement of their property, others, such as Clive and Mostyn, valued this grand publication as a status symbol. These individuals were the main funders of this work through subscription and their copies received impressive gilt bindings to show the wealth and status of a family before the book was even opened. The high price and nature of purchasing this book through subscription limited *British Zoology* to a subset of society

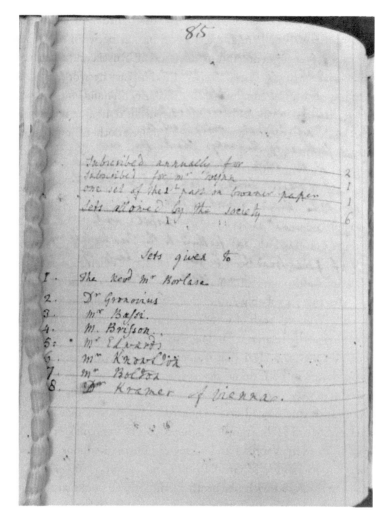

FIGURE 4.4. Thomas Pennant's list of naturalists who received presentation copies of the 1766 edition of *British Zoology*. By permission of Llyfrgell Genedlaethol Cymru/The National Library of Wales.

described by Lord Kames as "a few learned men who had money to spare, and to a few rich men who buy out of vanity as they buy a diamond or fine coat."[33]

Despite financial difficulties, Pennant retained eight copies of *British Zoology* to give away, as evidenced by his list of "sets given to" that he included at the end of his list of subscribers (figure 4.4). These presentation copies were crucial for allowing Pennant to maintain his standing with naturalists in Continental Europe and Britain. Four out of the eight sets were gifted to naturalists in Leiden, Bologna, Paris, and Vienna, cementing associations Pennant developed during his European "tour." These inspired foreign-language translations of *British Zoology*, such as that translated into German and Latin titled *Zoologia Britannica* (1771). Pennant described the plates as being "copied and coloured by the ingenious artists of that city [Nuremburg]," a product of the connections made during his continental journey.[34] Pennant gave copies to his most valued British correspondents, including the naturalist George Edwards, who produced several of the original drawings for *British Zoology*, and William Borlase, who had maintained an extensive correspondence on the natural history of Cornwall.[35]

In addition to reserving copies to give away in 1766, Pennant purchased twelve sets for himself and kept these as gifts for naturalists who contacted him as his reputation began to spread alongside the publication. In 1771 Linnaeus wrote to Pennant expressing an interest in his publications. Pennant responded: "You do me such honour in so earnestly desiring my works, that I beg the favour of you to accept all the remainder. viz. [of] my folio edition of the Br. Zool. 132 plates uncoloured. for I have never a set coloured."[36] The fact that Pennant did not have the sets he sent to naturalists colored shows a clear difference between copies reserved for the open market and those presented as gifts. In the advertisement leaf appended to the 1768–1770 edition of *British Zoology*, the 1766 edition was marketed as containing "132 plates beautifully coloured."[37] It seems that subscribers had a choice of having their copy colored, an addition that was often rejected by naturalists and accepted by aristocrats. This allowed the owner to have the images colored according to their specifications, or, in the case of Linnaeus, who had argued against the use of color as a defining character for species throughout his career, to leave plates in their uncoloured state. Linnaeus incorporated Pennant's uncolored plates into his system and broader approach to managing information in his library; at the foot of every plate from *British Zoology*, such as that which depicts thrushes and starlings, Linnaeus inscribed the binomial names published in the 1766 edition of *Systema Naturae*, linking Pennant's images with the Linnaean name and description (plate 8). This presents a very different means for interpreting these images when compared to the colored copies in aristocratic libraries and shows how images were read by an audience of naturalists.

The Economy of Books as Gifts

A result of continued financial difficulties with the grand folio, Thomas Pennant planned a different approach to publishing a new edition of *British Zoology*. This was outlined in a letter sent to John Murray, the Duke of Atholl, in 1769: "All I can propose is that I w^d join with any gentleman for engraving some of the most capital birds of the great plates; not for sale but to give away to friends."[38] In a similar manner to the copies of the folio edition he gave away, these books were designed to solidify Pennant's network of naturalists, gentry, aristocrats, and metropolitan elites who could send information from across Britain, Continental Europe, and North America. However, this plan changed when Benjamin White gave Pennant £100 to produce an octavo edition. To avoid tainting his genteel reputation by associating his publishing program with payments from a commercial (all be it genteel) publisher, Pennant promptly "vested [the money] in the Welsh charity school."[39] This was a charitable educational institution established to educate children and funded by private benefactors, although many were members of the Honourable and Loyal Society of Ancient Britons and the Honourable Society of Cymmrodorion.

Unlike the folio *British Zoology*, the vast majority of the five hundred copies of the new octavo edition published between 1768 and 1770 were designed to be sold on the open market. However, Pennant retained several copies for his "learned and ingenious friends," which included the lawyer and naturalist Daines Barrington; the collector William Constable; Joseph Banks; Benjamin Stillingfleet; William Borlase; the clergyman and classicist Thomas Falconer; the civil engineer and botanist Thomas Tofield; Richard Farrington, a Welsh Anglican priest; Owen Holland; Henry Seymer; Daniel Solander; Peter Collinson; Gilbert White; and George Edwards, whom Pennant acknowledged as the "Father of *British* Ornithologists."[40] This list has several crossovers with the subscribers to the 1766 edition and shows how Pennant sought to bond a specific network of naturalists through the ownership of his work while reducing criticism of a publication on which he was identified as the author. The octavo editions had a better reception among Pennant's circle than the folio *British Zoology*. For example, Peter Simon Pallas, who "had the good fortune of conversing with you [Pennant] in Holland" in 1765, commented that "I also saw the 8vo Edition of your British Zoology & prefer it to the atlantik

Edition."[41] Pallas's comments relate to the comparative ease he experienced when reading and carrying copies of the quarto edition on expeditions—a very different-sized text when compared to the folio.

As Geoffrey Turnovsky has suggested, it was common for those who had purchased a book to view its content as their personal property and subject it to public criticism. In comparison, those who received books as gifts from the author felt obliged to give favourable reviews and treat the content as the author's property.[42] There was also more of an incentive to recommunicate information and contribute to a network of natural history. The ability to maintain a reputation as a genteel author and patron of knowledge was of great importance to Pennant. Unlike the 1766 edition of *British Zoology*, in which Pennant remained anonymous, Pennant signed off the prefaces published in the 1768–1770 and 1776–1777 editions, identifying himself as the author. This shows his personal caution about being identified as the author: Pennant's name remains absent from the title pages of all the editions of *British Zoology* published during his lifetime. It was not until the posthumous 1812 edition, a book produced on commission, in which the title page identified Pennant as the author. Pennant's gifting of *British Zoology* combined with his hesitancy in claiming sole ownership of the content emphasizes the collaborative nature of his books. It also secured Pennant's defence against unfavorable reviews. Both the *Monthly Review* and *Critical Review* gave *British Zoology* (1777) and Pennant extensive praises, defining him as a genteel naturalist in the eyes of the reading public. After praising the quality of the plates and production, the *Critical Review* described both Pennant and his work:

> In the quotations from ancient authors, we still find the man of elegant taste, and the judicious critic, such as he [Pennant] has displayed himself throughout all his writings. His terminology is as useful, extremely apt, concise, and expressive; and the definitions and descriptions very perspicuous and just, Linnæus having been our author's chief guide in these lower links of the zoological system. Taking all the excellencies of the British Zoology together, our countrymen may now boast a more completely instructive account of the animal creation in this island, than any other country ever possessed. In regard to the historical part, the Fauna Suecica of Linnæus, however valuable, cannot be compared to it.[43]

Pennant's liberal presentation of British Zoology, then, had the simultaneous effect of solidifying his reputation as a genteel naturalist while promoting the intellectual and physical qualities of his books.

Pennant had several motivations for presenting his books to correspondents. The first was to repay collaborators, elevating their collective standing in the eyes of the reading public. Secondly, Pennant's work became a standard reference for naturalists in his network when conducting research in the field, consolidating notes, cataloguing specimens, constructing their letters, and compiling books. Many correspondents sent publications to reciprocate Pennant's gift. In Pennant's presentation copy of Anders Sparrman's account of his travels in South Africa and James Cook's second voyage, Sparrman addressed Pennant as "the English Plinius, Mr Thomas Pennannt F. R. S. Esqr" while Pallas ensured Pennant "will immediately receive a copy" of his new fauna of the Russian Empire.[44] Similar to Carl Linnaeus's use of Species Plantarum to unite an intercontinental network, Pennant created a "paper empire," bonding individuals through the gift economy and standardization of paper technologies established though circulating British Zoology.[45] Gifting British Zoology secured Pennant's learned networks while encouraging recipients to use the printed text as a template to observe and record the natural world, assessing the novelty of species and their distribution.

Reciprocal gifts and the communication of information across a network created a cyclical process, giving Pennant and his son, David, the opportunity to update British Zoology for new editions. Even by the 1830s, when David Pennant was preparing another edition of his father's work, the letters he received contained numerous references to specific pages and descriptions in British Zoology (1812). A typical example can be found in Pennant's correspondence with the local vicar of Llanasa, the Reverend Henry Parry, who, on September 21, 1837, wrote to describe a species of mackerel he had received from Sir Edward Mostyn that "answered exactly to its description in the Zoology except as to the tail which was more of a very obtuse angle" (figure 4.5).[46] British Zoology provided a baseline and important reference point for Pennant's correspondents to formulate the information they communicated through citing specific species, volume, and page numbers. Thomas and David Pennant then tipped these letters into the specified sections in their personal copies of British Zoology. This shows how correspondence was incorporated into new

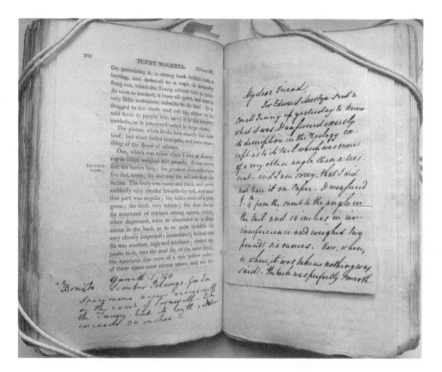

FIGURE 4.5. David Pennant's annotated copy of *British Zoology* (1812). Pennant annotated the pages, added flyleaves of notes, and tipped in letters from correspondents, including the letter shown sent by Henry Parry on September 21, 1837. Private collection.

editions, a network secured through those who received presentation copies, which shaped their means for practicing and observing nature through a series of species and page numbers.

Two surviving presentation copies of the 1768–1770 edition of *British Zoology* are those Pennant gave to Joseph Banks and Gilbert White. White added marginal notes on his frequent journeys around the parish of Selborne and Banks's marginalia relate to species observed during his voyage to Iceland in 1772.[47] After receiving his copy, White inscribed, "May 4: 1768. The gift of the Author," and his annotations relate to observations of species in and around Selborne. For example, White's marginalia appears next to Pennant's description of the season for the hatching of martin chicks, where he added the note, "In 1772 young martins continued in their nests 'til Octobr: 22."[48] This observation was communicated to Pennant in a letter, in which White remarked that "in 1772 they had nestlings on to *October* the twenty-first."[49] Pennant incorporated this observation into the 1776 edition of *British Zoology*, commenting that "Nestlings have been remarked in *Hampshire* as late as the 21st of *October* 1772."[50] This shows how Pennant's gifting of natural history books stimulated the observation, annotation, and redistribution of information that was incorporated into later editions.

Similar to White, Banks acknowledged that his copy of *British Zoology* was a gift, inscribing on the first end paper, "The Gift of my Esteemed Friend the author, Thos. Pennant Esq.[re]."[51] However, the geographical scale of Banks's use of these books and his approach to and motivations for recommunicating observations were very different to White's. Banks took these volumes to Iceland in 1772 when he and Daniel Solander used them to identify and compare the ornithology of the island to that of Britain. Nearly all of Banks's annotations relate to the variations between Icelandic and British birds, a typical example being Banks's note on Pennant's description of the Whimbrel; "× 2 shot in Iceland Sep[tr]. 1772 feeding upon berries of Empetrum & Vaccin. Uliginos." Banks recorded the locality of these birds, the weights of the specimens, and the date. This approach to recordkeeping is very similar to that Banks employed in his journal. For instance, on August 20, 1772, Banks described how "the lest Auk / Alca pica Linn./ was shot. It seemd to be a young bird & varied from Linnæus's & Pennants descriptions."[52] Banks's notes connect the printed text to the observation recorded in his narrative journal and the physical specimen, showing how print connected these different physical entities.

The layout of Banks's annotation is very similar to Pennant's description in *British Zoology*. When describing the whimbrel, Pennant noted the sizes and shapes of various physical characters and mentions that the bird never exceeds "the weight of twelve ounces." The specimens Banks observed in Iceland weighed 1 pound 3.5 ounces and 1 pound 3 ounces; both of these are significantly heavier than Pennant's claims in *British Zoology*, a weight Pennant continued to declare in *Arctic Zoology* (1784–1885).[53] Thus, Banks did not communicate this information to Pennant, using his copy of *British Zoology* to gather private observations whilst avoiding published references to his research. Banks's lack of interest in communicating this information was probably a result of the problems that arose during the publication of John Hawkesworth's and Sydney Parkinson's accounts of Cook's voyage in the early 1770s. Banks did not receive a presentation copy of the next edition of *British Zoology* due to his comparative lack of contributions to Pennant's enterprise.

As Pennant's network increased in scope, so did the content of each edition of *British Zoology* and the number of copies he distributed as gifts. The publication of the 1776–1777 edition in quarto and octavo formats emphasizes the different social ranks of correspondents Pennant intended to receive copies. The most exclusive copies were those printed in quarto. Out of the entire print run, 500 copies were printed in this format, and one was presented to Margaret Bentinck, the Duchess of Portland. On the dedication leaf Pennant wrote: "To the Duchess Dowager of Portland, this work is Dedicated as a Grateful Acknowledgement of the many favors conferred by her grace on her most Obedient Humble Servant, Downing, March 1, 1777. Thomas Pennant."[54] Pennant maintained a correspondence with the duchess throughout the 1770s, borrowed shells from her collection, and published several images and descriptions of these in *British Zoology*.[55] The Duchess of Portland represented a major aristocratic naturalist of the period. By the time of her death in 1785, the combined annual income from her estates and property totaled approximately £14,000. This allowed the duchess to expend vast amounts on her collection; some speculated this had cost up to £15,000.[56] The other 1000 copies of *British Zoology* were printed in octavo for purchase by the country gentry, clerks, and rural vicars. Many were also gifted to Pennant's correspondents as the smaller volumes were more practical to use—Pennant's own copies, which contain extensive annotations, interleaving, and original illustrations, are all in the octavo format.[57]

As he made clear to James Edward Smith in 1798, Pennant gave numerous copies of his books away. His list of "Br. Zoo. Given Away" shows the distribution of twenty-one copies of *British Zoology* (1776–1777) (figure 4.6).[58] Sixteen were circulated by Benjamin White and a "Mr Walton," showing the role of booksellers in distributing Pennant's books to genteel society. This list contains correspondents Pennant mentioned in the preface, although he seems to have concentrated his efforts to give copies to those who might not have been able to afford this book, individuals who contributed information to this edition, and acquaintances made on expeditions to Scotland and the Hebrides in 1769 and 1772. For example, Banks did not receive a presentation copy, although Gilbert White, to whom £2/8s would have been a substantial sum, did receive a copy.

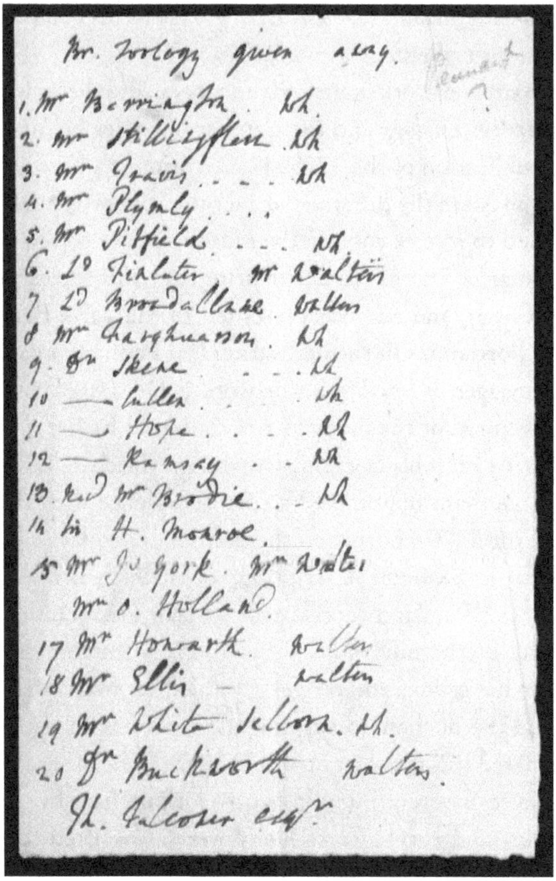

Br. Zoology given away.

Pennant ?

1. Mr Barrington. Wh.
2. Mr Stillingfleet. Wh
3. Mr Travis . . . Wh
4. Mr Plymly . . .
5. Mr Pitfield . . . Wh.
6. Ld Finlater Mr Walter
7 Ld Breadalbane Walter
8 Mr Farquharson. Wh
9. Dr Skene . . . Wh
10 — Cullen . . . Wh
11 — Hope . . . . Wh
12 — Ramsay . . Wh
13. Revd Mr Brodie. . . Wh
14 Sr H. Monroe.
15 Mr Jo York. Mr Walter
Mr O. Holland
17 Mr Howarth. Walter
18 Mr Ellis — Walter
19 Mr White Selborn. Wh.
20 Dr Duckworth Walter.
Jh. Falconer esqr

FIGURE 4.6. Thomas Pennant's list of "Br. Zoology given away." By permission of the Linnean Society of London.

Three major figures of the Scottish Enlightenment also received copies, these being John Hope, a founder of the Royal Society of Edinburgh and pioneer of the Linnaean system in Scotland; William Cullen, a professor at the Edinburgh Medical School who had particular interests in chemistry, agriculture, and natural history; and Robert Ramsay (1735–1778), the first holder of the Edinburgh chair in natural history and keeper of the University Museum. In the "Queries" Pennant circulated before his 1772 Scottish tour, he stated that the "object of my collecting is not selfish; whatever I may acquire will be at the service of the national repository, the Museum at Edinburgh, now under the care of my friend Dr Ramsay, Professor of Natural

History." The development of a national repository gave a direct motivation for local Scottish elites to respond to Pennant's questionnaires and assist with his expedition at a local level. [59] However, rather than being a reference to the physical specimens Pennant took back to his home at Downing Hall to construct *British Zoology*, Pennant's statement refers to gifts of his publications. Books were far more useful to correspondents than objects, whether these were individuals or institutions. *British Zoology* not only provided copperplate illustrations of specimens many naturalists believed to be second only to the real thing, they also gave a certain level of structure, providing examples of how to classify new material accessioned into the Edinburgh University Museum. Pennant's work could then be used to arrange information conveyed in correspondence and university teaching, bringing Scotland under the fold of the network Pennant governed in accordance with the descriptions, page references, and systems of classification in *British Zoology*.[60]

In comparison to the more practical application and distribution of octavo copies of *British Zoology* (1776–1777), Pennant intended for the quarto copies to find their way to the highest-ranking individuals in British society. This is immediately apparent from examining volumes that contain traces of past of ownership. Examples include Joseph Banks, who, by the late 1770s, placed himself at the head of an elite group of natural historians, aristocrats, and members of the government before his election to the presidency of the Royal Society of London in 1778; the Duchess of Portland, who gave Pennant access to specimens; and King George III (r. 1760–1820), whose copy has a gold tooled royal coat of arms embossed on the covers.[61] All these individuals owned vast estates, while landownership gave these figures a vested interest in the practical management of nature.

Banks consulted his copy of *British Zoology* when on trips around Britain and interleaved it with correspondence and flyleaves of notes. For example, Banks added a flyleaf next to the entry on the ring ouzel, on which he extended Pennant's original description by adding information from his own observations: "These birds breed in large quantities in the cliffs on the South side of the Isle of Wight where I saw them in the year 1777 & was informed that they arrive about the middle of april & stay the whole summer."[62] Although this information was not recommunicated to Pennant, it shows how *British*

Zoology brought a certain level of structure to observing the natural world. By the 1780s, this copy had been incorporated into Banks's library at 32 Soho Square.

King George III annotated his copy of *British Zoology* according to his interests in agriculture. A reputed enthusiasm for agricultural improvement gained the king some popularity early in his reign, earning the title of "Farmer George," before he succumbed to madness in the years around 1800.[63] George III's interests in natural history were probably inspired by his tutor, John Stuart, the Earl of Bute, a subscriber to the 1766 folio edition of *British Zoology* and a naturalist who was familiar with the processes of annotating publications. The king's use of these practices becomes apparent from a flyleaf tipped into the section on salmon that outlines the life cycle of the fish, the process of spawning, and experiments on young fish kept in freshwater ponds. George III's attempts to obtain information on these processes are apparent through comments such as "a Gentleman of my acquaintance (Mr. Dadosworth of Crakehill near Bedale) has frequently put them [samlets] into his pond where they grow to the size of 15 inches in two years but no larger. He has never had the opportunity of trying them in a good trout pond: his own is particularly unfavourable, being muddy & full of weeds."[64] The king described the weights of the different fish grown in the pond. These differed depending on the time of year they were caught and subjected to the experiment. George III's annotations show that *British Zoology* formed part of his organizational structure for agricultural information that he referred to as "that greatest of British manufactures" while his ownership of *British Zoology* (1776–1777) indicates the luxury audience of the quarto printing.[65]

Presenting Gifts from the Private Press

Thomas Pennant continued to present books as gifts throughout his career, compiling lists of recipients destined to receive complimentary copies for each new publication. This remained the case for his last book, *A View of Hindoostan*, published a few months before his death in 1798, a work covering the natural history and antiquities of India. Pennant compiled a list of seven individuals destined to receive copies as gifts. These included the Royal Society of London; the historian William Coxe; the ornithologist John Latham; John Douglas, Bishop of Salisbury; James Edward Smith, President of the Linnean Society of London; Thomas Maurice, keeper of manuscripts at the British

Museum; and the artist Thomas Daniell.[66] All had corresponded with Pennant and produced publications he had used to compile the text. Daniell gave Pennant permission to copy images from his *Oriental Scenery* and aquatints he produced for Alexander Allan's *Views of Mysore County* (1794). Examples include the plate Pennant described as "Sawen Droog." Pennant personally produced a reduced watercolor based on Daniell's aquatint in 1797 and commissioned Francis Chesham to produce an engraving for *A View of Hindoostan*.[67] In comparison, Maurice received a copy since he had maintained an extensive correspondence and facilitated Pennant's visits to the British Museum. Shortly after the publication of the third and fourth volumes of *Outlines of the Globe* Maurice sent a letter outlining the high scholarly value of Thomas Pennant's work while thanking his son David for a "very obliging and valuable present."[68] However, not all naturalists took Pennant's approach to covering expenses by maintaining a strict ratio between the copies given away and copies sold on the open market.

A typical example is the approach taken by William Jackson Hooker (1785–1865), who published two editions of his *Journal of a Tour in Iceland in the Summer of 1809* in 1811 and 1813. This was the main published product of Hooker's journey as "compensation for my not having it in my power, during that season, to put in execution a projected voyage to a tropical climate."[69] On the return journey, however, Hooker lost many of his specimens and records, describing how "the only things rescued from the flames were a portion of my journal." The loss of Hooker's collection caused Banks to offer the materials collected on his Icelandic expedition in 1772 to aid with the production of this work and some of these, such as the illustrations of the erupting geyser, were reproduced in Hooker's book.

Hooker's decision to privately publish the 1811 edition of his *Journal of a Tour in Iceland* was a direct result of his apprehension at presenting a book that had been cobbled together from memory, the scant remains of his own journal, Banks's collections, published accounts with no associated natural history collection. In a similar manner to Pennant's means for producing and distributing octavo books, such as *British Zoology* (1776–1777), this was done through commissioning a commercial printer. To do this, Hooker, who was originally from Norwich, relied on his local Norfolk network, exclaiming in the preface that "it is to Mr. Dawson Turner of Yarmouth that these sheets

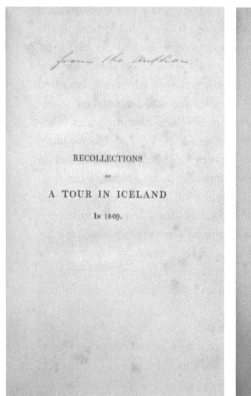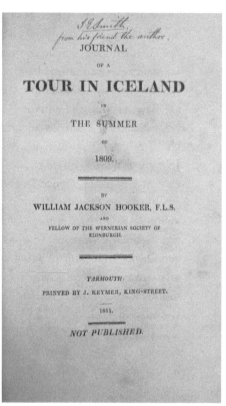

FIGURE 4.7. The title and half title pages from William Jackson Hooker's *Journal of a Tour in Iceland* (1811). The presentation inscription is at the top of the half title page (left) and the poignant italicized NOT PUBLISHED is at the foot of the title page (right). Private collection.

owe their existence."[70] Turner—a wealthy banker, traveler, and botanist, known in natural history circles for his *Fuci; Or Other Coloured Figures and Descriptions of the Plants, Referred to by the Botanists to the Genus Fucus* (1808–1819)—was responsible for combining his funds with Hooker's to employ the Yarmouth-based printer J. Keymer to print Hooker's *Journal of a Tour in Iceland* (1811). In the end, Keymer produced five hundred copies and the vast majority were printed in octavo, although ten copies were on large paper. These were designated for members of Hooker's family. For example, the large paper copy Hooker presented to his mother-in-law, Mary Turner, measures 26 × 17 centimeters, whereas the standard smaller copies measure 22.5 × 14 centimeters.[71] Thus, Hooker and Turner had no intention to break even financially when producing this work, reserving every single copy as a gift.

Hooker's distribution of the entire print run of this book as presentation copies was not just reflective of the standard genteel practice of avoiding commercial markets, but was essential for expanding Hooker's network of naturalists at an early stage of his career. A typical example includes the copy Hooker presented to James Edward Smith, on which a grateful recipient inscribed "from his friend the author." The desire to avoid commercial publishers was made clear on the title page, which, as opposed to the name and address of a publisher, retains the clear statement "NOT PUBLISHED" (figure 4.7). On the half title page of every copy, Hooker wrote "from the author," reemphasizing his personal interest in gifting this book while shielding his work from unfavourable reviews.[72]

A result of other naturalists' favourable reception of the 1811 edition, and, with Banks's encouragement, Hooker produced a new expanded edition of his *Journal of a Tour in Iceland*, published in two volumes in 1813. This edition was produced and distributed in a similar manner to Pennant's works. Numerous copies were kept aside from the conger of publishers listed on the title page, an example being the copy Hooker presented to John Brightwen who, after 1816, became a partner in Dawson Turner's Yarmouth bank. Brightwen's son married Hannah Turner, Hooker's sister-in-law (figure 4.8). However, Hooker continued to hold reservations about the content of his much-expanded work, fearing that "in the public I must expect to meet with less favourable judges" than those who received presentation copies. To defend it from criticism, Hooker dedicated this work to Banks, explaining that this was "to shield myself under the authority of a man, to whose judgement they are accustomed to pay the same deference that I do."[73] A similar line was taken by Turner in the dedication for his *Fuci* (1808), describing Banks as "the patron of science."[74] Hooker's faith in Banks's ability to defend this book from attacks in the popular press represents the power Banks held over the distribution and judgement of natural history publications by the early nineteenth century. The act of gifting books was central for maintaining Banks's "learned empire" and Hooker's emulation of this approach to the publication and distribution of natural knowledge associated him with the pinnacle of an elite social and scientific support network at the beginning of his career.

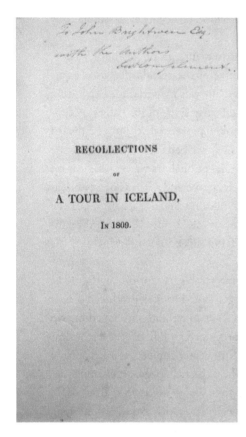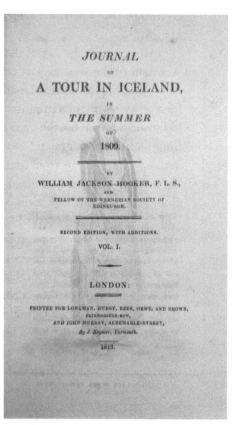

FIGURE 4.8. The half title, inscribed by William Jackson Hooker, and title page from John Brightwen's copy of Hooker's *Journal of a Tour in Iceland* (1813). Private collection.

Joseph Banks's "Monopoly" on the South Seas Plants

In contrast to publications by Thomas Pennant, Gilbert White, and William Jackson Hooker, Joseph Banks only ever intended to distribute the books he commissioned as gifts. This approach was inspired by Banks's poor experience with commercial publishing markets and the humiliation he received after the publication of Hawkesworth's *An Account of the Voyages Undertaken by the Order of His Present Majesty* (1773). Banks avoided the commercial publishing industry throughout production and booksellers during distribution. Banks had far stronger reservations than Pennant on the concept of marketing natural knowledge, believing it was improper etiquette for someone of his station to publish for financial gain, an attitude shared by other self-publishing aristocrats such as Horace Walpole. These views on authorship and publishing were spelled out in a letter Banks wrote to the Duchess of Somerset in 1816: "The Crier who is no Reporter of Loyal Subjects or the Truth has led your Grace into an Error in Suspecting me of having Published any thing whatever. I have hitherto avoided the lash of the Scotch Whippers in by Quietly Contenting myself with the Station I hold in life & not like the dog with his shoulder of Mutton hazarding what I feel to be Comfortable in seeking the reputation of an Author which in fact I do not consider a gentlemanly vocation. What your Grace is alluding to in an account communicated by me to the Journal of the Royal Institution of British Canoes found in Lincolnshire printed in the first No. of the Journal of the Royal Institution."[75] Thus, Banks believed gossip in polite circles and newspaper reports were responsible for circulating inaccurate information about his approach to producing publications. This letter spells out the fact that Banks was enthusiastic to act as a well-connected patron for those who published the results of their inquiries and communicate these accounts to journals—the *Philosophical Transactions* alone contains hundreds of accounts he communicated. Other journals that received accounts Banks communicated include *Annals of Botany, Transactions of the Linnean Society of London, Archaeologia, Communications of the Board of Agriculture, Transactions of the Horticultural Society of London,* among others. Many accounts Banks communicated were the product of research he patronised, that based on his collections or inquiries Banks stimulated—an example of Banks's approach to stimulating the production of published papers is discussed in chapter 5. It also reflects on the fact that, as Aileen Fyfe and Noah Moxham have suggested, Banks was concerned with gatekeeping his own collections, while playing a pivotal role in the production of numerous other works of natural history, antiquities, and travel.[76] Banks's desire to remain as the main patron of inquiries into natural knowledge is emphasized through his manner of distributing the books he commissioned and had produced at Soho Square. In response to the emergence of commercial publishing, Banks, one of the 250 or so wealthiest landowners in the country, sought to integrate his publishing program with an aristocratic gift economy that began in the sixteenth century, combining natural history publishing with the expectations of genteel society while building a diverse global network.[77]

Rather than turning to commercial booksellers, Banks's private production of books outlined in chapter 3 allowed him to control the distribution of the small quantity of copies he commissioned. These were presented to a select circle of naturalists, aristocrats, politicians, and institutions with interests in natural history or viewed these publications, which often contained large copperplate images, as symbols of Banks's position in society. This was central for bonding the diverse membership of institutions, such as the Royal Society, or interlinking Banks's collection at 32 Soho Square with the collections of other naturalists and emergent state repositories.[78] A typical example is Banks's intended means for distributing the copperplate images of plants based on the drawings Sydney Parkinson had produced on James Cook's first voyage. This was described with contempt by Georg Forster, who visited 32 Soho Square in 1790: "The great work of Banks is still a topic that keeps the conversation alive. He will, as he says and writes to his friends, never sell it, but only print a few copies and give them away.—Almost all 17 to 1800 plates are supposed to be ready. No one knows the reason for the further delay; Dryander himself does not seem to be able to say it or does not want to."[79] Thus, Banks only ever intended for a few copies to be produced that could then be given to fellow naturalists. The reasoning for this was again made clear by Forster, who, in a letter to the philosopher Christian Wilhelm von Dohm (1751–1820) described how Banks constructed a personal "monopoly" over the plants from the South Seas.[80] The limitations placed on the audience for this material are also alluded to by other commentators. For example, the American chemist and Yale professor Benjamin Silliman described the necessity of supplying a letter of introduction to visit Banks's collection, suggesting, "I had supposed myself to be precluded from calling on Sir Joseph Banks, as I had left a letter of introduction with my card, on my first arrival in London and had never heard anything further on the subject."[81] Although Silliman was granted access to Soho Square due to the intervention of James Watt (1736–1819), the necessity to supply letters of recommendation was an essential part of the process of managing the trustworthiness of those given access to Banks's collection. For example, it always ensured they looked at books and specimens on the terms dictated by Banks and his librarians—and only used them to produce publications that Banks approved of to ensure their continued access to the collection.

In the end, only two complete sets of impressions were taken from the 743 plates Banks had engraved to depict the plants from the South Seas. The motivation for this was to reduce exposure of this material to criticism, satire, and the literary pirates of Grub Street and overly critical reviews. Forster was not the only naturalist to criticize Banks for limiting access to those who visited his library. For example, in 1776 Anders Sparrman (1748–1820), who accompanied the Forsters on Cook's second voyage to the Pacific, warned against presenting Banks with original material. Sparrman commented: "I would not make an offer of any thing to this gentleman if he does not make first advances and proposals, and than he would get nothing but dupli-cates" preferring "to send somebody less avaricious of collections and the merit of them than he [Banks] is."[82] This is a direct reference to Banks's approach to printing and giving access to his own collections of plants from the South Seas and a warning to Forster—who had just sold the majority of his natural history drawings to Banks to alleviate the debts he accrued during the voyage—as to what would happen to his collection.[83]

Banks's regular instructions to his librarian, Jonas Dryander, clar-ify the fact that naturalists only viewed his natural history collection under his own terms. Those who did not stay on good terms with Banks and his curators or who developed approaches to publishing images without acknowledging the sources were either excluded or kept under close supervision. For example, Banks wrote to Dryander in 1786 to advise him on the forthcoming visit of the French naturalist Charles Louis L'Héritier de Brutelle (1746–1800). Banks instructed Dryander to "limit his visits to the Library" and "look out after speci-mens which are very scarce."[84] These comments are a result of L'Héri-tier's reputation for publishing materials from private natural history collections without the consent of the owner, reflecting the very differ-ent approach to natural history publishing and intellectual property held by French natural historians. Such activities posed a direct threat to Banks's efforts to control information published on the collections from Cook's first voyage, leading to its reservation for a select group of naturalists who were permitted to visit the inner sanctum.

Some naturalists received a far more favourable reception than L'Héritier. One was the Swedish naturalist Carl Peter Thunberg (1743–1828), who visited Banks in 1778 on his way back to Sweden from Japan. Thunberg described the process of examining Banks's collection: "I was permitted, previous to my departure, to view the Collection of Plants made from the islands of the Pacific Ocean, which were not as yet placed among the other plants [in the herbarium], and are not shown indiscriminately to every stranger."[85] Although Banks's library at 32 Soho Square was famed as a major meeting point for natural historians where, according to Thunberg, "several learned men daily assembled here, as though it were an Academy of Natural History," certain materials were kept aside.[86] To view these, one had to make a special request to Banks's curator and was kept under constant supervision. Both copies of the plates Banks commissioned for the South Seas plants were kept at 32 Soho Square—one was bound in two volumes and kept in the library; the other was kept alongside the materials Banks, Solander, and their team of artists, secretaries, and field assistants compiled and collected on the voyage. The latter could then be interlinked with Parkinson's original illustrations, Solander's notebooks, manuscript slips, and interleaved books (assembled by Banks, Solander, and Herman Spöring), and the specimens themselves.[87] This allowed users of the collection to read the printed images alongside Banks's wider Pacific collection, connecting each plate to the original illustration, description, and representative specimen.

Building Joseph Banks's Empire through Gifting Books

In comparison to the products of his own global voyage, Joseph Banks was more willing to publish information collected by other naturalists. This is apparent through the cases of *Reliquiæ Houstounianæ* (1781) and *Icones Selectæ Plantarum* (1791), the production, publication, and distribution of which were exclusively administered by Banks. However, neither of these books was designed to enter commercial markets and Banks presented every copy to a specific recipient. In comparison, other books Banks funded and organized the production of, such as William Roxburgh's *Plants of the Coast of Coromandel* (1795–1819) and William Aiton's *Delineations of Exotick Plants* (1796), were intended to be sold in small quantities. For example, a total of 210 copies of *Delineations of Exotick Plants* were produced in three printings, many of which were presented to notable individuals and only a small number were designated for sale.[88] These books followed a similar mode of production to that used by William Jackson Hooker and Thomas Pennant; both were privately commissioned from William Bulmer and neither were intended to make financial returns for their backers.

The bookseller George Nichol had a nominal role in the distribution process.[89] This is apparent from Roxburgh's *Plants of the Coast of Coromandel*. In 1798 Banks wrote to Roxburgh, commenting that production was slow because "the Court [of directors of the East India Company] find it expensive because they give away so many coloured copies."[90] The gifting of expensive colored copies to patrons and company shareholders reflects Banks's influence over the directors of the East India Company to ensure the majority of copies were presented to encourage the development of a powerful network and extend the patronage of natural history.

As seen in the previous chapter, Banks began his forays into private publishing with *Reliquiæ Houstounianæ*, a book he believed to be of great use to botanists throughout Europe and the West Indies. However, this work was not marketed, and no bookseller was involved in the distribution process. Rather, as reported by a reviewer in the *London Medical Journal*, this book was "printed at his [Banks's] expense, and liberally distributed to learned societies, public libraries and botanists in different parts of Europe." After employing Bulmer to produce the letterpress that was attached to the copperplate impressions printed on the rolling press in the room under the library of 32 Soho Square, Banks distributed *Reliquiæ Houstounianæ* to notable naturalists and institutions. This strategic distribution of presentation copies was central for securing Banks reputation in the already sophisticated correspondence networks of the North Atlantic.[91] Many of these were still dominated by the Spanish, resulting in one of the recipients of this book being the Madrid Academy of Sciences, the secretary of which, Casimiro Gómez Ortega (1740–1818), thanked Banks for his donation on February 17, 1785. Johann Reinhold Forster (1729–1798), a naturalist on James Cook's second voyage and by this time a professor at the University of Halle, also received a copy for which he thanked Banks on July 6, 1782, and the French physician and naturalist Johann Hermann (1738–1800) acknowledged his copy on May 3, 1788.

Jonas Dryander, Banks's librarian, presented several copies of *Reliquiæ Houstounianæ* to Johan Christian Fabricius (1745–1808), who had lodged at Soho Square for several weeks in 1782 when studying Banks's natural history collection. Fabricius was given instructions to distribute several copies to individuals and institutions around the Baltic and Central Europe, such as a "Professor Leike" at the Electoral

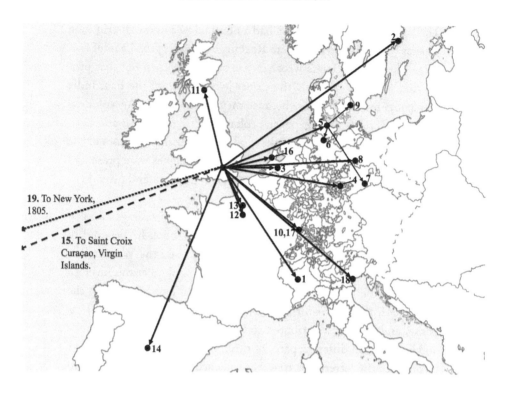

19. To New York, 1805.

15. To Saint Croix Curaçao, Virgin Islands.

FIGURE 4.9. A map and key showing the distribution of *Reliquiæ Houstounianæ* (1781) outlined in Joseph Banks's surviving correspondence.

Library in Dresden. Additionally, "Fabricius took also with him all the copies intended for Denmark, which he will distribute."[92] This shows how Banks not only circulated this book through his own extensive network of correspondents, but relied on those of other notable naturalists (figure 4.9).

Banks also gave copies to those who attended his "breakfast" meetings. These were held on a daily basis when he was resident at 32 Soho Square and were frequented by a range of naturalists, explorers, physicians, politicians, diplomats, and fellows of the Royal Society. The serving of a breakfast was of a secondary concern to attendees. Rather, these meetings were designed as forums for philosophical discussions while the unusual natural-historical food servings were the source of much ridicule and were immortalized in Thomas Rowlandson's satirical cartoon "Sir Joseph Banks about to Eat an Alligator," which served as the frontispiece for Peter Pindar's (a pseudonym

Key to Figure 4.9.

	Destination	Recipient	Method of acquisition and reference to Banks's correspondence	Page in Dawson (1958)
1	Turin	Carlo Alloni (1725–1804)	Thanks Banks for a copy of Houstoun's work, January 11, 1785.	16
2	Stockholm	Peter Jonas Bergius (1730–1790)	Thanks Banks for his copy of Houstoun's work on July 5, 1782.	47
3	Utrecht	Peter Boddaert (1730–1796)	Thanks Banks for a copy of Houstoun's work, May 1, 1783.	118
4	Halle	Johann Reinhold Forster (1727–1798)	Thanks Banks for a copy of Houston's work, July 6, 1782.	340
5	Kiel	Johann Christian Fabricius (1745–1798)	Dryander gave Fabricius several copies of *Reliquiæ Houstounianæ* to distribute in Denmark and Germany. From Dryander to Banks, September 24, 1782.	275
6	Hamburg	Dr. Ludwig	Distributed by Fabricius on Banks's behalf. "Dr. Ludwig" has not been identified. September 24, 1782.	275
7	Dresden	Professor Leike	Distributed by Fabricius on Banks's behalf. A copy for the electoral library, although "Professor Leike" has not been identified. September 24, 1782.	275
8	Berlin	Gesellschaft Naturforschender Freunde zu Berlin	The society thanked Banks for a copy of Houstoun's work on March 30, 1784.	47
9	Denmark	Unspecified	Distributed by Fabricius on Banks's behalf, who probably sent it to the Royal Danish Academy of Sciences, Copenhagen.	275
10	Strasbourg	Johann Hermann (1738–1800)	Thanks Banks for a copy of Houstoun's work, July 6, 1782.	408
11	Edinburgh	John Hope (1725–1726)	Thanks Banks for a copy of Houstoun's work, February 17, 1785.	424
12	Paris	Jean-Baptiste Lamarck (1744–1829)	Requested a copy of Houstoun's work, November 25, 1789.	519
13	Montdidier	Lendonny-Laucoug (dates not known)	Asked for a copy of Houstoun's work, August 24, 1784.	528
14	Madrid	Casimiro Gomez Ortega (1740–1818)	Returns the thanks of the Madrid Academy for a copy of Houstoun's work, February 17, 1785.	641
15	Curaçao, St Croix	John Ryan (F. R. S. 1798, d. 1808)	Thanks Banks for a copy of Houstoun's work, July 28, 1783.	726
16	Leiden	Edward Sandifort (1742–1814)	Requested a copy of Houstoun's work, September 21, 1784.	732
17	Strasbourg	Jakob Reinhold Spielmann (1722–1783)	Thanks Banks for a copy of Houstoun's work, November 9, 1782.	779
18	Venice	John Strange (1732–1799)	Thanked Banks for the "elegant botanical book" *Reliquiæ Houstounianæ*, May 29, 1783.	794
19	New York	Samuel L. Mitchill	Mitchill recorded Banks's gift in his journal, *Medical Repository* (1805), 413.	Cited in journal

for John Walcot) poem *Peter's Prophecy: Or, the President and Poet; An Important Epistle to Sir J. Banks* (1788).[93] According to Silliman, who visited in 1805, Banks was at the center of these events:

> We found Sir Joseph in his library, surrounded by a crowd of the litera-ti, politicians, and philosophers of London. These constitute his court, and would not dishonour the King himself . . . Among other distin-guished men who were present, was Dr. Wollaston, a chemical philos-opher of eminence, and Secretary of the Royal Society; Dr. Tooke, the historian of Catherine of Russia; Mr. Cavendish, who has done as much towards establishing the modern chemistry as any man living; Dalrym-ple, the marine geographer; Windham, the Parliamentary orator; and Lord Macartney, famous for his Embassy to China.[94]

These breakfasts were attended by individuals Banks deemed worthy to receive a copy of *Reliquiæ Houstounianæ*. This point was confirmed by the French traveler and naturalist Barthélemy Faujas de Saint-Fond (1741–1819), who remarked that Soho Square "is the rendezvous of those who cultivate the sciences . . . They assemble every morning in one of the apartments of a numerous library, which consists entire-ly of books on natural history."[95] Gifting books also gave the perfect opportunity and incentive for Banks to serve as a major patron for this circle who, in return, promoted Banks's interests in philosophical circles, Parliament, and on global expeditions. This gathering afford-ed Banks the ideal setting to distribute *Reliquiæ Houstounianæ* and explains why there were far more copies circulated amongst natural historians and philosophers than is made out in Banks's surviving correspondence. The conversations that took place among the select group at Soho Square often addressed the merits and drawbacks of the most recent publications on natural history and travel, providing the perfect backdrop for Banks to act as a supreme patron and circu-late copies of *Reliquiæ Houstounianæ*.[96] For example, when he visited Banks with Alexander von Humboldt in 1790, Georg Forster wit-nessed a conversation that covered the Scottish explorer of Ethiopic Africa, James Bruce's, *Travels to Discover the Source of the Nile* (1790). Forster described Soho Square as a place "where one judges sharply in general," adding that Bruce's work received "damning judgements" due to its author's utilization of Scottish commercial publishing markets to produce a "monstrous amount of copies" and secure a significant financial return.[97] Unlike authors who submitted their manuscripts to commercial printers, however, Bruce maintained authorial control over the production process, having the five large quarto volumes, containing numerous copperplate images, printed at his own expense in Edinburgh. The reason this work incensed the Banksian circle was that, to cover his expenses, Bruce contracted with the London bookseller George Robinson, who paid £6,666 for two thousand copies to be marketed at five guineas each. Bruce gained similar amounts from selling translation rights to booksellers in Paris and Leipzig. By 1794 Bruce was negotiating a further payment of £2,500 from the bookseller Thomas Cadell for a new edition of two thousand copies.[98] In total, Bruce gained a significant financial return.

Large payments from booksellers for the commercial production and distribution of publications went against all the notions of gentlemanly etiquette Banks held in esteem. For example, writing to the Duchess of Somerset, Banks described how he had avoided the "Scottish Whippers," a term used for the reviewers employed by various Edinburgh-based journals, perhaps the most notorious of being the *Edinburgh Review*, which rose to prominence in 1802.[99] Banks's means for removing his publications from commercial markets was central to the process of avoiding unfavorable reviews and assassinations of his character and scientific capabilities in the popular press. An example is the review of Banks's *A Short Account of the Causes of the Diseases in Corn* (1805), one of Banks's few publications to be distributed through a commercial bookseller, published in the Edinburgh periodical the *Farmer's Magazine*. In this review a typical "Scottish Whipper" described how the work was "perfectly unintelligible," adding that Banks was "imperfectly acquainted" with the subject.[100] As a result, *Reliquiæ Houstounianæ*—and the distribution of it within the circle who visited Soho Square—served as a reminder of a truly genteel means for publishing natural history that integrated wealth and leisure in the face of a rapidly expanding commercial publishing sector.

In addition to sending copies to prominent naturalists in Continental Europe and distributing them among his circle at 32 Soho Square, Banks ensured *Reliquiæ Houstounianæ* reached naturalists in the West Indies. On July 28, 1784, Banks received a letter from the Danish West Indian island of St. Croix, in which the plantation owner John Ryan returned his "most graceful acknowledgements" to Banks for his copy. Ryan was a correspondent and patron of the

Danish naturalist Julius von Rohr (1737–1793), founder of the botanic garden in Christiansted, St. Croix, and sent plants and descriptions to Banks.[101] Ryan, who had a medical degree from Leiden University and owned plantations on the island of Montserrat, collected botanical specimens throughout the West Indies and would have found the images in Houstoun's work useful for identifying species. The small size of *Reliquiæ Houstounianæ* ensured that this book could be used as a simple pocket field guide to assist in identifying the twenty-six species it described and depicted.

Banks's means for distributing privately printed books to a select network of correspondents seems to have been a standard approach used by aristocratic botanists during the 1780s. Four years after Banks published *Reliquiæ Houstounianæ*, John Stuart, the Earl of Bute, published his *Botanical Tables* (1785). Bute only had twelve copies of this book printed, and these were all gifted to individuals such as Queen Charlotte, wife of King George III; Empress Catherine II of Russia; Joseph Banks; Margaret Bentinck, the Duchess of Portland; George-Louis Leclerc, Comte de Buffon; and several important figures at the British court. Bute's careful distribution of his *Botanical Tables* was designed to strengthen his position in political circles and his relationship with royal and aristocratic patrons who had interests in botany. This was important given his fall from royal favor after a catastrophic premiership (1762–1763) and scandal in the 1770s.[102] In comparison, Banks's intercontinental distribution of *Reliquiæ Houstounianæ* shows how, from the early 1780s, he was starting to develop a sophisticated network of naturalists, using books as a cornerstone to solidify relationships and influence botanical practice across what became the Banksian Learned Empire.

Books donated to institutions often found a similar audience to the specific individuals Banks selected. Institutional libraries and societies that received copies of *Reliquiæ Houstounianæ*—such as the Royal Society, Madrid Academy of Sciences, and the Electoral Library in Dresden—were only accessible to fellows of those societies and those who carried letters of recommendation outlining their credentials. Similar precautions were taken to limit those who could access the collection of the British Museum, widely acknowledged as the world's first public museum, shortly after it opened in 1759. Access was only granted to study specimens in detail if they presented a letter of recommendation to prove their membership of the Republic of Letters.

Outlining the rules for entering the British Museum's reading room to its trustees in the early 1760s, Gowin Knight, the first principal librarian, described how "improper persons can be excluded," adding that this class was those "who have so little commerce with men of letters."[103] This confined the users of institutional repositories to those who already had access to private collections, limiting the audience of books such as *Reliquiæ Houstounianæ* to a close-knit group of naturalists who could be trusted to respect Banks's genteel approaches to the production and distribution of print.

Icones Selectæ Plantarum and the Publication of Empire

In comparison to *Reliquiæ Houstounianæ*, Joseph Banks spent considerably more money and time producing and distributing *Icones Selectæ Plantarum* (1791). This was a larger, more complex publication; it contains over double the number of copperplates, many of which were over ten times the physical size of those published in *Reliquiæ Houstounianæ*. Everything was produced at Banks's expense. As shown in the previous chapter, this book reflects the shift in Banks's interests from the botany of the South Pacific onto that of China and Japan, leading to his publication of a selection of illustrations from the British Museum's collection of Japanese material compiled by Engelbert Kaempfer. The increased size, quantity of plates, and fine quality paper used in *Icones Selectæ Plantarum* made it a far more powerful tool when forging relationships between Banks and naturalists throughout the globe. This book was central for Banks to emphasize his knowledge of East Asian botany and justify influence over the more frequent British expeditions to this region. The large plates, all of which were printed on Whatman's finest wove paper, not only produced high-quality illustrations suitable for botanical study but emphasized Banks's own great wealth and knowledge of producing complicated natural history books, positioning him at the pinnacle of an elite group of natural historians.

As with *Reliquiæ Houstounianæ*, Banks did not intend to sell a single copy of *Icones Selectæ Plantarum*. This is apparent from his response to the Tübingen-based publisher Christoph Friedrich Cotta (1750–1807), who asked for a copy in exchange for a selection of his firm's most recent publications. In the first instance, Banks refused Cotta's offer, commenting that "the book referred to in your letter

(Icones Kaempferianæ) was published for the Sake of my friends as a present for those who were so good as to assist me in Forming my Library ... I have determined therefore never to sell or exchange it in any way."[104] To receive *Icones Selectæ Plantarum*, Cotta had to prove his interests in natural history to Banks. As seen in the previous chapter, Banks was not concerned with the finances for this work, realizing that the expenditure would never be recuperated. Rather, Banks assessed that building relationships with naturalists throughout Europe and the world, along with providing a useful book on East Asian botany, was far more valuable than financial gain. Thus, this book served the same purpose as a diplomatic gift, maintaining the complex connections between Banks, other naturalists, and institutions.

Banks was prepared to offer free copies of *Icones Selectæ Plantarum* to botanists who made contact. This is apparent from his letter to Cotta: "If however any of your neighbouring botanists whose names are known to the list are without it & do me the honour of wishing to possess it I shall have great pleasure in supplying them."[105] Naturalists who contacted Banks included the Italian physician Francesco Antonio Notarianni (1759–1803) and the American physician Samuel Mitchill (1764–1831), who noted after receiving copies of *Icones Selectæ Plantarum* and *Reliquiæ Houstounianæ* in 1805 that "it is much to be regretted, that men of large fortunes do not more frequently expend a part of their income in promoting useful inquiry, end enlarging the bounds of knowledge, as this munificent and enterprising gentleman does." Another recipient included German botanist and pharmacist Carl Ludwig Willdenow (1765–1812), who was preparing a major new edition of *Species Plantarum* published between 1798 and 1826. Jonas Dryander started to use an interleaved copy of Willdenow's edition of *Species Plantarum* to order and record species present in Banks's herbarium during the first decade of the nineteenth century.[106]

William Roxburgh was another beneficiary who wrote to Banks requesting books. Roxburgh asked Banks to send *Icones Selectæ Plantarum* to Calcutta (now Kolkata) in 1801, believing it would be a useful resource for his botanical work. Although Roxburgh placed this request a decade after the initial publication of *Icones Selectæ Plantarum*, all Banks had to do was instruct an intaglio printer to run the plates through the rolling press in the basement of 32 Soho Square and append these to one of the letterpress components he had commissioned in 1791. This was one of many books Banks sent Roxburgh to assist with his fieldwork in Samalkota, deliveries that became routine after Roxburgh's appointment to his post in Calcutta in 1793. Banks saw Roxburgh's work as essential for expanding British trade and influence in the region, valuing it enough to commission Bulmer to print three hundred copperplate images as *Plants of the Coast of Coromandel* between 1795 and 1819. In the same letter, Roxburgh reported on his efforts to cultivate a new variety of Indian hemp. This material was essential for the production of ropes used in shipping, for which he was attempting to find a local substitute.[107] Banks recognized the importance of Roxburgh's work and the need for a new variety of hemp fiber to supply to British ships with ropes in India, probably a reason for his interests in the Calcutta Botanic Gardens and frequent gifts of expensive publications.

Not only did Banks respond to those who asked him for a copy of *Icones Selectæ Plantarum*, but he gave copies to recipients throughout Europe (figure 4.10). This was a strategic move, as these individuals consisted of notable and influential naturalists, such as Johann Friedrich Blumenbach of Göttingen (1752–1840), famed for conducting systematic studies of human beings as an aspect of natural history; Jacques-Julien Houtou de Labillardière (1755–1834), who had been appointed as the naturalist to accompany Bruni d'Entrecasteaux's expedition to the Pacific in 1791; and figures such as Carl Peter Thunberg, who succeeded Carl Linnaeus as professor of medicine and natural philosophy at Uppsala University in 1781. Many of these individuals had been central in allowing Banks to form his library by sending presentation copies of their books, an act Banks intended to reciprocate through sending copies of *Icones Selectæ Plantarum*.

After his visit to Soho Square, Thunberg viewed Banks as the ideal patron, not least because of the size of Banks's natural history collections. The gift of *Icones Selectæ Plantarum* reflects their good relationship. In a letter sent to Thunberg in 1787 Banks stated that "that Plates which Mr Dryander mentioned to you from kaempfer's Drawings are nearly finished had them for Presents in the same manner as the Reliquiæ Houstounianæ they will be elucidators of your Flora Japonica which is always Quoted you are welcome to as many sets as you please for your friends be so good as to let me know how many you chuse to have."[108] Banks regarded Thunberg as an authority on the plants of Japan and intended to print as many sets of plates

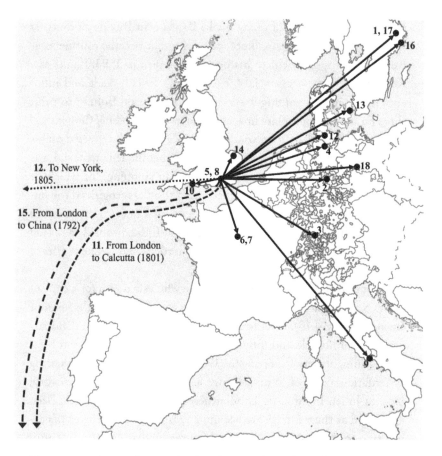

FIGURE 4.10. A map showing the distribution of Joseph Banks's *Icones Selectæ Plantarum* (1791).

for both *Icones Selectæ Plantarum* and *Reliquiæ Houstounianæ* necessary fulfill the demands of naturalists across Europe. Banks was prepared to print plates in response to the requests of fellow naturalists
and then append them to one of the spare sets of letterpress pages,
allowing him to distribute copies for decades after their initial publication as shown by the table and map in figure 4.10. Despite the serious political tensions between Britain and France in the early 1790s,
Banks attempted to maintain correspondence with naturalists on the
continent, an essential means for which was presenting valuable and
useful books. Banks's ability to print copies on demand proved useful; not only could he continue to distribute *Icones Selectæ Plantarum*
for the rest of his life (along with never having surplus copies), but he
could print new copies if those he sent out were lost or confiscated,

Key to Figure 4.10.

	Destination	Recipient	Method of acquisition	Page in Dawson (1958)
1	Uppsala	Peter Fabian Aurivllius (1756–1829)	Banks sent a copy for Uppsala University Library; Aurivllius, the librarian, acknowledged Banks on September 26, 1792.	25
2	Göttingen	Johann Friederich Blumanbach (1752–1840)	Thanked Banks in a letter dated June 5, 1791.	112
3	Tübingen	Christoph Frederick Cotta (1750–1807)	Banks initially refused to send Cotta a copy but seems to have done so in exchange for books.	233
4	Hamburg	Paul Dietrich Gieke (1745–1796)	Thanked Banks for a copy in a letter dated September 28, 1792.	n/a
5	London	Samuel Harper (d. 1803)	Thanked Banks on the behalf of the trustees of the British Museum in his capacity as secretary.	396
6	Paris	Jacques-Julien Houtou de Labillardière (1755–1804)	Thanked Banks in a letter dated July 25, 1791.	514
7	Paris	Jean Baptiste Lamarck (1744–1829)	Banks wrote that he would soon send a copy to Lamarck in a letter sent January 12, 1789.	519
8	London	Linnean Society of London	Copy presented and inscribed by Jonas Dryander, fixed vice president of the Linnean Society of London.	n/a
9	Fondi	Francesco Antonio Notarianni (1759–1803)	Wrote to ask Banks to send a copy, January 5, 1794.	640
10	Blandford	Richard Pulteney (1730–1801)	Thanked Banks in a letter dated February 4, 1794.	690
11	Calcutta	William Roxburgh (1755–1815)	Asked Banks to send a copy in a letter dated September 27, 1801.	735
12	Rendsburg	P. Schlanbusch (Theodor Georg Schlanbusch, 1756–1829)	Norwegian judge in Rendsburg, thanked Banks for a copy in a letter dated November 1, 1791 (Dawson, *Banks Letters*, 735).	735
13	Copenhagen	Heinrich Schlanbusch (1757–1830)	Thanked Banks for a copy in a letter dated May 28, 1803.	737
14	Norwich	James Edward Smith (1759–1828)	Presented by Joseph Banks, c. 1791. Copy at Linnean Society of London, CF.536/4.	n/a
15	China	George Leonard Staunton (1737–1801)	Copy sent by Banks on August 18, 1792 for use on the Macartney Embassy.	784
16	Stockholm	Olaf Swartz (1760–1818)	Thanked Banks for a copy in a letter dated November 27, 1791.	799
17	Uppsala	Carl Peter Thunberg (1743–1828)	Thanks Banks for a copy in a letter dated February 24, 1792.	821
18	Berlin	Carl Ludwig Willdenow (1765–1812)	Asked Banks to send a copy in a letter dated September 24, 1794.	872

a consistent problem when corresponding with European naturalists during the Revolutionary Wars.[109]

Distributing Books in East Asia

Along with ensuring *Icones Selectæ Plantarum* was distributed to major institutions and notable naturalists in Europe, America and India, Joseph Banks made a point of giving copies to British naturalists who participated in the more frequent expeditions to East Asia. This is implied in a letter he wrote to George Leonard Staunton on August 18, 1792: "Herewith you will receive a Copy of the Icones Kaempferianæ which I beg you to accept & Enclosed in it such loose hints relative to Horticulture & Botany as I could Conveniently throw together, I beg you will Present them with my heartiest Good Wishes to Ld Macartney & request that his Lordship will in the Course of the Voyage put them in such hands as he may find the most willing & the most able to make use of them."[110] Banks sent this letter and an annotated copy of *Icones Selectæ Plantarum* to Staunton when the latter was preparing for the Macartney Embassy to China, a British diplomatic mission financed by the East India Company from 1792 to 1793 and headed by George Macartney (1737–1806). This expedition had the joint purpose of improving the conditions of British trade with the Qing Empire and surveying the economy, social customs, and natural history of China. Central to Macartney's purpose was an entourage of artists, musicians, and "gentlemen of Science capable of making Philosophical Experiments."[111] Staunton was appointed as Macartney's secretary and naturalist on the expedition; along with gifting Staunton a copy of *Icones Selectæ Plantarum*, Banks lent a further seventy-eight volumes from his library so Staunton could survey the natural history of China.[112]

In Banks's view, books were central to undertaking a survey of the natural history of China. In the "instructions for collecting" he wrote for Staunton shortly before the voyage departed, Banks included a list of plants "described in the following books that are likely to be more acceptable to his Majesties Botanic Garden [at Kew]." These books included Carl Peter Thunberg's *Flora Japonica*, Engelbert Kaempfer's *Amoenitatum Exoticarum*, and *Icones Selectæ Plantarum*.[113] Banks copies of these books from his library and his requests for particular species, some of which were depicted in *Icones Selectæ Plantarum*, made these works useful for identifying plants. For example, one species Banks "recommended to Ld Macartney to Procure . . . for Kew" that had been depicted in *Icones Selectæ Plantarum* was *Limodorum striatum*. This species of orchid (plate 7) represents a genus that was becoming increasingly desirable for the nobility to cultivate in newly built greenhouses. Banks's and his librarians' annotations in the volumes Staunton received were central for providing a unified repository to aid in identifying plants, establishing the foundations for the processes of collecting and identifying new species before the journey even began.

In addition to advising Staunton on books, Banks recommended the best means for collecting, recording, and illustrating plants in the field. These notes were influenced by the hierarchic structure Banks had imposed on his own team of natural history staff during James Cook's voyage, a means for organizing information and people Banks intended for Staunton to emulate in China. Banks sent Staunton a botanical illustration "to shew the most Commodious manner of making Scetches," emphasizing the need to have "the Parts of Fruitification drawn separately in their natural size." These were for classifying plants according to the Linnaean system. For the specimens themselves, Banks advised that "every Species Especially in China & Japan are indiscriminately dried and preserved for Examination in Europe . . . to be able to distinguish with absolute Certainty whether a plant is /Absolutely/ really /new."[114] These processes were central for allowing botanists in Europe to identify new species from Staunton's collection, classifying the plants of China under the Linnaean system.

Banks gave large portions of his instructions to Staunton over to obtaining information on Chinese agriculture and horticulture. This reflects the interests of earlier travelers in East Asia, including Kaempfer, who gave an extended account of the processes of cultivating and refining Japanese tea in his *History of Japan* (1727).[115] Banks included a section titled "Hints on the Subject of Gardening Suggested to the Gentlemen who attend the Embassy to China," under which he listed the approaches used by the Chinese for growing and training plants that had been reported in *Mémoires concernant l'histoire, les sciences, les mœurs, les usages, &c. des Chinois: par les Missionaires de Pekin* (1776–1791). Banks expected those who attended the Macartney Embassy to obtain information on the Chinese practice of producing bonsai trees, processes for improving soil in gardens, and "the method of accelerating the blossoming of Plants."[116] Although the immediate application of these practices is obvious for administering and caring for institutions such as Kew Gardens, they also relate to Banks's more pressing concerns relating to agricultural improvement in Britain. The introduction of new horticultural practices was of interest to educational institutions and Banks, as a major landowner, after successive European famines in the decades around 1800. In response to the "alarming state of the Harvest in August 1804" Banks published *A Short Account of the Causes of the Disease in Corn* (1805). This was distributed to "intelligent Agriculturalists whose residence in the country enables them daily to examine, not only the progress of their crops, but the origin and advances also of all those obstacles which nature has opposed to the success of agricultural labours."[117] Banks's pamphlet was an unusual foray into commercial publishing—resulting from his wish to circulate it to educated farmers across Britain—and influence teaching in new agricultural and horticultural schools.[118]

A main outcome of Staunton's journey through China to Peking was *An Authentic Account of an Embassy from the King of Great Britain to the Emperor of China* (1797). This was the first detailed account of the topography, social customs, antiquities, and natural history of China that resulted from a British expedition to the region and the first substantial account published since Jean-Baptiste Du Halde's *The General History of China* (1737), a book compiled from the reports of seventeen French Jesuit missionaries (Du Halde had never visited China). By the 1790s, this sort of account had ceased following the suppression of the Jesuit order in 1773, making the information collected by Staunton even more valuable. Large portions of Staunton's final published work were given over to matters of natural history and plant cultivation. For example, Staunton gave several lists of the plants found in different regions while providing detailed descriptions of those he found to be new or interesting, an example being "a species of epidendron, that is capable of vegetating alone." Species such as tea also received considerable attention. The second volume of Staunton's work contains a long description of the process that went into tea production, the different kinds of tea, how it could be used in medicine and the various climatic zones suitable for tea cultivation. Tea was of such importance to readers of Staunton's *Account of An Embassy* that an expensive copperplate image was produced to accompany the description (figure 4.11). Staunton's emphasis on

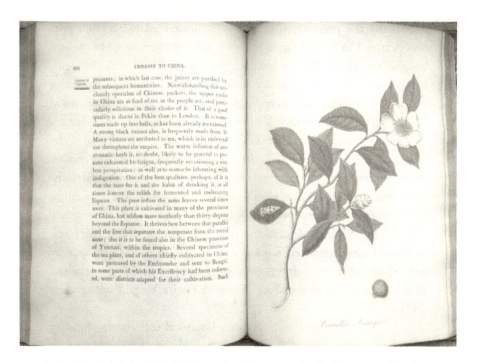

FIGURE 4.11. The expensive copperplate image opposite the letterpress description of the tea plant from George Leonard Staunton's *Embassy to China* (1797). Private collection.

natural history was a result of Banks's involvement in the publication process. Owing to Staunton's ill health after his return, the main task of editing and producing this work was left to Banks. In addition to financing its production, Banks selected the images he wished to publish and was responsible for funding and organizing their production. This involved organising sixteen engravers to produce forty-one large plates for the final publication.[119] These were of two different sizes; each of the twenty-four largest plates measured 18 × 12 inches (45.72 × 30.48 centimeters) and cost Banks eighty guineas each. The seventeen smaller plates each cost fifteen guineas. These added a total £2,283/15 to Banks's expenditure on this book, which, in addition to purchasing enough fine wove paper to print two thousand copies, amounted to £4,111/2/6, enough money a skilled artisan, such as the engraver Daniel Mackenzie, could expect to earn in a lifetime.[120]

Banks's involvement in the publication process of Staunton's *Account of an Embassy* shows that in addition to knowing how to manage and publish his own privately printed books, he was capable of organizing the production of a heavily illustrated publication that was destined to be circulated to polite society thorough commercial booksellers. This utilized a similar approach to that Thomas Pennant used to publish *A View of Hindoostan*, although the fact that Staunton was listed as the author gave Banks a certain degree of anonymity, insulating him from unfavorable reviews in the popular press. Bad reviews and satire, including James Gilray's famous "the reception of the diplomatique and his suite, at the court of Pekin," published shortly after the return of the Macartney Embassy and news of its catastrophic failure, encouraged Staunton to employ "a consciousness of talent and literary attainments which might enable him to defy the severity of criticism."[121] The wide distribution and desire to insulate this book from criticism influenced Banks's decision to select the best quality materials. The use of Bulmer's print shop and fine wove paper resulted in one reviewer recommending this work for its "elegant typography and paper."[122]

Staunton satisfied Banks with his botanical collections, showing that the achievements of the Macartney Embassy in natural history were far more successful than the diplomatic outcomes. In his letter to King George III, Emperor Qianlong (1711–1799) famously remarked, "As your Ambassador can see for himself, we possess all things. I set no value on objects strange or ingenious, and have no use for your country's manufactures." This ensured that British trade with China was confined to the port of Canton until the First Opium War of 1839–1842.[123] Banks's interests in publishing are reflected through his distribution of *Icones Selectæ Plantarum*; he only gave copies to members of the Republic of Letters whom he could rely on to use it to further natural knowledge and British imperial ambitions. This is reflected by Banks's distribution of *Icones Selectæ Plantarum* in Europe and his insistence that it accompanied the Macartney Embassy. Although the diplomatic aspects of the mission was a failure, Banks's aims were met: enough information was collected to successfully produce a new natural history of China.

Changing Markets and an End to Genteel Natural Histories?

The different routes taken to distribute natural history books were shaped by individual interpretations of genteel publishing. Authors adapted the physical content of their works to suit a specific branch of society, reflecting a lack of reliance on commercial publishing and booksellers for distributing publications. For example, Thomas Pennant utilized the commercial publishing industry to distribute his work to elites throughout Britain and Continental Europe, encouraging a grand program of national improvement while bonding a network of naturalists. In comparison, Joseph Banks associated himself with the aristocratic publishing practices of the Earl of Bute and Horace Walpole to secure a global network of naturalists while integrating his publishing program with the aims of a growing imperial state. This genteel vision avoided any commercial associations throughout the production and distribution processes. A lack of reliance on commercial markets while maintaining a distinct understanding of how they worked shaped content such as images, described by James Raven as being "greatly conditioned by the demands of the market place."[124] Genteel naturalists removed images from commercial conditions to maximize their quality and usefulness in taxonomy. It was only after the production process that some sought to cover their expenses through introducing a set number of copies to commercial markets.

The personal incomes of naturalists and the extent of their interactions with the commercial publishing industry shaped the longevity of these publications. For example, Banks's production of private press books relied on his personal ambition and income—a process that ceased after his death in 1820. In comparison, the structural makeup of Pennant's *British Zoology*—combined with his means for utilizing commercial markets to cover the costs of production and his research expenses—inspired a constant stream of communications from loyal correspondents, ensuring that the collection at Downing Hall continued to grow after his death in 1798. This led Thomas Pennant's son, David, to edit and produce a new edition of *British Zoology* published in 1812. However, in comparison to the previous editions that were meticulously managed and funded by his father, David produced this book in collaboration with a conger of twelve publishers who managed the finances and production process. This is similar to many on-commission books and reflects a consistent trend in publishers taking on far more financial responsibility in the early nineteenth century.[125]

The publishing house with the largest share was White and Cochrane, the successors to Benjamin White's business, who continued to operate from their Fleet Street premises and publish expensive natural history books by genteel authors. On December 2, 1813, John Cochrane (1781–1852), a partner at the firm, wrote to David Pennant:

> I beg you will not attribute it to any inattention that I have not before written to you on the subject of new editions of the Tour in Scotland or any other of Mr Pennant's works. The subject has been more than once discussed among the patrons since the publication of the Zoology, & the London, but there was so strong a feeling amongst the whole of them that the sale of the works hitherto published had been so slow, and so large a capital yet remained locked up in them (which last argument pressed most strongly upon us who hold the largest share) and that it was therefore imprudent to print any more at present.[126]

Cochrane's comments reflect the changing state of the British book trade by the early nineteenth century, when books compiled from single octavo or duodecimo volumes became more popular than large, expensive, heavily illustrated, multivolume works.[127] Books such as *British Zoology* became untenable to commercial publishers without significant funding from their authors. Even after White's refusal to publish new editions of Pennant's works David still persevered with funding the production of a new edition of *British Zoology* printed in Dublin in 1818.[128] Financial troubles caused by changes in the publishing market and the post–Napoleonic depression were coupled with the new approaches to classification based on the anatomical differences between species proposed by figures such as Georges Cuvier (1769–1832), whose *Animal Kingdom* was published in English between 1827 and 1835. Outdated content combined with the serious financial difficulties faced by many natural history publishers from the mid-1810s brought a sudden end to Pennant's *British Zoology*.

Distributing natural histories as gifts set a precedent for the strategic distribution of books for the next century. Despite the major changes in the commercial publishing industry, Banks continued to administer publications. These were concluded with *Strelitzia Depicta: Or Coloured Figures of the Genus Strelitzia from the Drawings in the Banksian Library* by Ferdinand Bauer (1818) and the final volume of Roxburgh's *Plants of the Coast of Coromandel* (1819).[129] The plates for *Strelitzia Depicta* were printed by Moser and Harris, an early lithographic press, and copies were retained for private distribution. This legacy of avoiding commercial markets continued to be expressed by scientists throughout the nineteenth century. When petitioning

Robert Peel for Richard Owen's Civil List pension in 1847, William Buckland expressed concern that a lack of private income would cause Owen "to descend to the Condition of a bookseller's Hack," risking a severe decline in the quality and content of Owen's work.[130] Similar networks for distributing complimentary copies of publications were retained by institutions over which genteel naturalists—such as Banks—had held considerable influence. Examples include the British Museum and Royal Society, both of which maintained "free lists" of the recipients of gratis copies until the mid-twentieth century. This followed a model initially developed by Banks and his associates in the 1780s. By 1954 this ran to 276 free copies of the Royal Society's *Philosophical Transactions*, designed to connect individuals, foreign scientific societies, and institutions across the British Empire.[131]

The success of natural history books, then, pivoted on their individual creators, who intertwined approaches of strategic distribution with the practice of natural history. Books gave naturalists the opportunity to depict, describe, and distribute rigorously classified selections from their collections to a specified audience, shaping their networks and unifying the approaches used to collect and communicate information. Thus, these portable collections shaped natural knowledge in a variety of different spaces and settings, from the comfort of European libraries, to voyages in Asia and the Americas, allowing their creators to practice and shape natural history on a global scale.

Chapter 5

The Use of Books

Practices of engaging with print and natural history extended across national and global domains by the late eighteenth century. Moving beyond the previous chapter's exploration of Joseph Banks's, Thomas Pennant's, and others' specific distribution of books, the question remains as to how print is used within these elite circles and when it escaped to become exposed to broader sectors of society. Although genteel naturalists aimed to create standardized working practices to gather and communicate information, books and the practices they promoted started permeating beyond select networks. However, rather than adhering to the more general "diffusionist models" that, since the 1980s, have integrated the distribution of books with commercial interests and since been applied to studies of distributing natural history books,[1] this chapter shows how those who engaged with these books remained specific, emanating from certain ranks of the genteel elite or those encouraged to engage with print to meet the intentions of this subset. Genteel readers had the motivation to engage with natural histories in addition to the education, time, and resources to receive, analyze, exchange, and contribute information. Exploring authors' and recipients' uses of natural history books, it becomes possible to see how printed matter was fuelled by global exchanges to stimulate the cyclical process of natural history to grow collections, publications, and networks of correspondence.

The objective here is to close the cyclical process of natural history outlined in the introductory diagram, showing how these enterprises relied on the continual build-up of new information. Specified individual correspondents were only capable of transmitting material across their own established networks, creating a system of redistribution that only facilitated a limited level of growth. In comparison, this chapter explores how printed materials integrated with a broader global society. This took place in a transitional period that witnessed significant changes in the book trade. Economic problems caused by the French Revolutionary and Napoleonic Wars resulted in challenges for those who produced natural history books around 1800. As William St Clair has suggested, problems in book production originated from gradual changes in established protocols of supply and demand. Emerging in the 1750s, the more general book trade saw a reduction in higher income groups' interest in expensive, large copperplate books, which began to compete with works published in smaller formats with fewer overheads. These changes influenced the stock offered by booksellers, resulting in the emergence of a range of specialist sellers such as Benjamin White (c. 1725–1794) catering for different branches of the reading public.[2]

When compared to more general publishing markets, the demand for expensive copperplate books remained strong among elite naturalists. Thomas Pennant and Joseph Banks saw themselves as gatekeepers of natural knowledge, taking a central role in commissioning, producing, and distributing natural history. In most cases books did not increase the wealth of their authors. Many genteel naturalists had a limited interest in profits, viewing books as tools for expanding a network that served to enrich their natural history collections. This presents an interesting paradox with more general publishing and distribution practices in an age that witnessed what St Clair referred to as an "explosion of reading."[3] Natural histories remained marginal to the distribution networks of major publishers while authors, compilers, and funders maintained significant authority over their dissemination. Power exerted over the production processes ensured these works were distributed to a specific group, identifying sectors of society receptive to national and global enterprises of knowledge production.

Practices of distributing books built new connections between members of the genteel elite who naturalists did not automatically include in their networks. Many had to be alerted to the publication of new books through correspondence and advertisements printed in newspapers and on the preliminary leaves of books and journals. Naturalists often relied on others in their personal networks to facilitate the further distribution of knowledge in addition to employing publishers with a genteel reputation. For much of the late eighteenth century, Benjamin White's firm dominated this sector. White's reputation and familial ties played a central role in connecting his brother, Gilbert, with Thomas Pennant and Daines Barrington, both of whom distributed their books through White's firm. White's shop served as a meeting point. Thomas Martyn, professor of botany at the University of Cambridge, first met the noted botanist Richard Pulteney at White's bookshop in 1786—a chance in-person meeting between two naturalists who had been corresponding for twenty-six years.[4] A downside to distributing books, albeit through a genteel publisher, was that naturalists started to lose control over the information they compiled. Perhaps the best example is Banks's experience with the publication of John Hawkesworth's *Account of the Voyages* (1773), which placed his moral character under public scrutiny. As a result, many naturalists were selective about the material they distributed. Some attempted to maintain absolute authority over certain aspects of their collections, limiting their accessibility to genteel visitors and only distributing aspects they viewed as crucial for serving their broader imperial interests. Others used print to stimulate the continual growth of their collections and correspondence network, building a loyal group who recommunicated information to facilitate the growth of these collections and publications through multiple editions.

Through exploring how broader distribution networks established continual streams of incoming information, it is necessary to examine the design, adaptation, and use of books in localities ranging from the parishes and country estates of Great Britain through to the South Pacific. It shows how print was used in the process of collecting to create standards of observation, initiate information exchange, and grow global networks of correspondence. Participants in these networks contributed specimens and information through transmitting objects, letters, and printed ephemera, much of which was interleaved within existent published works to create repositories of all known species. Through distributing print across diverse networks, naturalists gathered new information, facilitated the expansion of collections while updating publications that were produced, reedited, and extended as information was accessioned to extend networks alongside print.

Images, Order, and the Identification of Useful Plants

Joseph Banks commissioned books with the intent for them to be used across the globe, from a transnational network of European correspondents to naturalists who participated in voyages to Asia, the Pacific, and the Americas and the various Indigenous groups they encountered. Banks's books comprised compilations of copperplate illustrations, reflecting his belief that images were the most effective means for communicating information on species. In addition to producing finished books, Banks mobilized his in-house production line of clerks, artists, and engravers to print copperplate images designed to be distributed separately. Images not bound within books were more accessible to diverse audiences: in Europe they could be viewed in the windows of printsellers' shops, whereas global travelers distributed examples to Indigenous people to communicate information, providing a cheaper and more efficient approach when compared to giving away bound volumes. Illustrations played a greater role in global voyages, overcoming language barriers to facilitate the collection of specimens and information.

Typical examples include a plate Banks commissioned depicting *Cinchona officinalis*. In comparison to plates bound in books designed to be read as a series of images or alongside a related letterpress description, the content of stand-alone plates was structured to broadcast the physical properties of specific species to different cultural and ethnic groups. Banks believed the distribution of information on specific species was central for generally improving British ventures in the tropics. Some plates were designed to protect crews from dangerous animals, such as one Banks commissioned that depicts a species of puffer fish.[5] Others were geared toward discovering substitutes for important species used in tropical medicine.

Concentrating on the plate Banks commissioned to depict *Cinchona officinalis*, a plant many believed essential for British economic and political expansion, the following pages show how illustrations overcame linguistic barriers when communicating and gathering information. Inspired by his own experience of communicating through images in the Pacific, Banks concluded that images of species with a high economic potential ought to be distributed to facilitate British imperial expansion. Examples include the breadfruit, an image Banks commissioned from Sydney Parkinson and permitted to be published in Hawkesworth's *Account of the Voyages* due to its perceived economic potential as a food source. Cinchona served a similar broad purpose: preventing disease in European armies.

Banks's interests in cinchona followed the long-standing interests of British governments, physicians, apothecaries, and natural historians who had attempted to examine specimens of this species since the late seventeenth century. A chief export from the Spanish territories of South America, cinchona cured various intermittent fevers (since diagnosed as malaria) and was essential for facilitating European attempts to maintain a significant mercantile and military presence in the tropics.[6] By 1706 the apothecary James Petiver was promising "a *Guinea* to the first Person that brings or sends him a fair Specimen [of cinchona], or Branch of its Leaves, with Flower and Fruit on it."[7]

From the 1750s, the monopolies of the Spanish Crown made importing cinchona far more expensive for British merchants. As such, many wealthy individuals with connections to the interests of the British state, such as Banks, mobilized their resources to seek a viable alternative. Trading restrictions impacted domestic medical markets and the East India Company, which viewed cinchona as essential for sustaining armies in India. By the early 1760s the East India Company had a continual military presence in the Indian subcontinent and by 1800 maintained an army of one hundred thousand. To avoid Spanish monopolies and the high prices on Continental European markets, Banks decided to diversify his empire.[8] This involved seeking a plant with the same medical properties as cinchona that could be cultivated in Asian and American territories under British control.

For Banks, this meant mobilizing his growing correspondence network and in-house production line to engrave, print, and distribute copperplate images in response to the British government's need to obtain a substitute for cinchona. The distribution of this image correlates with Banks's global, but selective, distribution of his other books, *Reliquiæ Houstounianæ* (1781) and *Icones Selectæ Plantarum* (1791). After production, Banks used these books to solidify his reputation across European correspondence networks, using them as diplomatic gifts that had the added benefit of influencing practices of collecting and British trade ventures. Unlike many contemporary copperplate images of romantic views and classical scenes,[9] the prints depicting cinchona were never designed to be sold. Rather, this image was distributed to a range of individuals Banks deemed to have the necessary skills to identify and compare the taxonomic properties outlined on the plate with potential substitutes.

The physical construction of the plate depicting *Cinchona officinalis* is very similar to the thousand or so other Linnaean botanical prints

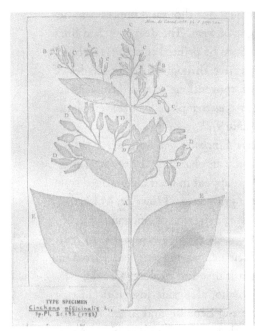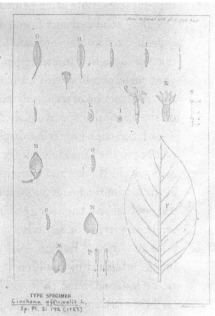

Figure 5.1. The two plates that depict *Cinchona officinalis* published by Charles Marie de la Condamine in *Mémoires de l'Académie royale des sciences de Paris* (1738), which Joseph Banks inserted into his herbarium collection at 32 Soho Square. Courtesy of the Trustees of the Natural History Museum, London.

Banks commissioned between 1771 and 1800 either as part of his privately printed published works or the botanical illustrations designed to depict species collected on the *Endeavour* voyage. The process of production is similar to that used for Banks's other plates. Richard Jones supplied the copperplate, as evidenced by Jones's stamp on the verso, and Banks's staff printed copies on a small rolling press located in the rooms under the library at 32 Soho Square on Whatman's paper. This gave Banks the ability to have new impressions printed on demand, updating the paper used in the production process as the decades progressed. Earlier impressions were printed on steam-pressed laid paper, although by the 1790s Banks's printer used fine wove paper.[10] The engraver was Daniel Mackenzie, who was simultaneously employed to produce the plates that depict the plants of the South Pacific. This engraving is characteristic of a "private plate" commissioned by Banks, who positioned himself outside the commercial print trade. Similar to Banks's books, private plates had a low print run; engraved copper plates could often print around one thousand impressions before they exhibited significant signs of wear. The deeply etched lines suggest Banks never intended for it to be printed in color. The black ink was designed to create perspective through different shades of gray, corresponding with Linnaean conventions intended to facilitate the construction of standardized images that placed the sole emphasis on the physical shape of a plant (plate 9).[11]

Mackenzie's engraving displays *Cinchona officinalis* arranged according to the Linnaean system while emphasizing the physical characters of the flowers and fruit. These appear in their anatomized state at the foot of the image above the engraved text. However, Mackenzie did not base this engraving on a recent depiction of cinchona by a contemporary artist or a specimen from Banks's collection due to their poor condition.[12] Rather, Mackenzie's engraving originated from an image published by the French naturalist and explorer Charles Marie de la Condamine (1701–1774) who, alongside Joseph de Jussieu (1704–1779), spent a decade exploring Ecuador and Brazil from the mid-1730s. In 1738 Condamine published a paper titled "Sur L'Arbre du Quinquina" in the *Mémoires de l'Académie royale des sciences*, for which he had two plates engraved that depict the cinchona plant (figure 5.1).[13]

Banks's interests in this image were twofold. Initially, Condamine gave a clear representation of the species that distinguished it from other members of the same genus, conforming to the Linnaean convention of emphasizing characters that are "constant, certain and organic."[14] This is similar to other contemporary illustrations of South American plants that Banks received from the Spanish botanist and physician Casmiro Gómez Ortega (1740–1818), a recipient of *Reliquiæ Houstounianæ*. These included materials sent by José Celestino Mutis (1732–1808), who was careful to ensure that the images he commissioned during the Royal Botanical Expedition to New Granada (1783–1816) gave a general representation of each specific species, instructing artists to employ standardized iconographic strategies to combine features gathered from multiple specimens. Mutis made sure the anatomized parts of the plants were present at the foot of every illustration, arranging each species according to Linnaean conventions and giving essential information for its visual representation, identification, and classification.[15] However, Condamine's images were split between two different plates in the journal—the first depicting the leaves and flowers, the second showing the anatomized parts of the flowers and fruit. This is representative of many herbarium specimens sent back to Europe. A typical example is a specimen of cinchona Mutis sent to Carl Linnaeus, which traveled to London with the rest of the Linnaean collection in 1784. Mutis's specimen is divided between two sheets, one containing the stalk and leaves and another the flowers and fruits. Following similar artistic conventions to those used by Mutis's artists in South America, Banks ordered Mackenzie to combine Condamine's two plates in a single sheet. This ensured features central to the classification of this species according to the Linnaean system were obvious while following standard conventions used in other botanical illustrations.[16]

Banks's interest in Condamine's illustration of cinchona derived from the taxonomic authority placed on this specific image by Linnaeus, who used it as the main source for the entry *Cinchona officinalis* published in *Species Plantarum* (1762–1763). Linnaeus cited Condamine's article and used the image to construct his diagnosis. This was a result of Linnaeus not possessing a representative specimen until he received one from Mutis in September 1764.[17] Throughout his botanical career, Banks regarded the Linnaean collection as "the real standard to prove the meaning of old Linnaeus's works," a body of material that incorporated specimens, descriptions, and images, such as that produced by Condamine. Linnaeus's use of this image made it an essential standard for establishing the name *Cinchona officinalis*, creating an indispensable resource in Banks's campaign to stabilise botanical nomenclature and facilitate global communication.[18] Copies of the plates were cut out from Condamine's article and inserted next to the species of cinchona in Banks's herbarium.[19] By basing his copperplate on Condamine's illustration and Linnaeus's diagnosis, Banks not only distributed an image, but circulated a perfected "classification type" anchored in earlier published accounts, reproducing the specimen on which Linnaeus had based his description whilst updating it according to Linnaean conventions of botanical depiction.[20] Central to the purpose of the plate, however, was releasing this image from the bound volumes of an old journal only accessible to those who owned copies or had access to private and emergent public library collections.

Descriptive captions formed a standard part in stand-alone prints such as portraits, views, and scenes from classical antiquity. In comparison to having a mere descriptive function, the engraved text at the foot of Banks's plate showing *Cinchona officinalis* served to instruct the viewer on how to use the plate to identify cinchona and communicate their findings. Those who received copies were told: "This Tree is to be sought for on Mountains and if found will produce a valuable Trade. If any dried Specimens (with flowers & fruit) of Trees resembling this are sent to Mr Banks at New Burlington Street, London, the person sending them will be acquainted whether they are of the true sort, or likely to become a usefull substitute for it."[21] This description followed the Latin, French, English, and Spanish names for this species, the most common European languages in the Americas, showing that Banks desired for the image to be viewed and identified by as many different people as possible. This goes beyond the names cited in *Species Plantarum*, showing how Banks's plate was designed to engage with more people than this Latin botanical book. Thus, unlike the close control Banks exerted over books, this plate was designed to be viewed, interpreted, and stimulate a response from a variety of different peoples, gathering information on potential substitutes.

Banks's plate represents an example of how print was employed in a broader global hunt for cinchona from the mid-eighteenth century, a global hunt participated in by numerous emergent European nations—in addition to Great Britain, there was significant interest in France, the Dutch Republic, Portugal, the German states, and Scandinavia. Banks initially sent copies to naturalists, physicians, and surgeons in the Americas who monitored different groups' use of these plants. These included the Scottish military physician and naturalist William Wright (1735–1819), a partner in a medical practice at the Hampden estate in Jamaica, who was responsible for treating over 1,200 enslaved persons before his appointment as surgeon general of Jamaica in 1774.[22] Wright was in correspondence with Banks from the early 1770s and received one of the plates depicting *Cinchona officianalis*. Evidence for Wright's possession of this plate can be found in an article Banks communicated to the *Philosophical Transactions* titled "Description of the Jesuits Bark Tree of Jamaica and the Caribbees." Wright cited Banks's plate, describing the specimens he collected as being "in every respect similar [to] those of the *Cinchona officinalis* as depicted in a plate sent out by MR. BANKS." Wright often experimented on the medicinal properties of cinchona and similar plants; in addition to himself, Wright tested new treatments on enslaved Africans. Comparison of this plate with Wright's specimens was essential for

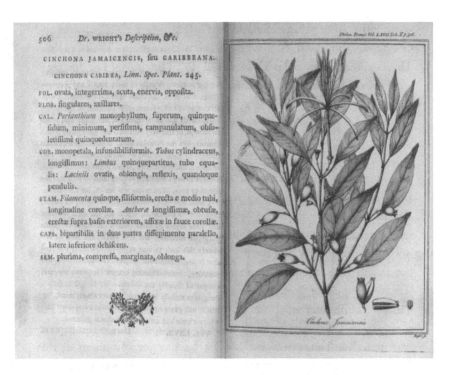

Figure 5.2. William Wright's description and image of *Cinchona jamaicensis* published in the *Philosophical Transactions* (1778). Royal Society of London.

assigning *Cinchona jamaicensis* a place within the Linnaean system, as evidenced by the final page of Wright's article, in which he named, described, depicted, and classified this species according to the arrangement of the 1762–1763 edition of *Species Plantarum*. *Cinchona jamaicensis* proved a viable substitute for *Cinchona officinalis*, as evidenced by Wright's comments on how the drinking of a serum derived from this plant "conquered speedily the disease." "My success in such a dangerous malady," he continued, "leaves not a doubt in my mind, but that it will prove equally efficacious in every other case where tonic and antiseptic medicine is indicated." Wright's results were almost certainly derived from the experimental use of this species on the enslaved population.[23] Wright's use of Banks's plate to compare the different physical characters of cinchona species shows it served its purpose of finding a viable substitute for *Cinchona officianalis* with a potential for cultivation in British territories (figure 5.2).

After publishing in the *Philosophical Transactions*, Wright received offprints of his article, including the copperplate depicting *Cinchona jamaicensis*, asking Daniel Solander to "get Prints of the Cinchona and Geoffroya from the R. S. Secretary: I should be glad of 100 of each."[24] This became a standard practice of the Royal Society of London from the 1760s, which allowed authors to take up to one hundred offprints of their articles after making a special application, so they could distribute these to correspondents and increase the author's reputation. One recipient was William Roxburgh, who was attempting to find local substitutes of cinchona in India during the 1790s.[25] Wright's distribution of offprints stimulated further exchanges on cinchona, showing how Banks's copperplate image stimulated evermore extensive programs of publishing, collecting, experimentation, and correspondence. This placed Wright within Banks's network, ensuring his continued loyalty as a botanical collector. Writing to Banks from Jamaica in 1783, the physician Everard Home described how Wright was "very much devoted to your service."[26] Banks's collection quickly became a main reference point for those with interest in cinchona. For example, in 1780 Alexander Anderson mentioned to William Forsyth that "Dr Young tells me Mr. Banks has got specimens of this Cinchona sent him from Jamaica & Dominica."[27] Many of the letters, illustrations, printed pamphlets, and manuscript descriptions Banks received were incorporated into his collection and bound into albums. One example, titled "Officinal plants," contains manuscripts and prints relating to cinchona dating between the mid-seventeenth and early nineteenth centuries.[28]

Other correspondents within the Banksian cinchona network included George Davidson, a physician on the island of St. Lucia, who published "An Account of a New Species of the Bark-Tree, Found in the Island of St. Lucia" in the *Philosophical Transactions*. The physician George Wilson added a prefatory note, discussing how he compared Davidson's specimens with those "in the possession of Joseph Banks."[29] Davidson's interest in finding a substitute for *Cinchona officianalis* in St. Lucia was inspired by the circulation of Banks's image and Wright's article. Banks was responsible for organising the production of Davidson's article, enacting a similar strategy to that he used to manage publications in the 1790s, employing James Basire (1730–1802), who had previously engraved Wright's plate, to produce the engraving. Banks's circulation of this copperplate maintained his authority on cinchona for several decades while Banks's collection became the main reference point for those intending to examine cinchona specimens.

In addition to gifting the plate depicting *Cinchona officinalis* to individuals who traveled to the West Indies, Banks gave copies to global travelers. One example is Banks's former secretary and assistant naturalist Sigismund Bacstrom (c. 1750–1805). Banks wrote to Bacstrom on August 20, 1791, setting out his terms and instructions for botanical collecting on global voyages: "You may be of use to me in collecting specimens of plants; my terms, you know are sixpence for each species of which there is either flowers or fruits, & a shilling when there are both if you chuse to accept these terms & employ your leisure hours in my service I am ready to engage with you. . . . I have enclosed you a draught on my stationer for a Ream of Cartridge Paper & one of packing paper which if you undertake my buisness you will call for. I enclose you also a Bank note of £10."[30] After being employed in Banks's library between 1771 and 1775, Bacstrom joined a series of international voyages as a ship's surgeon. In 1791 Bacstrom embarked on a global circumnavigation and trade venture known as the Butterworth Squadron, organized by Alderman William Curtis; Theophilus Pritzler, a London ship owner, and the shipbuilder John Perry.[31] This voyage was destined for China and the East Indies via Cape Horn and Nootka Sound, the aim of which, according to Bacstrom, was "to bring home Cortex perur [cinchona bark]: and Whatever valuable Druggs or natural productions we can meet with."[32] For Bacstrom, Banks's offer of payment for specimens was an incentive for collecting; he had been in financial trouble since 1789 and had joined the Butterworth Squadron as a paid employee. Banks saw this far-flung commercial venture as an opportunity to circulate images of *Cinchona officinalis* and obtain specimens from South America, which he regarded as one of "the places whence I posess the fewest plants."

Since traveling aboard James Cook's *Endeavour*, Banks had viewed the Indigenous peoples of the Americas, Pacific, and Asia as an essential source of plant knowledge for global travelers. Many Indigenous groups were far more receptive to illustrations than verbal and written communication. As such, Banks instructed Bacstrom to distribute images to Indigenous peoples encountered on this voyage in case they knew of a plant with similar physical characters, stating that "I have ordered Mackenzie to give you as many Prints as you chuse of the tree that bears the peruvian bark, in order that you may put them into the hands of the natives where you touch, to encourage them to bring you plants that resemble it."[33] Although no evidence survives for how these images were interpreted by their recipients, it is clear that by the 1790s remote communities in Tierra del Fuego and the Pacific Northwest were becoming more familiar with this sort of imagery. When Bacstrom passed through the region around Cape Horn it had been over twenty years since Banks had visited when traveling aboard the *Endeavour*, during which time European naturalists and traders paid evermore frequent visits while producing and distributing botanical illustrations. The prominent image of *Cinchona officinalis* and singular nature of this plate meant it could be displayed in a less intimate manner than a printed book, either being held up by Bacstrom or passed around the local community. This shows how Banks's interests in discovering new species were intertwined with his aim to entrench Linnaean practices of identification, description, and communication in a network ranging between European naturalists, "go-betweens" and "knowledge brokers," such as Bacstrom, and Indigenous communities.

Banks's offer of a financial reward was an incentive for Bacstrom to compile a botanical collection that contained the necessary features for the identification and classification of each species according to the Linnaean system.[34] This is represented in the illustrations Bacstrom produced during this voyage. Typical examples include an image titled "American Tea," which Bacstrom observed on April 3, 1793, in Cross Sound, now in the Alexander Archipelago, Alaska. Bacstrom's illustration reflects long-standing interests in the cultivation of tea as an economic resource although, due to the limited time the ship made port and the nature of the season, he was unable to produce an image of the flowers.[35]

Bacstrom produced illustrations of the people he encountered in the Pacific Northwest with whom members of the Butterworth Squadron intended to trade furs for European goods and probable recipients of the plate depicting *Cinchona officinalis*. Among these is Cunnyha of the Haida people, whom Bacstrom met on March 18, 1793, probably one of the best-known chiefs of the region whose lands extended around the Parry Passage area between Langara and Graham Islands. Bacstrom described Cunnyha's village as being "on the North-Side of Queen Charlotte's Island, N. W. coast of America." Such was its strategic importance for trade and military dominance in the region, Bacstrom produced a detailed chart of "Queen Charlotte Islands," marking "Cunnih[a]'s village" and drawing a traditional longhouse. Such details were crucial for stimulating the European fur trade in the region, identifying tribes open to trading, their geographical location, and the most important individuals. In addition to trading sea otter pelts for iron, the former with the intent to be traded across the Pacific to China, it seems that Bacstrom's search for botanical species were integrated with these negotiations. In addition to recording people, the views Bacstrom depicted of places like the Spanish settlement in Nootka Sound on February 16, 1793, paid special attention to detailing gardens and land used in agriculture. Cunnyha's interests in European trade is represented on Bacstrom's image: many of the clothes he is wearing are European in origin, traded in exchange for furs presenting a stark contrast to the Mowachaht people Bacstrom encountered on the main continental coast who resisted European trade.[36] In spite of Bacstrom's failure to record the essential aspects of botanical species, his images of Indigenous people, their settlements, Spanish military bases, a whole range of views ranging from the western coasts of the Americas to southern China, and maps provide a valuable insight into the people who engaged with botanical images in the Pacific.

Banks's plate was essential for standardizing knowledge on cinchona to streamline communication. This allowed visitors to Banks's library, such as Aylmer Bourke Lambert, to utilize the information generated through the distribution of Banks's plate to produce books such as *A Description of the Genus Cinchona* (1797). Lambert acknowledged his use of Banks's collection, noting that the work was "accompanied by figures taken from the specimens themselves, preserved in the herbarium of Joseph Banks, and assisted by drawings in his possession."[37] Similar to the elite group who visited Banks's library, this expensive copperplate book had a limited audience, a factor solidified by its distribution through the genteel publishers Benjamin and John White. However, Banks's influence and collections were impacted by continual misfortunes during global voyages. In 1796 Bacstrom summarized his journey, describing how "I had collected 4 large Quires full of plants and Mosses at Staten Island and On the N:º W:ʳ Coast of America, and a greater quantity of Seeds from Nanking, which I had purchased at Canton but during my long and disagreable Voyage, Obliged to be in 6 different Ships, taken prisoner 3 times, detained 6 months in the Island Mauritius, I lost your Valuable Collection which had cost me a great deal of Trouble to preserve so long."[38] Banks had become used to these setbacks. When returning from New Holland

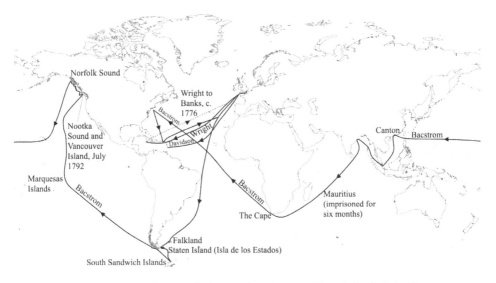

FIGURE 5.3. A map showing the known distribution of Joseph Banks's *Cinchona officinalis*, L. plate, showing the journeys of William Wright, George Davidson, and Sigismund Bacstrom.

in 1803, Captain Matthew Flinders was imprisoned in Mauritius for seven years—a consequence of the French ownership of the island during the French Revolutionary and the Napoleonic Wars. A result of the multiple storms, incarcerations, and ships boarded, Bacstrom returned to London empty-handed, although copies of Banks's plate were now distributed on a global scale (figure 5.3).

Accumulating and Ordering in Successive Editions of *Species Plantarum*

While the desired purpose of distributing printed material was to stimulate the continual influx of information, the recipients of these objects often adapted them for their own reasons. As we have seen in the first two chapters, many books destined to be taken on expeditions were interleaved with blank pages to facilitate the systematic accumulation of information. Others were ripped apart, bound into separate sections, and the plates serve a similar purpose to Banks's image of *Cinchona officianalis*. Others were adapted for use within a natural history collection, recording the products of global correspondence networks stimulated and standardized by printed matter, creating links between large private and institutional repositories.[39] Similar to the practices used when collecting information on expeditions, collections were governed by a whole range of paper technologies intended to integrate, order, classify, and cross-reference information on species from different sources. Paper technologies bridged between specimens, illustrations, and manuscript descriptions, while printed images and descriptions circulated as individual plates, offprints, and printed books. Although Joseph Banks and Daniel Solander performed these tasks in the field, individuals who incorporated references to new species, descriptions, and images into broader collections, often under the instruction of a curator or owner, were relegated to mere clerical assistants by the 1790s.[40] This reflects a continuation between hierarchies established on expeditions in metropolitan natural history collections.

The distribution of books created fixed bonds between collections through standardizing information, a practice exemplified through Banks's relationship with the Prussian botanist and mentor of Alexander von Humboldt, Carl Ludwig Willdenow (1765–1812). Willdenow corresponded with Banks throughout the 1790s, and in 1799 mentioned that "I read in the proceedings of the Linnean Society of London about a certain book, which was titled Kaempfer's Icones, but in spite of the greatest effort by the booksellers, I could not get hold of this book from the English. I therefore ask you, illustrious man, if you could tell my friend Hannemann, who lives in London, where this book can be bought."[41] Willdenow's request for a copy of *Icones Selectæ Plantarum*, a book commissioned and distributed by Banks in 1791, was one of several requests he made to Banks in the 1790s who both provided Willdenow with books and information on the botany of the South Pacific. In 1792 Willdenow described Banks's gift of seeds: "The most welcome among the rest of the seeds were those from New Holland, especially the genus Banksia."[42] Willdenow needed *Icones Selectæ Plantarum* to edit his new edition of Carl Linnaeus's *Species Plantarum*, published between 1797 and 1825, citing it alongside Thunberg's *Flora Japonica* (1784) in the descriptions of species native to Japan. In return for Banks's gift, Willdenow sent volumes of *Species Plantarum* to Soho Square. The first arrived in 1797: "I send you an edition, supervised by me, of the first part of the *Species Plantarum*. It contains the first three classes of Linnaeus."[43]

Willdenow's edition of *Species Plantarum* soon became the main book used by Banks's librarian, Jonas Dryander, to classify and record information on the botanical collection held at 32 Soho Square. It succeeded two earlier editions, both of which had been interleaved and annotated by Banks's secretaries, librarians, and assistants. The first, published between 1762 and 1763, was that used aboard the *Endeavour* and annotated by Solander and Herman Spöring. This remained in use until the late 1770s, when it was replaced with a more updated version edited by the German physician and botanist Johann Jacob Reichard (1743–1782), a decision made after many of the species represented by annotations in Banks's earlier edition had been published.[44] By 1797 this was being replaced by the new edition edited by Willdenow for similar reasons. These three titles formed keystones for organizing information on species held within Banks's botanical collection between 1771 and 1820, showing how books were adapted to facilitate the constant process of updating and accumulating information on the botanical world.

Many annotations in *Systema Plantarum* relate to species and publications held in Banks's collection that had not been recorded in the printed text. A large proportion originated in Banks's Pacific collection, in addition to specimens, images, and descriptions in published works accessioned into the Banksian library between 1780 and 1797.[45] It was the responsibility of Dryander's secretaries, especially Samuel Törner (1762–1822), to undertake the laborious task of transcribing this new information onto the interleaved pages, aligning this with the Linnaean arrangement of *Systema Plantarum*.

Since the interleaved annotated copy of *Species Plantarum* Banks's party took on the *Endeavour* was discussed at length in chapter 2, the following concentrates on the books that succeeded it. Replacing this book in the early 1780s, the majority of annotations in the interleaved copy of Reichard's *Systema Plantarum* relate to species held in Banks's collection. Törner transcribed several references relating to species depicted in *Icones Selectæ Plantarum* into Banks's copy of Reichard's *Systema Plantarum* and added these alongside citations to Thunberg's *Flora Japonica* and an article titled "Botanical Observations on the Flora Japonica" (1794). Manuscript references have been added below the plates in Banks's copy of *Icones* to allow the illustrations to be cross-referenced with the relevant descriptions.[46] Triangulating illustrations and descriptions, Banks's librarians unified entities to define the unique qualities of each species. These refer to the successive interleaved copies of Linnaeus's works that provided a central repository of information in the Banksian collection, referencing specimens in the herbarium, illustrations arranged in a series of portfolios, and published descriptions.

Willdenow's correspondence on specimens collected in New Holland, during the Macartney Embassy to China, and in South America extended his edited edition of *Species Plantarum* into six volumes. It is often described as the last detailed attempt to present a complete picture of all plant species.[47] Willdenow added living examples of these species to the collections of the Berlin Botanical Garden. Writing to Banks in 1802, Willdenow described: "This summer I have already added more than 3000 species to the garden; among them many new ones. Since I possess only a few of the plants of New Holland, I ask you fervently, most illustrious man, to share the fresh seeds from New Holland, especially of the genus Banksia, also the seeds from the East and West Indies & the Cape of Good Hope."[48] Willdenow's success in obtaining plants from across multiple imperial networks ensured his new edited edition of *Species Plantarum* became the main reference point and means for classifying collections used by naturalists in Britain. For example, James Edward Smith believed it was necessary to donate volumes of Willdenow's work to the Linnean Society from 1799 to give a main point of reference for fellows visiting the library.[49] Others, such as Dryander, regarded Willdenow's edition of *Species Plantarum* as an essential tool to support the Linnaean classification of the Banksian herbarium and library. As he received each volume, Dryander accessioned it into the library at 32 Soho Square, commissioning a binder to interleave it with blank pages. Dryander anticipated that these volumes would be needed to incorporate even more new information than Banks's earlier interleaved copies of *Species Plantarum* (1762–1763) and *Systema Plantarum* (1779–1780). Eventually, the interleaving of Banks's copy of Willdenow's *Species Plantarum* extended this work from six to twenty volumes. This contained almost triple the original page count due to there being two blank leaves inserted for every printed page.[50] The integration of Willdenow's *Species Plantarum* with physical collections was seen as a necessity for many British naturalists; for example, Robert Brown, Banks's librarian between 1810 and 1820, had his copy interleaved and extended to sixteen volumes.[51]

Willdenow's publication of numerous species before only named and described in the manuscript annotations in Banks's interleaved copies of Linnaeus's works reduced the need for Dryander and his

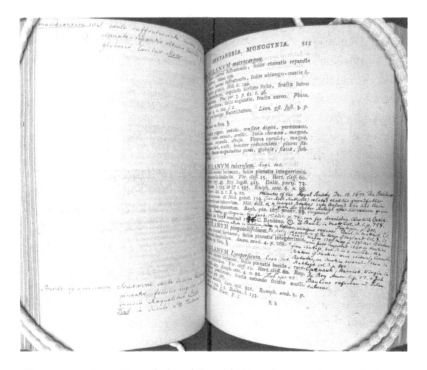

FIGURE 5.4A. Jonas Dryander's and Samuel Törner's annotations on the inter-leaved page and within the text in their copy of Johann Jacob Reichard's *Systema Plantarum*. The faint vertical pencil line through the name *Psudo lycopersium* is by Frederick Schulzen. By kind permission of the Trustees of the Natural History Museum, London.

secretaries to transfer information into *Species Plantarum*. Howev-er, some information still had to be transferred within the system-atic order. A result of the declining status of such tasks by the late 1790s, Dryander relegated this task to his assistant naturalist Fred-erick Schulzen (1770–1848) who worked at 32 Soho Square between 1797 and 1801, and the secretary John Swan, who was employed for a period between approximately 1799 and 1805.[52] As new volumes of Willdenow's *Species Plantarum* were interleaved and accessioned into the collection, Schulzen and Swan checked the printed descriptions against the manuscript and printed entries in Banks's interleaved copy of Reichard's *Systema Plantarum*. Those annotations containing information Willdenow had not published were transferred onto the interleaved pages in Banks's copy of *Species Plantarum*, showing that information on several important plants had not been communicated. As they went through the earlier editions, Schulzen and Swan either

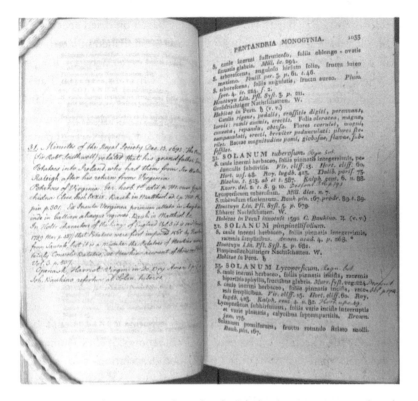

FIGURE 5.4B. John Swan's and Frederick Schulzen's annotations in Joseph Banks's copy of Carl Ludwig Willdenow's *Species Plantarum*, which transfer the annotated information from Reichard's *Systema Plantarum* (fig.5.4a). The ticks and annotations in the right-hand margin are by Schulzen and indicate that the species in the printed text appear as manuscript additions in the earlier edition. By kind permission of the Trustees of the Natural History Museum, London.

crossed through or underlined the generic name of the species, indicating that the information had either been transferred as an annotation or published in Willdenow's edition of *Species Plantarum*.

A typical example of this activity is apparent from the species *Solanum lycopersicum*, commonly known as the tomato, an annotated name and diagnosis of which Törner copied from Banks's interleaved copy of *Species Plantarum* (1762–1763), which had been annotated by Herman Spöring aboard the *Endeavour*, into the interleaved copy of Reichard's *Systema Plantarum*.[53] It seems probable that this annotated name and diagnoses were transcribed since they added information on a common species: the name and descriptive terms were based on Banks's, Solander's, and their team of assistants' observations in the field, whereas Reichard's were most likely based on dried specimens. Thus, it was necessary to identify the species published in Reichard's work with the related manuscripts in Banks's collection. Törner cited these through adding the note "Mscr*" while noting that Banks and Solander had observed this species in the South Atlantic island of Saint Helena. To show Swan had transferred this description into Willdenow's *Species Plantarum*, Schulzen placed a faint vertical pencil line through the annotated specific name of the binomial in the interleaved copy of Reichard's *Systema Plantarum*. Next to the printed description in Willdenow's work, one reader, most likely Dryander, has added a tick checking the information had been transferred in the form of a published name and diagnoses (figures, 5.4a and 5.4b). Since Willdenow's published name and diagnoses did not require extensive comparison with the manuscript descriptions in Banks's collection, the annotations concerning *Solanum lycopersicum* in *Systema Plantarum* were not transferred into Willdenow's work.

In addition to linking two different editions of Linnaeus's botanical work, the interleaved copy of Willdenow's *Species Plantarum* served as a checklist that united published descriptions with species represented in Banks's herbarium. Unlike Swan's and Schulzen's task of transferring the information, it was Dryander's responsibility to make sure the descriptions given in Willdenow's work were supported by the physical characters of the specimens. Although the description of *Solanum lycopersicum* did not need updating, some of Willdenow's other descriptions required major revisions. For example, Dryander erased the species *Pancratium caribaeum*, justifying this decision on the interleaved page opposite: "When Linné published the species plantarum, first edition, he had no other Pancratum in his herbarium, than caribaeum, but from the differentia specifica it is impossible to tell if it was fragramus, speciosum or amœnum. The synonyms are so uncertain to determine any thing from them, as the old figures cannot be depended upon in such uicatier as distinguish these 3 species."[54] Dryander's ability to define and edit descriptions while erasing synonyms represents a very different role to that of the assistant naturalists and secretaries. Dryander had the power to make these additions and corrections to Linnaeus's work; his access to Banks's collection allowed for the examination and comparison of specimens from Linnaeus's own collection that had been purchased by James Edward Smith in 1784. Dryander was a close associate of Smith, earning the title of "fixed vice president" of the Linnean Society, becoming the de facto president after Smith's move to Norwich in 1796.[55] After the Linnaean collection had been brought to London, Banks described how he, Dryander, and Smith "are masters of the definitions in Species Plantarum."[56] As shown in the Soho Square copy of Willdenow's *Species Plantarum*, Dryander's combined access to Banks's and Linnaeus's collections gave him the authority to define species and identify synonyms, comparing multiple examples, including the specimens Linnaeus identified as "typical," to correct later editions of Linnaeus's works. For example, when editing the description for *Aspergo ægyptiana* L., Dryander noted his observation of "the specimen from Linné's Herbarium."[57] Supressing synonyms required extensive experience, employing what Joeri Witteveen has described as the "requisite tacit knowledge" for identifying typical species, showing how Dryander came to define his expertise as a botanist.[58]

Dryander's actions reflect a distinct hierarchy of knowledge when managing Banks's collection. Dryander added and refined content on references and new species across editions of *Species Plantarum* while the secretaries and assistant naturalists duplicated information that had not been included in the printed text. As Bettina Dietz has shown, such collaborative working practices were essential for reducing synonyms for species across collections.[59] For example, in the interleaved copy of Reichard's *Systema Plantarum*, Dryander added a long list of references to published works and manuscripts next to the printed description of *Solanum tuberosum* (figure 5.4a). These include a reference to the minutes taken during a meeting of the Royal Society in 1693, manuscripts Dryander had unlimited access to through his position as librarian of the Royal Society, that describe when Sir Robert Southwell "related that his grandfather brought Potatoes into Ireland who had them from Sir Walter Raleigh after his return from Virginia."[60] This is followed by a string of references to books such as Linnaeus's *Genera Plantarum* and Richard Hakluyt's works, all of which were cited in an article titled "An Attempt to Ascertain the Time when the Potato (*Solanum tuberosum*) was First Introduced to the United Kingdom," published under Banks's name in the *Transactions of the Horticultural Society of London*.[61] In the first lines Banks noted that that the content "was chiefly collected by my learned friend Mr. Dryander," reflecting Dryander's consistent updating and editing of different editions of Linnaeus's works.

In Banks's copy of Willdenow's *Species Plantarum*, Swan has transcribed Dryander's annotations onto the interleaved page opposite the printed description.[62] A keystone in this process of updating *Species Plantarum* was the acquisition of books that remained central to the referencing process. The ever-expanding library was catalogued in an interleaved copy of Dryander's catalogue of Banks's library. The acquisition of books was facilitated by Banks's continual orders from booksellers. For example, Banks had worked with Benjamin White while other booksellers took a more direct role in acquiring for Banks and Dryander. For example, William Sancho (1775–1810)—son of the notable African abolitionist, writer, and composer Ignatius Sancho (1729–1780)—is noted as being "librarian to Sir Joseph Banks, and collects literary curiosities with the most unrivalled diligence."[63] Acquiring published works represents a very different role to Dryander, a process central for connecting descriptions of species represented in Banks's herbarium to bibliographical records in the library while placing special emphasis on Banks's and Dryander's own literary productions.

The activities of Dryander and his assistants were not the only occasions when this information was transferred between interleaved copies of Linnaeus's works. For example, Robert Brown, Banks's librarian between 1810 and 1820, had also commissioned an interleaved copy of Willdenow's *Species Plantarum*. Brown subscribed to this work as it was published, considering it a valuable tool to assist with compiling and ordering botanical descriptions. After leaving Madeira on July 25, 1801, when traveling with Matthew Flinders aboard the HMS *Investigator*, the first voyage to circumnavigate Australia, Brown described how he was "occupied upwards of an hour in copying specific characters from Wil[l]denow's species plant[arum]: into my Florula Maderae."[64] Brown's interest in identifying new botanical species made having the most complete edition of Linnaeus's work aboard the *Investigator* a useful tool during a voyage to the South Pacific, using it to compile descriptions in a manuscript ledger and assess how novel the species he encountered were in relation to accounts by other authors. Transcribing the annotations from Banks's copy of Willdenow's *Species Plantarum* augmented the content of Brown's copy, updating descriptions of unpublished species represented by the specimens Banks and Solander had collected on James Cook's voyage between 1768 and 1771.

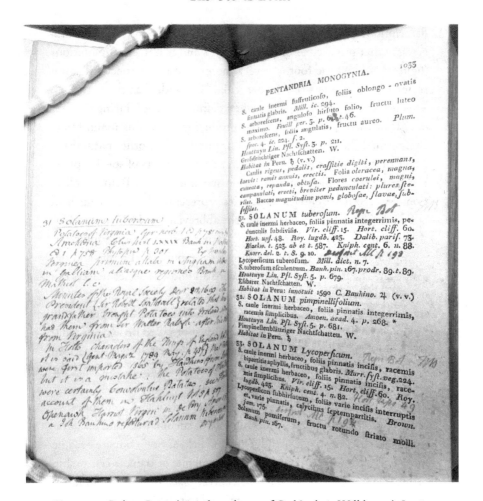

FIGURE 5.5. Robert Brown's interleaved copy of Carl Ludwig Willdenow's *Species Plantarum*. The annotations shown have been transcribed from those by Frederick Schulzen in Joseph Banks's copy of this work. By permission of the Linnean Society of London.

Brown's use of Willdenow's *Species Plantarum* shows how it started to unite the collections amassed during the *Investigator* voyage with the Banksian herbarium, two repositories brought together when Brown was appointed as Banks's curator in 1810. It also reflects on how the Banksian library became more open to visits from other naturalists in the decades around 1800, giving Brown extensive access to these materials while providing him with enough time to transcribe Schulzen's and Dryander's annotations from Banks's copy of Willdenow's work before embarking on the *Investigator*. An example is the description Brown added to the interleaved page opposite *Solanum tu-berosum* that transcribes both Dryander's and Schulzen's annotations (figure 5.5).[65] At the foot of several notes that revise species diagnoses or erase them entirely Brown cited "Dryander" as the authority. This reflects Dryander's power when defining species during the first decade of the nineteenth century, producing numerous manuscripts that formed the basis for descriptions published by other naturalists. Many of these notes updated references with aspects of species' physical characters while basing these on specimens from Banks's collection. Next to most entries for species represented by specimens held at Soho Square, Brown has added the abbreviation "HB" to stand for "Herb Banks," maintaining a complex network of bibliographic references and physical objects to support the published name and description. The use of this book to record all known species reflects the necessity of an updated copy of Willdenow's *Species Plantarum* at the start of Brown's voyage, allowing Brown to identify species previously unknown to European naturalists. This went toward Brown's foundation of his own collection and the use of this work to establish certain descriptive and botanical standards across collections, allowing for the exchange of information and specimens.

Willdenow's *Species Plantarum* created a standard point of reference for British botanists around 1800. As such, Banks and his curators had to use this book as the collection held at Soho Square became more open. However, it also becomes symbolic of Banks's censorship of his South Seas collection. The central place of these volumes in collections reflects on how they were used to transfer information between Banks's, Brown's, and the Linnaean collections while serving as a unified repository that listed and classified all known plants. In addition to facilitating the transfer, collection, and addition of information, these books made these collections accessible.

Interleaving, Annotating, and Updating the Natural History of Britain

In comparison to Joseph Banks, whose publications served as a form of patronage for a global network, Thomas Pennant's careful distribution of published works aligned the practices used by a network of observers across Britain while stimulating the communication of new information. As Anne Secord suggests, the unification of natural history practices created a "commonwealth of observers," who, by the early nineteenth century, were united in the task of defining a distinct British natural history in the face of Napoleonic invasion.[66] Pennant's distribution of successive editions of *British Zoology* was central for formalizing links between those who received copies of different editions as gifts and those who purchased copies from Benjamin White's Fleet Street shop. Both these approaches to distributing information were designed to bond a group of elites, all of whom had sufficient wealth to afford copies of these books and leisure to observe the natural history of their surroundings and communicate their findings to Pennant.

As shown in the previous chapter, the distribution of *British Zoology* in different formats and editions provided central reference points to standardize correspondence through the consistent citation of Pennant's work. Citations to the *British Zoology* allowed correspondents who already contributed to Pennant's network to reference published descriptions. As Dániel Margócsy has shown, this standardized information across collections, allowing for engagement with the descriptions in different copies of the same printed book and allow for the interpretation of brief references given in letters.[67] The consistent use of *British Zoology* throughout Britain, Europe, and the wider world secured Pennant's position as a focal point in a network, communication across which was structured through the standards of description, depiction, and citation outlined in *British Zoology*. In addition to citing *British Zoology* in correspondence, those who owned copies interleaved them for notes or added these to the margins. Marginalia was then reordered to construct correspondence, some of which was communicated to Pennant, providing the basis for new editions of *British Zoology* as the decades progressed.

Pennant's own copies of *British Zoology* are at the centre of this intricate web of information. They contain extensive annotations, flyleaves of additional notes, original correspondence, and specimens representing the cyclical process of gathering information from across Britain. These books relied on the continual provision of information for their improvement, a practice applied to successive editions of Pennant's works. Examples include the final annotated copy of *British Zoology* (1812), discussed at the end of chapter 4. Correspondents' careful citation of precise page numbers allowed Pennant and his assistants, who included family members, Moses Griffith (who acted as Pennant's secretary while also producing artworks), and the son of the local postmaster, Thomas Jones, who was employed as Pennant's scribe, to insert this information into these bespoke copies of *British Zoology*. This created a manuscript used to print new editions. For example, Pennant's copies of the 1768–1770 edition contain numerous additions, including a flyleaf opposite the description of the scaup duck on which Pennant described a specimen shot on January 13, 1776. Pennant noted the precise dimensions of the animal alongside descriptions of its color, inserting a selection of feathers taken from this bird to provide a reference point for his description. Inserted specimens became more prominent in Pennant's own copies of the *British Zoology*.[68]

Pennant's copy of the 1776–1777 edition contains far more annotations than his copies of the earlier edition of this book, outlining the contributions of Pennant, his son, several amanuensis, and reviewers. Pennant used these copies of *British Zoology* to serve as the manuscript for a new edition, a fact apparent from his annotations on the first endpaper of the first volume that designated a set of "Directions to Mr. white." Under this heading Pennant specified for White, and the printer he employed, "to alter the running number of species where one is added or supressed. The same for the numbers of the plates & the list of them."[69] This reflects the numerous additions and amendments Pennant and several other people made throughout these volumes, revising sections, omitting descriptions that duplicate species, and adding new descriptions communicated by a national network. This is very similar to the activity Jonas Dryander was commencing in Banks's copies of *Systema Plantarum* and *Species Plantarum*, exercising significant authority over the inclusion, description, and definition of species while leaving tasks, such as the numbering of individual species, to those who had less authority over philosophical aspects. Pennant left the task of numbering species in *British Zoology* to Benjamin White, who, as a genteel natural history publisher, was already familiar with these practices through corresponding with his brothers, the naturalists John and Gilbert White, and Pennant.[70]

A typical example of Pennant's annotations can be found in the new description of the crested titmouse. This consists of a single folio sheet; on one side Pennant has added the name, list of references, and description of this species and on the other he has extended the description of the bearded titmouse's nesting habits. Pennant's addition of a species to the running order of *British Zoology* ensured that

Benjamin White had to realign the numbers for all the descriptions published in the new edition. In addition to describing the physical attributes of the crested titmouse, Pennant added information on its geographical distribution in Europe, describing how it is "found in *Schonen* in *Sweden*, and the western and more temperate parts of *Russia*, but does not reach *Sibiria*."[71] These notes appeared in print in the 1812 edition of *British Zoology* and reflect Pennant's continued correspondence with naturalists who lived and traveled in these regions. For example, in the preface to *Arctic Zoology* (1787), a book Pennant was writing when adding these annotations to his copy of *British Zoology*, Pennant thanked the Stockholm-based naturalist Samuel Ödeman for sending "valuable remarks on the quadrupeds of Sweden." Pennant's description of the distribution of the crested titmouse reflects his consultation of recent published descriptions and correspondence when constructing *Arctic Zoology*. These include Peter Simon Pallas in St. Petersburg, Alexander Garden in Carolina, and those who had traveled with James Cook into the polar regions of the Pacific.[72]

In addition to Pennant's hand, the manuscript insertion describing the crested titmouse contains notes by Pennant's son, David. These are compiled from annotated references, including page references to Carl Linnaeus's *Systema Naturae*, reflecting the distribution of labor within Pennant's collecting and publishing enterprise. David Pennant's notes are present throughout these volumes and reflect his takeover as the main editor of the new edition after his father's death in 1798. After this point, the editing process became fundamentally collaborative. At the beginning of the first volume, David Pennant noted: "Part 1. Quadrupeds p. 1 to 151 sent by Mr White to Mr. Hamilton. Jan 19 1811. Rest of vol. to Rev. H. Davies per mail, same day—returned & forwarded soon after to Mr White who sent it to Dr Latham. Vol. II sent to rev. H. Davies per mail. Feb. 7. Returned march 5 sent to Dr. Latham, & mail to White. March 7."[73] These notes present a rare insight into the process of reviewing and printing natural history books in the first years of the nineteenth century. The fact that Pennant and White sent specific parts—such as those on quadrupeds, birds, fish, shells, and crustacea—to different reviewers before sending them to Samuel Hamilton, the Weybridge-based printer commissioned to produce this work, reflects the levels of expertise these individuals had over different branches of natural history. For example, Hugh Davies (1739–1821) was sent the sections on birds, fish, and shells, whereas

John Latham (1740–1837), one of the most notable British ornithologists who also published with Benjamin White, only received the sections on birds. Pennant's and White's division of these volumes based on different branches of zoology also attests to the condition of the bindings; many, such as the first volume, were ripped apart so sections could be sent to different reviewers.

Although significant attention has been paid to the processes of reviewing scientific journals,[74] the procedures used to review printed books have received minimal attention. The notes in *British Zoology* demonstrate the complexity of reviewing books; not only were different parts sent to specialists in their respective fields, but each individual held a position of independent authority while annotating and updating the respective volumes, notes used by the printer to set the type for the new edition. This was maintained through the specific roles of the publisher and editor. David Pennant personally sent Hugh Davies the sections and volumes on birds, fish, shells, and crustacea to annotate. In comparison, Latham was approached by Benjamin White, who sent him the respective ornithological sections. After reviewers had added their notes White forwarded the relevant annotated pages to Weybridge so Hamilton could set the type.

Latham's annotations are often quite straightforward, adding references to his publications printed since 1776. Perhaps the most numerous refer to Latham's *A General Synopsis of Birds* (1781–1785), updating each of Pennant's descriptions with the relevant page references. Latham's other notes relate British birds to a broader European fauna, comparing aspects of Pennant's descriptions with accounts of certain species' feeding, migratory, and breeding habits. For example, under the description of the raven, Latham added, "In Italy [the Raven] quits the subalpine woods in Octo[ber] and spreads over the lower countries; returns to the woods in April."[75] Many of Latham's notes are drawn from the personal observations of grand tourists and works by naturalists such as Giovanni Antonio Scopoli (1723–1788), who had a profound influence on other naturalists in Pennant's circle, such as Gilbert White. This reflects on an increasing interest in the geographical distribution of species by the 1780s, with Latham, Pennant, and other naturalists paying far more attention to debates over seasonal migration, contributing to the emergence of biogeography in the decades around 1800.[76]

Despite his efforts in the process of editing, Latham was

dissatisfied with the final publication. Writing to David Pennant in 1812, Latham offered his thanks for a copy of the *British Zoology* on large paper. However, he noted how "I have looked through it and join with you in lamenting ye most gross & careless errors in misspelling names on ye plates of the last vol.—shewing that no one could have looked over them." Latham then went on to attack White and Cochrane, describing how they "have always appeared to me true superstitious Booksellers, not caring for any thing beyond emolument."[77] This represents a distinct change from the values shared by Pennant and Benjamin White. White and Cochrane lost the trust of numerous elite naturalists in addition to those who purchased these books. This reflects a significant change in natural history publishing during the early nineteenth century, social and economic pressures that led to White and Cochrane's bankruptcy in 1815.

Hugh Davies of Beaumaris on the Isle of Anglesey was another main contributor to and reviewer of *British Zoology*. Davies was a long-term correspondent of the Pennants, communicating new information on specific species to assist with the editing and production of several editions of *British Zoology*. However, the information Davies communicated did not only take the form of conventional letters and annotations. This is apparent from a small package Davies sent to David Pennant in February 1802 on which Pennant inscribed "Wings of Buntings" (plate 10).[78] In comparison to more mainstream letters, the physical proof of specific physical characters displayed by a specimen designed to represent a species gave Pennant sufficient evidence to make substantial revisions to *British Zoology*. These took the form of alterations to the printed text and the number of distinct species of bunting. Initially, Pennant changed the name of the "tawny bunting" to the "snow bunting" for part of the description, reflecting on how these are not distinct species but "against the rigorous season [winter], they become white on their head, neck, and whole under side."[79] Davies's specimen shows the bunting at the point of seasonal transition, with a strong element of brown gradually turning to white. As such, David Pennant erased the separate page-long description that defined the snow bunting as a distinct species, combining elements of this with the section on the tawny bunting. To unite these descriptions and justify that the snow and tawny buntings are the same species, Pennant added an interleaved page with notes designed to assist those responsible for setting the type with merging these descriptions. The specimens contained within these volumes show how the material culture of natural history played an active part in the editorial process, supporting decisions to make revisions and define distinct species.

The snow bunting's wings represents just one example of the addition of physical specimens to Pennant's printed books and manuscripts. This practice represents the conjoined nature of text and specimen collections in Pennant's and numerous other contemporary natural history collections. For example, in his copy of Engelbert Kaempfer's posthumously published *The History of Japan* (1727) John Martyn (1699–1768) inserted leaves of plants described in the printed text and identified these with handwritten labels.[80] The insertion of specimens emphasizes that descriptions did not always provide an adequate representation of a species and a specimen was needed to supplement this. Other examples can be found in Pennant's copy of the 1768–1770 edition of *British Zoology* in which Pennant inserted several feathers from a snipe opposite the description of this bird on page 358, labeling them as "four coverts from the tail longer than the rest" and "four greater coverts" in addition to "four quill feathers." These were inserted next to an illustration and detailed description of a specimen "shot at Sandwich Jany 1779," describing its weight and anatomy.[81]

Similar practices are exhibited in Pennant's other books and manuscripts. These include the butterfly pressed between the pages of his copy of August Johann Rösel von Rosenhof's *Insecten-Belustigung* (1746–1761), discussed in chapter 1. Specimens between the pages of Pennant's manuscript notebooks are plentiful and represent a similar practice to his extra-illustration of many of the printed books in his library. Examples include samples of the fur of cattle, describing their "horns 10 ½ inches long/ round at the base 5 ½ / distance between the tips 12 1/3 / at the bottom ½ original color brown. range large." Pennant's specimen was designed to preserve a record of the color of this species in his notebook for future reference.[82] Other specimens can be found in a notebook titled in Pennant's hand "Plants of Barbary, Senegal, Guinea, Aethiopia, Canary Isles, Madeira, Azores" in which he aims to catalogue all known species in known regions of Africa and the Atlantic islands. This notebook was a collaborative enterprise. Pennant's handwriting is interspersed throughout with that of Hugh Davies. Toward the middle of the notebook Davies has written, "The following is a list of some few of the plants I had selected to compose a probable Flora of Senegal and Guinea from considering the Latitudes in which they are known to grow. In doing this I have expressed myself in general terms to avoid deception. S. stands for Senegal G. for Guinea. S. G. for both." Under Davies's note Pennant has written "copied By the revd. Mr Hugh Davies. This was the last act of friendship (out of numbers) which was done me by that man able & universal Botanist the revd. John Lightfoot of Uxbridge: This with his last letter dated Decr 27 1787."[83] Leaves of plants Pennant, Davies, and Lightfoot received from Senegal are then inserted between the pages of the catalogue, representing their connections in West Africa. Many were associated with the slave trade and it is probable that the information in this manuscript went toward the content used in the volumes covering Africa for Pennant's vast *Outlines of the Globe*. Pennant's and his wider family's insertion of numerous images into these volumes has a similarity to the addition of specimens to printed texts and manuscripts throughout Pennant's collection that was, for the most part, a collaborative enterprise relying on contributions from a wide range of individuals.

Examples of Davies's input can be seen throughout the third and fourth volumes of *British Zoology*, which concentrate on fish and molluscs. Davies's notes all expand on his observations of specimens collected on foreshores in Anglesey and North Wales. A typical example is a note Davies inserted between the pages concerning the dolphin, stating that "Mr P[ennant] is mistaken in the number of teeth of the Dolphin Delphis—in the year 1793 I had an opportunity of examining at least half a score fish catch ashore near Caernarfon when I found that each fish had in the upper jaw from 24 to 30 in each side, that is from 48 to sixty in the whole."[84] Davies signed this note "H.D." to emphasize his contribution while elements of this description have been added to the printed text in the new edition. Pennant noted that the specimens "were so much mutilated when the reverend Hugh Davies had an opportunity of seeing them, that he was unable to form any particular description."[85] The account of a firsthand observation of a specimen serves a similar purpose to the physical specimen of the bunting Davies posted to Pennant—using objects observed in his local vicinity to back up and correct specific elements of a description. Davies's interest in teeth is important, given that the quantity and type of teeth a particular species possessed was central to its placement in the Linnaean system. Linnaeus regarded the artificial digits devised from the number and structure of teeth to be essential for classifying and grouping species of mammals, a view that generated severe criticism of the Linnaean system.[86]

Through emphasising the number and kind of teeth possessed by dolphins, Davies was engaging with ongoing debates surrounding the definition of cetaceans from fish. To address this, David Pennant consulted Bernard-Germain-Étienne de la Ville-sur Illon, Comte de Lacépède's *Histoire Naturelle des Cétacées* (1804), in which Lacépède suggested whales were more closely related to quadrupeds than fish, emphasizing their areal respiration, warm blood, and lactation. Lacépède cited Joseph Banks's "handwritten notes" as a main source of information concerning a sighting of six gladiator dolphins (now known as killer whales) in the River Thames.[87] A result of Lacépède's citation of Banks's manuscript notes, David Pennant wrote to Banks in 1811 on the subject of these cetaceans. Banks sent a curt response: "On my Return to London I shall be happy to give you all the information in my power Respecting whales . . . I have no Recollections of having given Le Cepede any information Respecting the Beaked Whale The Bottle nose I always believed to be a Porpoise with a long thin snout . . ."[88] Thus, rather than corresponding with Banks, it seems Lacépède received his information secondhand. One source was the numerous handbills and newspapers that cited Banks as an authority when describing "miraculous" sites of nature. Many were used by David Pennant to add new information to his annotated copies of *British Zoology*, filling them with newspaper cuttings and advertisements—ephemeral items given out and sold on the street.[89] A typical example is a broadside advertising the "Monodon Monecerous. Linnæus. Unicornum Marinum" that Pennant tipped into *British Zoology* describing "The Sea Unicorn" that could be viewed at "Richard Pollard's, the Swan & Salmon, St John's Street, Stamford." Banks was cited as a verifier of the authenticity of this spectacle and was quoted to have described it as one of "the greatest" natural curiosities ever "seen in this Kingdom."[90] This is a typical example of Banks's name being added to a specimen to verify its uniqueness. Other examples include a broadside issued to advertise "The Original Stone Eater," whose performances of "Stone eating and Stone swallowing" could be viewed on the Strand, events verified by "Sir Joseph Banks, Dr. Hunter and Dr. Monro &c. &c. &c."[91]

Banks's offer of supporting Pennant's efforts to edit *British Zoology* went further than just communicating information. This involved allowing access to his London library at 32 Soho Square. Although predominantly a botanical collection, Banks did possess some zoological books and manuscripts—many of which Thomas Pennant examined in the early 1770s after Banks's return from the South Pacific. An example is Banks's quarto copy of the 1776–1777 edition of *British Zoology*. This was incorporated into Banks's main library and remained accessible to those who visited his collection.[92] One of these visitors was David Pennant, who consulted Banks's copy of *British Zoology*, annotating a flyleaf that he loosely inserted between the pages of his own copy of this work concerning the fish his father named "Morris" "in memory of our worthy friend" William Morris of Angelsy. The strange translucent fish Pennant named "Morris" is now known as the larvae of a marine eel. David Pennant noted:

> In the Br. Zoo. in Sir Jos. Banks library with a bad sketch of the fish "This fish I found in the month of January 1745 in Penrhyn Dyfr just left by the hole & alive. It was in length 5 inches and about 3/10 of an inch wide as transparent almost as glass the thickness at the neck & tillg about 1/6 of its breadth, it had no fins, and all its braces appeared as in the cut with small black spots from one end of the other.
>
> L. Morris.
>
> Copied from a blank page in Louis Morris's Ray Synopsis g. to mr Lloyd of Aberystwyth. 1786.[93]

Pennant's notes reflect his access to books and manuscripts when visiting London as a member of the elite circles who associated with Banks. Morris's annotated copy of John Ray's *Synopsis Methodica Avium & Piscium* (1713) had recently been acquired by the British Museum and Banks's copy of *British Zoology* was held in a private library of natural history—the qualification to access books held by these collections was to be known to Banks. This reflects yet another source of information Pennant used to edit *British Zoology*, additions that became so substantial that Hamilton wrote from Weybridge to comment that "the charge for Cancelling two pages will be about 21/-" if Pennant decided to remove sections after the new pages had been printed, although Hamilton added that "if the cancel be necessary from any blunder of mine I shall be happy to rectify it without any charge."[94]

David Pennant's use of annotated books in Banks's library reflects practices of reading and sourcing information in the more accessible collections of the early nineteenth century. Davies and Latham, in addition to a network of correspondents, including Banks, provides an insight into the collaborative methods of gathering information and organizing it into a logical format. Firsthand observations gathered from across Pennant's network were supplemented by information sourced from newspapers and street literature such as broadsides, in addition to the books Pennant read in private and public libraries.

Defining the "Reading Public" for Natural History

In a direct appeal to polite society, the advertisement printed at the end of the third volume of the 1776–1777 edition of *British Zoology* promoted the "NINETY ELEGANT PLATES of the Shell and Crustaceous Animals of Great Britain, with Descriptions" in the forthcoming volume.[95] The much-delayed fourth volume on Mollusca and Crustacea had been held back by the cost and time it took to engrave the plates. The Warrington-based printer William Eyres wrote to Thomas Pennant in 1776, commenting that "I have just got a letter from Mr. White who says the Quarto Zoology is to make two volumes & the octavo three volumes which I Imagine you also intend them. He desires me to write to you about the division of the volumes of which you'll please acquaint me in time."[96] This shows Pennant's direct reliance on sales to recuperate his personal expenditure on producing books, distributing these through Benjamin White's Fleet Street shop.

By the late 1770s genteel society could be relied on to purchase Pennant's books; sales that ensured he made no financial losses while maximizing distribution. White's success in circulating these books was a direct result of the popularity of natural history among parish priests, physicians, lawyers, and aristocratic women throughout the 1770s. For example, writing to John Strange (1732–1799), the British envoy in Venice, Thomas Martyn, botany professor at the University of Cambridge, reflected on the increased class sizes and popularity of natural history when compared to their time at the university two decades before: "In our time, you know it [natural history] was a study scarce heard of among us; we were looked upon as no better than cockle-shell pickers; butterfly hunters and weed gatherers; and I can remember very well that when I walked forth now & then, with a little

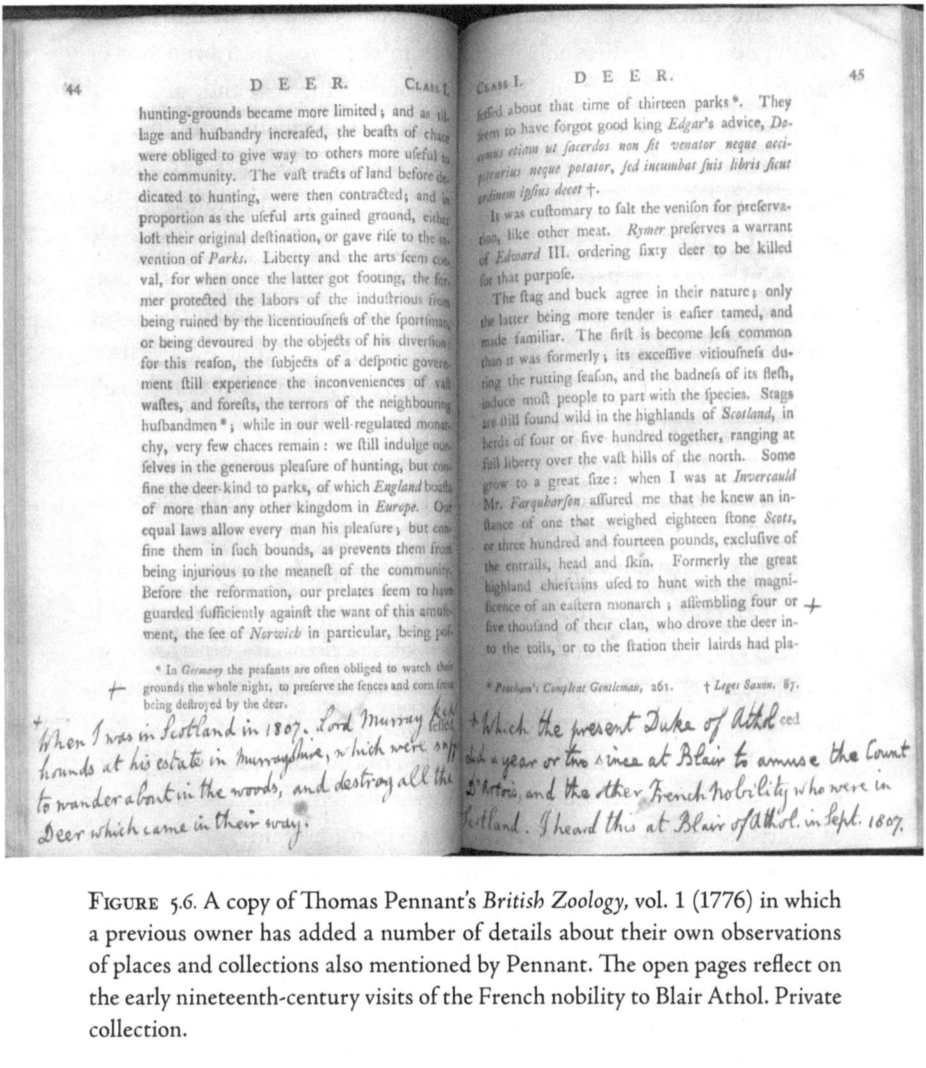

44 D E E R. CLASS I.

hunting-grounds became more limited; and as tillage and huſbandry increaſed, the beaſts of chace were obliged to give way to others more uſeful to the community. The vaſt tracts of land before dedicated to hunting, were then contracted; and in proportion as the uſeful arts gained ground, either loſt their original deſtination, or gave riſe to the invention of *Parks*. Liberty and the arts ſeem coeval, for when once the latter got footing, the former protected the labors of the induſtrious from being ruined by the licentiouſneſs of the ſportſman, or being devoured by the objects of his diverſion: for this reaſon, the ſubjects of a deſpotic government ſtill experience the inconveniences of vaſt waſtes, and foreſts, the terrors of the neighbouring huſbandmen*; while in our well-regulated monarchy, very few chaces remain: we ſtill indulge ourſelves in the generous pleaſure of hunting, but confine the deer-kind to parks, of which *England* boaſts of more than any other kingdom in *Europe*. Our equal laws allow every man his pleaſure; but confine them in ſuch bounds, as prevents them from being injurious to the meaneſt of the community. Before the reformation, our prelates ſeem to have guarded ſufficiently againſt the want of this amuſement, the ſee of *Norwich* in particular, being poſ-

* In *Germany* the peaſants are often obliged to watch their grounds the whole night, to preſerve the fences and corn from being deſtroyed by the deer.

+ When I was in Scotland in 1807. Lord Murray kept hounds at his estate in Murrayshire, which were suff to wander about in the woods, and destroy all the Deer which came in their way.

CLASS I. D E E R. 45

ſeſſed about that time of thirteen parks*. They ſeem to have forgot good king *Edgar*'s advice, *Docemus etiam ut ſacerdos non ſit venator neque acceptrarius neque potator, ſed incumbat ſuis libris ſicut ordinem ipſius decet* †.

It was cuſtomary to ſalt the veniſon for preſervation, like other meat. *Rymer* preſerves a warrant of *Edward* III. ordering ſixty deer to be killed for that purpoſe.

The ſtag and buck agree in their nature; only the latter being more tender is eaſier tamed, and made familiar. The firſt is become leſs common than it was formerly; its exceſſive vitiouſneſs during the rutting ſeaſon, and the badneſs of its fleſh, induce moſt people to part with the ſpecies. Stags are ſtill found wild in the highlands of *Scotland*, in herds of four or five hundred together, ranging at full liberty over the vaſt hills of the north. Some grow to a great ſize: when I was at *Invercauld* Mr. *Farquharſon* aſſured me that he knew an inſtance of one that weighed eighteen ſtone *Scots*, or three hundred and fourteen pounds, excluſive of the entrails, head and ſkin. Formerly the great highland chieftains uſed to hunt with the magnificence of an eaſtern monarch; aſſembling four or + five thouſand of their clan, who drove the deer into the toils, or to the ſtation their lairds had pla-

* *Peacham's Compleat Gentleman*, 261. † *Leges Saxon.* 87.

+ Which the present Duke of Athol did a year or two since at Blair to amuse the Count D'Artois, and the other French Nobility who were in Scotland. I heard this at Blair of Athol. in Sept. 1807.

FIGURE 5.6. A copy of Thomas Pennant's *British Zoology*, vol. 1 (1776) in which a previous owner has added a number of details about their own observations of places and collections also mentioned by Pennant. The open pages reflect on the early nineteenth-century visits of the French nobility to Blair Athol. Private collection.

hammer concealed under my coat, I looked carefully round me, lest I should be detected in the ridiculous fact of knocking a poor stone to pieces."[97] Martyn's observation, as someone aimed to promote natural history prospective priests in Cambridge, marks a distinct rise in the popularity of the subject by the 1780s, something reflected by the increased numbers of people visiting the Cambridge University Botanic Garden and buying botany books. Examples include Martyn's translation of Jean-Jacques Rousseau's *Letters on the Elements of Botany Addressed to a Lady* (1785), a book written for those who "wished to amuse themselves with Natural History."[98] The initial publication of the octavo editions of Pennant's *British Zoology* was central for stimulating educated individuals' interests in the zoological branches of natural history throughout Britain, North America, and, after the publication of French and German editions, Continental Europe.

Purchasers of *British Zoology* used these books to record observations, adding to descriptions of specific species mentioned in the printed text. For example, when traveling around Scotland, one early nineteenth-century user of Pennant's works added several annotations relating to the practices of hunting deer, describing the large parades of clansmen, "Which the present Duke of Athol did a year or two since at Blair to amuse the Count D'Artois, and the other French nobility who were in Scotland, I heard this at Blair of Athol in Sept. 1807." This note relates to one of the numerous events orchestrated for Charles-Philippe, Comte d'Artois (1757–1836), brother of the executed French king Louis XVI, who took refuge in Scotland during the French Revolution and the Napoleonic Wars (figure 5.6).

Other notes relate to the localities of specimens Pennant cited. When Pennant described a specimen of the field lark "in the magnificent and elegant museum of Ashton Lever, Esq," the annotator has added a note to the margin, stating that "it is now sold by auction."[99] This refers to the famous auction of the contents of the Leverian Museum in 1806. Lever's museum attracted numerous members of genteel society who could afford the half-guinea entrance fee. One of these was Pennant who, on January 24, 1780, purchased an admission ticket from Lever's curator and secretary Thomas Waring.[100] This reflects Pennant's, and many other visitors,' interests in viewing a diverse natural history and ethnographic collection from around the world, including material collected during James Cook's voyages. The wealthy, learned individuals who could afford to visit the Leverian Museum, housed in the Blackfriars Rotunda after 1786, represent Pennant's desired readership for *British Zoology*. For example, writing to his brother, Gilbert, on April 6, 1774, John White described how "Mr. Lever, I find, continues to be wonderfully generous . . . In my monography I have made honourable mention of his museum."[101] Those who engaged with Pennant's books and institutions, such as the Leverian Museum, are representative of the higher accolades of the "reading public" who had specific interests in natural history, a pastime that became integrated with devotions to reading, going to the theater, and writing sermons and diaries.

The emergent genteel groups of parish priests, lawyers, country squires, and wealthy yeomen farmers were among the most common users of Pennant's works, using these to add a systematic structure to their observations and natural history collections. Many had access to specimens, often being the younger children of local gentry and educated at either Oxford, Cambridge, or one of the Scottish universities, and developed interests in natural history through their rural upbringings that gave them access to a wide array of different flora and fauna. For example, Pennant's second son, also called Thomas (1780–1846), became the rector of Weston Turville in Buckinghamshire. Many others, such as Anna Blackburne, compiled and had access to collections held in their relatives' country houses. *British Zoology* became a central tool when recording and ordering observations and collections, a typical example being Blackburne's notes on shells and birds in her copy of Pennant's *British Zoology*.[102] Copies of *British Zoology* were often purchased from Benjamin White during visits to London, a typical example being that acquired by Gregory Lewis Way (1756–1799), a former lawyer of the Inner Temple and owner of the small estate of Spencer Grange, Essex, who had his copy of the 1768–1770 edition of *British Zoology* interleaved with blank pages to accommodate notes. Interleaving was a common service supplied by London stationers and was something White offered to his customers from the 1750s.

Way added copious notes throughout his copy of *British Zoology* recording detailed observations of specific species. Many compare specimens observed at different times. An example is the great crested grebe. Way described how he "examined a fine male bird of this species, shot during the Great Flood near Eton, March 1774," adding a description of the bird's plumage. Close observation of this bird allowed Way to add specific details on its physical characters not listed in the printed text while describing how "I found a little fish in its mouth, which I suppose it has just caught before it was shot." Another example relates to Way's observation of the goldeneye duck, writing at the foot of the printed text that in 1773 "my brother BW shot one of these on his water at Denham," adding that its physical characters matched Pennant's description, except for the legs, which "were of a blackish or lead colour." This relates to Benjamin Way, his older brother and the main inheritor of the family estate in Denham, who supplied his brother with information gathered through observing the natural history of his private land.

In 1776 Way added another description of male and female examples of the goldeneye duck, paying special attention to the color of the legs. In comparison to his brother's specimen, both examples conformed to Pennant's description on the opposite page, where Way underlined "the legs are of an orange colour."[103] Way's notes reflect the systematic updating of the printed descriptions, using an interleaved copy of Pennant's work to add structure to his observations. Interleaved books were far more useful than blank notebooks since they provided incentives to incorporate observations from local surroundings into a defined systematic order.

Evidence for the use of *British Zoology* by individuals who were not presented with a copy can be found through examining its citation. Use of Pennant's *British Zoology* reflects a growing interest in natural theology by the early nineteenth century. This was manifested by the work of William Paley, whose *Natural Theology* (1802) was widely read by naturalists and philosophers. Paley's work was soon connected to natural history by its readers, many of whom included prominent genteel naturalists and those who worked in universities.[104] Pennant's *British Zoology* supported Paley's interest in commencing a program to explore nature and "discover" divine intervention. Every edition of Pennant's work retained the prefatory passage, suggesting that the purpose of these volumes was to explore and spread information on the zoology of Britain, a study designed to uncover "the wisdom of our Creator, which his divine munificence has so liberally, and so immediately placed before us."[105] A typical example of a natural theological publication that engaged with Pennant's *British Zoology* when describing animal species is Edward Polehampton's *Gallery of Nature and Art; Or, a Tour through Creation and Science* (1815). A similar use of Pennant's work as an authority on natural history can be found in an edition of Augustin Calment's *Dictionary of the Holy Bible* edited by Charles Taylor (1800). This shows how the sale of Pennant's work to an elite subset of society facilitated broader engagement with those interested in natural theology. Ownership of Taylor and Polehampton's works, large quarto and folio books made up from fine wove paper and containing numerous copperplates, would have been seen as an important purchase for aspirational parish priests and wealthy yeomen, groups who used Pennant's work as a point of reference to support their own interest in exploring the extent of God's creation.

A main user of Pennant's *British Zoology* was Gilbert White, who was presented with copies of successive editions in 1768 and 1777. However, in comparison to Pennant, who relied on the income from his estates to fund his publication, White's limited finances as a country curate meant that he had to sell copies of his *Natural History and Antiquities of Selborne* to recuperate the expenditure on the plates, paper, and charges of printing, posing a significant risk to his genteel reputation. Unlike Pennant, White's expenditure on the quality his book made its production unaffordable, resulting in White having to agree a contract with his brother's firm. Brokering a contract allowed White to afford fine paper and copperplates, describing how "my brother and nephew have spared no expense about it [*Selborne*], and particularly on the engravings, that have cost a considerable sum."[106] White lost his independence when distributing this book, only presenting a few copies to his main correspondents, Daines Barrington and Thomas Pennant, in addition to members of his family. On December 3, 1788, Henry White, Gilbert's White's brother, wrote that he had received a "hamper from London containing ye Natural History of Selborne, presented by ye author. A very elegant 4to with splendid engravings & curious invests [investigations]."[107] Pennant inscribed "The Gift of the Author" on the first endpaper of his copy.[108] Contracting with a commercial publisher was a worrying prospect for White. Shortly before the publication of *The Natural History and Antiquities of Selborne*, White wrote to his friend and one of the main proponents for the book, Richard Churton, commenting that "as you were accessory to making me an author, you must defend me if I am attacked unreasonably."[109] White's reservations came as a direct result of risking his genteel reputation through becoming associated with a commercial venture.

Contrary to White's concerns, the financial risk of purchasing the highest quality materials for *The Natural History and Antiquities of Selborne* paid off in genteel circles. White's correspondent, John Mulso (1720–1791), another Hampshire cleric, commented that "I have hardly ever seen a book so well attended to, and so happily finished off."[110] In addition to receiving praise from his established network, *The Natural History and Antiquities of Selborne* brought White several new correspondents. One example is the naturalist and army officer George Montagu (1753–1815), who published his *Ornithological Dictionary* with Benjamin White's firm in 1802. Striking up a correspondence with White in May 1789, Montagu praised *The Natural History and Antiquities of Selborne*, saying, "I have been greatly entertained by your 'Natural History of Selborn,' in the ornithological part."[111] The good reception of White's *Natural History and Antiquities of Selborne* led to praise in the *Monthly Review*, in which the author acclaimed White as an "ingenious writer," adding that the writing style reflects on the good "TASTE of the author."[112] By May 1789, the reception of White's work had been so promising that he noted that "the world has been so indulgent on my book, that I begin to hope that the editors will be paid for the trouble and expense they have bestowed in the publication."[113] The reception of White's *Natural History and Antiquities of Selborne* among genteel naturalists was central for its immediate success. Its long-term success, being published in numerous editions throughout the nineteenth century, was a direct result of the sale of the publication rights to commercial publishers. This meant *The Natural History and Antiquities of Selborne* adapted in accordance with the general changes taking place in the commercial publishing industry. For example, although the first three editions were published in large quarto volumes with wide margins, the 1822 edition was published in two small octavo volumes by a consortium of publishers. This book continued to be used by a genteel elite. For example, the copy of the 1822 edition owned by Henry Norman, a landowner in Bromley, Kent, and a neighbor of the naturalist Charles Darwin, has been interleaved with blank pages annotated with observations relating to the printed text.[114] Despite White's reservations, the use of commercial publishing houses ensured the longevity of *The Natural History and Antiquities of Selborne*, and its continued influence over practices of observing the natural world.

Networks, Systems, and Continued Exchange

Books and stand-alone prints became integrated in the late eighteenth century, demonstrating the flexibility of printed matter when collecting, ordering, and producing knowledge on nature. Practices of distributing and using printed materials were shaped by their physical format. Examples include Joseph Banks's plate depicting *Cinchona officinalis* produced to stimulate a personal network, the distribution of printed items as presentation copies of books or by publishers at the author's request, and genteel natural history books sold on the open market. The use of these items by individuals ranging from global travelers to librarians employed to catalogue private collections, those editing new editions and the reading public of late eighteenth-century Britain emphasize a wide social and geographical diversity of users who all had different motivations for reading such items. Copies were taken to the other side of the world while some, such as the interleaved copies of *Species Plantarum* kept at 32 Soho Square, were designed to record the products of these expeditions. Others still were designed to accumulate information from specimens, correspondence, observations, and materials viewed in other libraries. Similar practices of reading, annotation, travel, and observation were further impressed upon the reading public. Few copies of these books, perhaps most significantly Thomas Pennant's *British Zoology*, were left clean of annotations, reflecting how they inspired an intense engagement with the natural world.

The maintenance of a continual cycle of gathering information relied on a fine balance that rested on the motivations of individual naturalists, commercial publishing markets' ability to cater for genteel groups, and the standardization of knowledge. Early nineteenth-century changes in the British publishing industry represent the first cracks that started to permeate this genteel process of generating information.[115] These were followed by new systems of classification. For example, writing to David Pennant in 1822 the aged naturalist and reviewer of *British Zoology*, John Latham, complained that he now found the new names and classifications ascribed to species unintelligible, adding that "I therefore am resolved to follow my former Plan, except in Particular cases, whereby I hope to be better understood, & I trust you will approve of my conduct." Latham attributed these problems to the "French ornithologists [who] wish to bury the name of Linnaeus, by creating new systems, but ultimately, these System-makers do not agree with each other, so that out of them all no perfect one will be found—how far I shall be blamed for my obstinacy by these systematists, we shall herewith see."[116] For many genteel naturalists, the Linnaean system became essential for distributing information across diverse networks in an intelligible format—practices central for the continual buildup of information on the natural world. In many cases, the lack of unity between the so-called natural systems of the early nineteenth century and practices of communicating information ensured Linnaean systematics remained dominant across the British Empire, with Latham suggesting that the Linnaean system "may be mended in Parts, as new lights occur, The Basis of the Structure will I trust remain firm."[117]

Complaints about new classification systems reflect a lack of unity when gathering and synthesizing information. However, practices employed when reading and interpreting this information remained independent of particular systems. As emphasized in the introductory diagram, these depended on the social conditions associated with the genteel practices of collecting information used to assemble and fund books containing numerous copperplate images. Rather than falling apart with new classificatory approaches, systems for managing information facilitated the continued exchange of information, growing collections, and new editions of published works.

Conclusion

Publishing, Markets, and the End of Genteel Natural History?

In 1816 an advertisement in the *London Morning Post* notified its readers of the bankruptcy of the Fleet Street booksellers White, Cochrane & Co., necessitating for any with a claim on "Books &c. deposited on sale or Return, are requested by the Assignees immediately to send an Account of the same."[1] The firm's genteel authors were not impressed. One of these—the Anglican priest, Oriental scholar, and keeper of manuscripts at the British Museum, Thomas Maurice (1754–1824)—complained that White and Cochrane's bankruptcy had "thrown upon my hands all remaining unsold copies of my works." Maurice, who had corresponded with Thomas Pennant on the latter's *View of Hindoostan*, had been "compelled to become my own BIBLIOPOLIST."[2] Maurice was forced to sell his books, risking his reputation in polite circles. By April 17, 1816, White and Cochrane's books were being sold "at reduced prices, with a further deduction of 10 percent, for ready money, until 30 April 1816, when the sale will positively close, and the lease of the house disposed of."[3]

Economic problems in the book trade and changing classificatory systems caused a distinct schism in the complex relationships between practices of natural history, the construction of books, and commercial publishing houses. Founded in 1759, the fifty-six-year lifespan of Benjamin White's firm encapsulates a major epoch in the practice of natural history. As exemplified by the introductory diagram, the use and production of books was intertwined with diverse paper technologies that governed the collections compiled by genteel authors.

Central to shaping the emergence of the Linnaean system in Great Britain, genteel naturalists used books to influence voyages of discovery, the development of publishing markets, and the take-up of natural history as a fashionable pursuit. The diverse processes that went into collecting, organizing, producing, and distributing knowledge of natural history in Britain between 1760 and 1820 reflect the extent naturalists aligned themselves with the growing commercial world and global circulation of knowledge. Many naturalists—including Joseph Banks, Thomas Pennant, and Gilbert White—sought to combine their interests with plans for national improvement, integrating natural history with specific economic, social, and imperial ambitions. These are apparent throughout the practices, ranging from collecting information in the field through to the production, distribution, and use of books—approaches to natural history that mirrored the workings of elite households while interacting with the new nation state and expanding empire.

The constant gathering of information with a view to its publication, distribution, and reuse remained central to the practices of natural history in the late eighteenth century. As represented in the diagram in the introduction (figure I.1), the practices of natural history connected the processes of producing and distributing books within a genteel framework. Working around this diagram in a clockwise direction, this book has surveyed the process of using books to gather information on local, national, and global voyages; the practices of constructing books and the different routes "genteel" naturalists took to produce these; their approaches to distributing books as a means to maintain their genteel social standing; and how these books were actively used by their recipients. Paper technologies surrounded these processes and became integrated with books that served as platforms for annotation used to gather information and facilitate the cyclical process of building collections and publishing editions. This was a practical process and paper technologies both formed a central role in organizing and linking between diverse records of the natural world while guiding the process behind finishing field sketches, producing copperplate engravings, binding books, and distributing these across the globe.

Paper had the ability to combine the results of different expeditions to create the foundations for all-encompassing general natural histories. General programs were embodied through philosophical systems such as those devised by John Ray, which Pennant used to bond a national network, and Carl Linnaeus, the standardizing nature of which made it applicable on global expeditions and as a tool for communicating information to a specific group across an empire. Ray's system bonded a diverse network of naturalists, gentry, aristocrats, and clergy throughout the newly defined nation: it developed in England and was a system that combined physical resemblances with the major uses of animals, aligning these with the long-standing interests of these groups to "improve" their private landholdings. In contrast, the Linnaean system offered a standardized means for identifying species and communicating information. The Anglophone nature of Ray's system proved essential in the 1820s and 1830s for overcoming national prejudice when Robert Brown and John Lindley started to develop and introduce "natural systems."[4]

Connecting scholarship on the development of processes for managing information through assembling printed books and paper technologies in libraries with the literature on the emergence of field science,[5] this book has revealed how these elements integrated with the active process of collecting. Paper technologies were formulated to overcome the scales of time and geography traveling naturalists had to work across. Larger expeditions, such as Pennant's tour of Scotland and Banks's voyage with James Cook to the Pacific, show how these paper technologies governed hierarchies surrounding the activities of naturalists, artisans, secretaries, and assistants.

In contrast to previous research on eighteenth-century field science, which tends to emphasize the role of single individuals in processes of collecting and recording information, this book has emphasized how these processes relied on collaborative teams.[6] Different paper technologies defined specific tasks, emphasizing a close collaborative approach to recordkeeping and collecting specimens, facilitating a distinct chain of command for the collection and systematization of empirical information obtained from a diverse variety of sources. These extended from a naturalists' observations through to information supplied by Indigenous peoples. For example, when on Tahiti, Banks's artist, Sydney Parkinson, showed one Polynesian "some of my drawings, which he greatly admired, and pronounced their names as soon as he saw them."[7] Although Anne Salmond and others have suggested that Parkinson showed images depicting Indigenous Tahitians, it is more probable that Parkinson was showing images of plants to gather information on their names that Daniel Solander incorporated into his notebook.[8] Parkinson produced more botanical illustrations on Tahiti than any other subject and—unlike the images he produced of Indigenous Tahitians who supplied their own names—Parkinson needed to communicate with local people to find the names and uses for plants. Solander's close collaboration with Parkinson facilitated the systematic integration of Indigenous names for the rest of the voyage, relying on Tupaia's role as a bridge between different cultures. These descriptions were then refined by Herman Spöring, who transcribed them into interleaved copies of Linnaeus's works and the "Primitæ Floræ" manuscripts. This has certain similarities with the interleaved pocket book Pennant used to construct a trilingual vocabulary of species names in the English, Welsh, and Gaelic languages. All these interleaved books provided the structures needed to produce books containing information gathered on these voyages. Book production and distribution was an important step used to define naturalists such as Banks and Pennant as the owners of important collections and the intellectual outputs produced by the hierarchic teams of naturalists and artisans they assembled. Ownership of information was of central importance to figures such as Banks and Pennant and their respective artisans had limited rights over the artworks and prints they commissioned, hence Banks's conflict with John Frederick Miller and Pennant discussed in chapter 3.

Social hierarchy remains prominent throughout all the practices of natural history outlined in the introductory diagram and correlate with the workings of elite households in the late eighteenth century. Central to this was the establishment of clear social hierarchies transposed onto the practices of natural history. Although these genteel views have been prominent throughout the general historiography since the work of Roy Porter in the 1970s, since the early 1990s attention has concentrated on expanding our understandings of a diverse array of actors.[9] Through analyzing the collaborative practices used to gather, produce, distribute, and create new information, this book has brought these two literatures into conversation. It shows how genteel interests shaped a diverse global array of actors who participated in natural history, incorporating knowledge systems originating in places ranging from Polynesia to the Scottish Highlands, while drawing the practical knowledge of artisans into objects designed to be integrated with a genteel network. Typical examples are apparent from the teams Banks and Pennant took on expeditions, emphasized by the pyramidal structure outlined in figure 2.2. Clearly defined roles allowed for the rapid assimilation of information. Large numbers of travelers, including artisans and servants, all funded by one naturalist, emphasize the general lack of interest in profiting from these expeditions—contrasting with other British travelers of the period who relied on publishing to support their incomes. Similar social hierarchies influenced genteel naturalists' approach to the production and distribution of books. This was defined by the tasks they employed specific artisans to complete, including the production of illustrations, copperplate engravings, typesetting, and intaglio printing.

Social hierarchies were further enforced by the distribution of books, objects designed to provide readers with a lens through which they could view a private collection, elevating these objects and the information surrounding them in the eyes of genteel naturalists. The increasing dominance of a single elite naturalist over hierarchic teams, who was often accredited with the final published products, resulted in the declining status of those who kept records on expeditions and in collections from the late 1770s. By 1800 such tasks had been relegated to the position of mere secretarial work and manual labor, a result of numerous naturalists turning their attentions to broader philosophical problems, including biogeography and the development of so-called natural systems. For example, Pennant employed "as secretary, Thomas, the son of Roger Jones, our parish-clerk," and Banks employed several Linnaean naturalists to manage his library from the 1770s until 1820.[10] However, hierarchies continued even among those employed by elite naturalists. For example, Jonas Dryander was employed as Banks's curator at 32 Soho Square from 1782 to 1810. His main tasks included managing the team of secretaries responsible for accessioning new material into the collection, facilitating access for visitors during Banks's long absences when visiting his country estates, and managing the acquisition of the most up-to-date reference works. Central to Dryander's role was enforcing Banks's will in relation to who should use the collection and for what purpose. Banks viewed Dryander as irreplaceable, describing shortly after the latter's death that "I have lost my right-hand man. My chief pleasure, that of my library, is reduced almost to a shadow."[11]

Scholarship on the late eighteenth century relegates the practices of collecting information, producing, distributing, and using books to their commercial viability, examining the growing influence of publishers and commercial markets on book production and distribution.[12] Commercial aspects certainly apply to more common publications such as novels, almanacs, tables of calculations, grammatical guides, poetry anthologies, and popular natural histories, books often associated with the growth of the publishing industry and the "print revolution." In contrast, this study has shown the variety of new ways concerns about commerce informed elite natural history publishing programs. Many naturalists distanced themselves from commercial modes of book production and booksellers when distributing their works. This was partly a result of individual reactions to the increasing dominance of commerce in society and its interactions with natural history while naturalists realized that the more commercial approaches to publishing reduced their ability to control the quality and content of a book. Typical episodes include John Hawkesworth's catastrophic editing of Banks's journal for *An Account of the Voyages Undertaken by Order of His Present Majesty* (1773). Although some naturalists such as Pennant funded and organized the production of their published works to make sure they aligned with collections and met particular production standards, the distribution of a certain number of books, albeit through a genteel publisher, posed certain risks to the reputation of genteel naturalists. In spite of attempts to maintain absolute control over their published works to ensure their alignment with specimen collections and paper technologies, the use of commercial publishers to distribute books and recuperate expenditure exposed these works to scrutiny, risking the reputations of their respective authors. Consequences included a loss of control over the content of these works, risking its use by literary hacks such as Oliver Goldsmith, critical reviewers who had the potential to damage naturalists' reputations in polite circles, and accusations of "authorship" as a profession, which Banks and many others did "not consider a gentlemanly vocation."[13] However, booksellers' networks had the advantage of securing more loyal correspondents from across respectable society, allowing for the continual expansion and revision of collections and the publication of new editions to accommodate the new information this growing group supplied. Connections between books and natural history collections ensured they carried all the social connotations elite naturalists associated with a specimen collection, characterised by Krzysztof Pomian as "a set of natural or artificial objects kept temporarily or permanently out of the economic circuit, afforded special protection in enclosed places adapted specifically for that purpose and put on display."[14] Books served to extend these objects and surrounding social connotations into the world of print.

Naturalists' consistent attempts to avoid commercial markets presents a very different set of priorities to those outlined in Robert Darnton's "communication circuit." More recently, Jonathan Topham has stressed the essential role of booksellers in distributing publications, overlooking the large proportion of books presented for free, sometimes via booksellers on an author's behalf, to solidify and expand networks.[15] Although later renditions of Darnton's diagram have gone toward displacing overtly commercial practices,[16] the chart depicted in the introduction (figure I.1) and outlined in this book integrates the practices of natural history with book production. Through combining these processes, it has been necessary to base the diagram around the beliefs and activities of individual practitioners, defining how they continued to avoid commercial connotations through self-funding these projects to maintain a genteel social standing. This is reflected by the large numbers of presentation copies, sometimes amounting to the entire print run of a book. For other natural history books, such as those managed and financed by their author and published under the auspices of a genteel publisher, gifts constituted a substantial percentage of the copies produced. These were calculated into the revenue authors accrued from sales to ensure they broke even with their production costs. Presentation copies were essential for bonding naturalists' global correspondence networks where books and prints presented as gifts had a higher intellectual value than those purchased on the open market.[17]

Presentation copies were valuable to both the authors and their recipients, inspiring a degree of civility towards the work and its compiler and more favourable reviews. However, unlike the French philosophes whose rejection of payments was associated with a desire to stimulate the wide-scale production and distribution of their works,[18] British natural historians attempted to maintain control over intellectual content, production processes and print runs. This careful regulation of the content and number of copies defined a select readership for a specific range of books, representing how the "closed systems" described by James Raven dominated the production and distribution of many natural history books. All readers were selected for a specific purpose. For example, in addition to producing specific books such as *Icones Selectæ Plantarum* (1791) to assist with knowledge production on global voyages in East Asia, Banks distributed this work to repay correspondents and those naturalists who had sent presentation copies of their publications. Similarly, Pennant's strategic distribution of his books can be seen as a kind of intellectual repayment that consolidated, stimulated, and streamlined the communication of information. Although some authors, such as Pennant and William Jackson Hooker, utilized booksellers' resources, the production of these works remained the singular concern of the authors who employed provincial printers to produce books to occupy the shelves of a London bookseller.

Partnerships between natural historians and booksellers regulated the number of natural history books exposed to the open market. This ensured books found their way to a select group who could be trusted to derive some benefit from the information they contained. Bookshops became important meeting points for defined groups, perhaps the most noted example in natural history being that of Benjamin White, a kind of philosophical coffeehouse. The elite audiences and the high price commanded by specialist natural history booksellers ensured natural histories were only accessible to a specific sector of society. The limited audience often led to copies remaining unsold for years after publication. In the case of Pennant's *British Zoology* (1776–1777), Benjamin White was still attempting to sell new unbound sets in 1786.[19] The diverse approaches to distributing natural history publications reformulates current interpretations of the practices associated with the redistribution of knowledge in late eighteenth-century Britain, presenting a further move away from scholarship on a "print revolution" that emphasizes the wide-scale arbitrary dispersal of numerous books on the open market.[20]

Practices of producing natural history books began to change in the early nineteenth century, a period that witnessed the onset of on-commission publishing when, although authors supplied the funding, it was publishers who managed the production process. A result of these being author commissions managed by publishers who had an incentive to maximise profits from sales, the books printed remained the author's legal property. White and Cochrane employed this approach in the 1810s for Maurice's works, including his guidebook *Westminster Abbey* (1813), the remaining sets of his *Indian*

Antiquities (1793), and *The Fall of the Mogul* (1806), the remaining copies of which Maurice had to collect after the firm's bankruptcy in 1816. When compared to his father, who meticulously managed the production of earlier editions of *British Zoology*, David Pennant was dissatisfied with the on-commission production of the 1812 edition. In the preface David Pennant complained that he was "sorry to find, that owing to his not having seen the impressions of the Plates to the *fourth* volume (and a few of the third,) after the inscriptions and references had been engraved upon them, a number of mistakes have been committed."[21] This reflects a wider trend of booksellers having a far more active role in elite publishing practices by the early nineteenth century—reducing authorial control over the process of engraving and proofing the plates. Instead of directly commissioning an intaglio engraver and printer, David Pennant left this task to the publisher. In spite of these problems, it seems David had a hand in the production of the Dublin edition of *British Zoology* (1818). A result of the decline in quality of such books, many publishers who catered for a genteel audience faced financial difficulties and bankruptcies. The case of natural history shows how the character of the publishing industry was changing a decade before the crash of 1826.[22] Natural history books became smaller and cheaper, as is apparent from the reduction in size of White's *Natural History and Antiquities of Selborne* between 1789 and 1822. Large books containing numerous copperplates designed to extend and distribute a collection—such as Pennant's *British Zoology*, George Leonard Staunton's *Authentic Account of an Embassy*, and William Roxburgh's *Plants of the Coast of Coromandel*—started to disappear from commercial markets.

In response to the changing relationship between publishing and the practice of natural history, the middle decades of the nineteenth century witnessed the foundation of societies with a specific remit to publish natural history books. Societies were instituted by the genteel elite and perhaps the most notable was the Ray Society, founded in 1844 by the physician George Johnston. The Ray Society was established to follow a specific agenda of printing useful books on botany and zoology that were not viable for commercial publishers. This was a direct result of the costs associated with the production of high-quality illustrations, the funding for which came from membership subscriptions, developing a financial structure reminiscent of the subscription volumes organized by Pennant in the 1760s. By 1845 the Ray Society had 650 subscribing members, each of whom paid an annual fee of one guinea to receive publications.[23] Ray Society books, such as Joshua Adler and Albany Hancock's *British Nudibranchiate Mollusca* (1845), which contained seventy-five plates based on specimens from private and institutional collections, were subscribed to by an elite membership. Subscribers comprised wealthy naturalists, those who received institutional support and aristocrats. Examples include the Earl of Derby and Prince Albert, who had interests in and collections of natural history. The Paleontographical Society, founded in 1847, was modeled along similar lines, with a view to publish figures of all undescribed British fossils. Images and descriptions contained within these books were based on specimens held in private and the increasingly prominent institutional collections, many of which were being rigorously catalogued to form national repositories of natural history.[24] The foundation of these societies allowed natural historians to continue the cyclical process of publishing the materials they compiled from expeditions, correspondence networks, and institutional repositories. Books received a similar treatment to many private and institutional collections, keeping them off the open market while ensuring copies were distributed to select groups who would derive a certain benefit from the material they contained.

In addition to producing and distributing books, the paper technologies developed by eighteenth-century naturalists continued to be used in natural history collections. Typical examples include that of the British Museum, the main repository for Banks's collection after it was donated by Robert Brown to form the Banksian Department in 1827. Pennant's descendants donated his collection and several associated manuscripts to the British Museum (Natural History) (now the Natural History Museum, London) between 1911 and 1913.[25] From the 1830s John Edward Gray (1800–1875), curator of the Zoological Departments, began publishing catalogues of the British Museum's collections. The printing of these catalogues was funded by the institution and many were interleaved with blank pages—to encourage their recipients to add information to a predetermined systematic order. Interleaved copies of Gray's catalogue, many of which were presented to individuals and institutions throughout the British Empire, served to structure the information communicated back to the museum. This could then be fed into a central catalogue where "every species [was] to have a leaf devoted to it, so that in any future time the leaves may be separated and bound in any other form."[26] Gray's distribution of interleaved catalogues is a legacy of the practices developed by Banks, Pennant, and Solander. Interleaved books play a central role, serving as tools to collect information, feeding into a central repository of more "flexible paper technologies" while the distribution of printed catalogs served to extend standardized working practices across the British Empire at the dawn of the Victorian age.

Combining the practices of natural history with the history of books has allowed for a reassessment of knowledge production between 1760 and 1820. The distancing of natural history collections and associated books from the market shows how elite naturalists combined their collections and means for distributing knowledge with the social standing they gained through property ownership. Natural history libraries served as meeting points and workshops of knowledge production, developing the standards for formulating and distributing information. Independent incomes and notions of leisure became ingrained into the practices of natural history, playing into a distinct anti-commercial approach to book production. High-ranking British naturalists—ranging from parish clergy, gentry, and wealthy aristocrats—used their interests in natural history as an avenue to maintain their power and places in society. By combining the cyclical processes of natural history collecting, book production, and distribution with notions of gentility, luxury, and the idea of maintaining an income based on wealth from landed estates, many attempted to preserve their dominance in the face of rapid commercialization and political changes.

The defined intellectual niche provided by natural history and associated practices allowed genteel practitioners to retain a firm grip over the sciences even when their more general social positions were threatened. This becomes apparent from their reactions to the French Revolution—an event that instilled absolute terror in the hearts of many high-ranking British naturalists. Thomas Pennant formed "an association for the defence of our religion, constitution, and property" in response to "the dangerous designs of the *French*." Writing to William Hamilton in 1792, Joseph Banks expressed his "terror from the active pains those who wish for a scramble are taking to raise the lower orders into a wish for Equality." Six months before his death in 1793, Gilbert White wrote that "you cannot abhor the dangerous doctrines of levellers and republicans more than I do. I was born and bred a Gentleman, and hope I shall be allowed to die such."[27]

Appendix

A List of Those Who Subscribed to Thomas Pennant's *British Zoology* (1766)

Individuals who are marked on the map of transatlantic and European recipients (figure 4. 3) are marked with an asterisks (*). The rest are located on the map of Pennant's British subscribers (figure 4.2). The numbers of those who received presentation copies of this work have been placed in square brackets ([]).

	Name	Location
1	All Souls College Library	Oxford
2	James Pitt Andrews (1737–1797)	Donnington Grove, Berkshire
3	Rev. Dr. Richard Newcome (1701–1769)	St Asaph
4	Edward Astly (1729–1802)	Melton Constable
5	Sir John St. Aubin (1758–1839)	Clowance, Crowan, Cornwall
6*	The Prince of Brunswick, Charles William Ferdinand (1735–1806)	Brunswick Palace, Braunschweig
7	Frederick Calvert Lord Baltimore (1731–1771)	Woodcote Park, Epsom, Surrey
8	Mrs. Ball	Chester
9	Bishop John Egerton (1721–1787)	Bangor
10*	Pierre-François Basan (1723–1797)	Paris
[11]*	Dr Laura Bassi (1711–1778)	Bologna
12*	Varenne de Beost (1700–1791)	Paris
13*	Gasper Beels	Amsterdam

14	Thomas Bennett	?
15	Hugh Bethel (1727–1772)	Yorkshire (Probably York)
16*	Ashton Blackborne (Blackburn)	Hempstead, Queens County, New York
17	John Blackborne (Blackburn) (1694–1786)	Orford Hall, Warrington
18	Sir Edward Blackett (1719–1804)	Maften Hall, Maften
19	John Blayney	Gregynog, Montgomeryshire
20	Richard Wilbraham Bootle (1725–1796)	Lathom House, Lancashire
[21]	Rev. William Borlase (1696–1772)	Ludgvan, Cornwall
22	Sir Griffith Boynton (1745–1778)	Beverley, Yorkshire
23	Owen Sailsbury Brereton, Esq. (1715–1798)	Chester
[24]*	Mathurin Jacques Brisson (1723–1806)	Paris
25	John Peyto-Verney, Baron Willoughby de Broke (1738–1816)	Compton Verney House, Warwickshire
26	Pusey Brook (d. 1768)	Chester
27*	Georges-Louis Leclerc, Comte de Buffon (1707–1788)	Paris
28	John Stuart, Ear of Bute, (1713–1792)	Highcliffe, Hampshire
29	Lord Robert Clive (1725–1774)	Claremont, Esher
30	Peter Collinson (1694–1768)	Mill Hill, London
31	William Constable (1721–1791)	Burton Constable Hall
32	William Cook	London
33	Sir Richard Corbett, Bart (1696–1774)	Longnor, Shropshire
34	Cymmrodorion Society	London
35*	M. Davits	Paris
36	John Davies Esq.	Denbighshire
37	John Dehainy Esq	?
38*	Paul de Demidoffe (Pavel Grigoryevich Demidov) (1738–1821)	Moscow, Russia
39	James Dickson (?)	London
40	Jonathan Disney (?)	London
41	William Dixon (?)	?
42	James Dickson, Dean of Downe	Downpatrick, Ireland
[43]	George Edwards (1694–1773)	London
44*	John Ellis	Jamaica
45	Dr. England (?)	Dorsetshire
46	Rev. Mr. Falconer,	Lichfield
47	Thomas Falconer (1738–1792)	Chester

48	Rev. Mr. Farrington	Dynas, Caernarvonshire
49*	Christian Fleischer (1713–1768)	Copenhagen
50	John Fothergill (1712–1780)	London
51	Rev. Mr. Green	Yorkshire
52	John Grimston	Grimston Garth, Yorkshire
[53]*	Laurens Theodoore Gronovius (1730–1777)	Leiden
54	Mr. Hauliter, Painter	Chester
55	John Hill	Yorkshire
56	Owen Holland (1720–1795)	Conway
57	Timothy Hollis (1709–1790)	London
58	William Hudson (1730–1793)	London
59	John Hunter (1728–1793)	London
60	Jesus College Library	Oxford
61	Mrs. Inge	Thorpe Constantine Hall, Staffordshire
62	Mr. Joliffe, Bookseller	London
[63]	Thomas Knowlton (1692–1782)	Londesborough Hall, Yorkshire
[64]*	Dr. Kramer (?)	Vienna
65	Thomas Kyffin (c. 1710–1776)	Maenan Hall, near Conway
66	Roger Kynaston (?)	Foundling Hospital, London
67	Mr. Lanton, Bookseller	Chester
68	Edward Leigh (1742–1786)	Stoneleigh Abbey, Warwickshire
69	Ashton Lever (1729–1788)	Manchester (Later London)
70	John Gideon Loten (1710–1789)	London
71	Daniel Lysons (1721–1800)	Gloucester
72	Manchester Library	Manchester
73	John Major, Engraver (1720–1799)	London
74	William Marshal (?)	Lincolnshire
75	Rev. Richard Pocock (1704–1765)	Charleville Castle, Meath, Ireland
76	Owen Meyrick (?)	Bodorgan Hall
77	Francis Willoughby, Right and Honourable Lord Middleton (1692–1758)	Woollaton Hall, Nottingham
78	John Milan, Bookseller (?)	London
79	Richard Morris (1703–1779)	London
80	William Morris (1705–1763)	Holyhead

81	Rev. Mr. Morton (?)	Shropshire
82	Sir Roger Mostyn (1734–1796)	Mostyn Hall
83	Rev. Mr. Roger Mostyn (?)	Chrisleton, Cheshire
84*	Georg Christian Oeder (1728–1791)	Copenhagen
85	Paul Panton (1731–1797)	Baglit, Flintshire
86*	Marc Antoine René de Voyer (1722–1787)	Paris
87	Thomas Pennant (1726–1798) (twelve sets)	Downing Hall, Flintshire.
88	William Pennington Esq. (?)	Bodmyn
89	Robert Edward Petre (1742–1801)	Ingatestone Hall
90	Joseph Plymley (?)	Longnor, Shropshire
91	William Portman (?)	Bryanston, Dorset.
92	Richard Parry Price (d. 1785)	Bryn y Pys, Flintshire
93	William St Quentin (1729–1795)	Scampston, Yorkshire
94*	John Ray (?)	Leiden
95	Charles Lennox, Duke of Richmond (1735–1806)	Goodwood House, Sussex.
96	Rev. Mr. Roberts (?)	Derbyshire
97	John Ker, Duke of Roxburghe (1740–1804)	London
97	Henry Seymer (1755–1783)	Hanford, Blandford Forum
98	William Fayne Sharpe (?)	London
99	William Spencer (?)	Yorkshire
100	Henrietta Grey, Countess of Stamford (?)	Enville Hall
101	Benjamin Stillingfleet (1702–1771)	Felbrigg Hall
102	Humphrey Sturt (1724–1786)	Crichel House, Dorset
103	Captain John Thomas (1698–1791)	London
104*	Otto Thot (1703–1785)	Copenhagen
105	Thomas Tindal (?)	?
106	Thomas Tofield (1730–1779)	Wilsic Hall
107	John Tomlinson (?)	London
108*	Christoph Jacob Trew (1695–1769)	Nuremberg
109	Dr. Thomas Tylston (d. 1766)	Chester
110	Miss Tylstons (?)	Chester
111	William Vaughan (c. 1707–1775)	Corsygedol, Merioneth
112	Richard Varlo (?)	Portsmouth
113	Horace Walpole (1717–1797)	Strawberry Hill, London
114	John Walters, Bookseller (1738–1812)	London
115	Lady Waters (?)	?

116	Philip Carteret Webb (1702–1770)	London
117	Benjamin White (1725–1794)	London
118	Ralph Willet (1719–1795)	Merley, Canford Manor, Dorset
119	Rev. Mr. Williams (?)	Fron, Flintshire
120	Rev. Mr Worth (?)	Norfolk
121	William Wright (?)	Cheshire
122	Pierce Wynn (?)	Dyffryn Aled, Denbighshire

Notes

Manuscript Sources Abbreviations

BL	British Library, London
BM	British Museum, London
CUL	Cambridge University Library
HLH	Houghton Library, Harvard University, Cambridge, MA
LHL	Linda Hall Library, Kansas City, MO
LSL	Linnean Society of London
LWL	Lewis Walpole Library, Yale University, Farmington, CT
MUL	McGill University Library, Montreal
NHM	Natural History Museum, London
NLA	National Library of Australia, Canberra
NLS	National Library of Scotland, Edinburgh
NLW	National Library of Wales, Aberystwyth
RBGK	Royal Botanic Gardens, Kew, London
SLNSW	State Library of New South Wales, Sydney
UUL	Uppsala University Library
WRO	Warwickshire Records Office, Warwick

Introduction: Natural History and the History of Books

1. Turner, *Avium, Quarum Apud Plinium*. This is often regarded as one of the first printed books to exclusively concentrate on birds.

2. Pennant, *Literary Life of the Late Thomas Pennant*, 9.

3. See Brener, Kunstmann, Mukhapadhyay, and Rogers, *Global Histories of Books*, esp. chap. 2. On Bacon, see H. Cook, *Matters of Exchange*, 40; Müller-Wille and Charmantier, "Natural History and Information Overload," 4.

4. On information, see Eddy, *Media of the Mind*, and Blair, *Too Much to Know*. For scientific journals, see Csiszar, *Scientific Journal*; Dawson et al., *Sci-*

ence *Periodicals in Nineteenth-Century Britain*; Fyfe et al., *History of Scientific Journals*.

5. Yale, *Sociable Knowledge*. See also Kühn, *Wissen, Arbeit, Freundschaft*.

6. See, for example, Lack, *Bauers*; Roos, *Martin Lister and His Remarkable Daughters*; Nickelsen, *Draughtsmen, Botanists and Nature*.

7. Larson, *Interpreting Nature*; Stevens, *Development of Biological Systematics*.

8. See Pomian, *Collectors and Curiosities*; Hickman, *Doctor's Garden*.

9. Darnton, "What Is the History of Books?," 81.

10. Quoted in Bayly, *Imperial Meridian*, 2; Bayly, *Birth of the Modern World*, 11. See also Fullagar and McDonnell, *Facing Empire*; Manning and Rood, *Global Scientific Practice in an Age of Revolutions*.

11. Porter, *Making of Geology*; Rousseau and Porter, *Ferment of Knowledge*; Rose, "Natural History Collections and the Book," 16.

12. Porter, "Gentlemen and Geology"; Rudwick, *Great Devonian Controversy*, 440; Shapin, *Social History of Truth*, 45.

13. Lindenfeld, *Practical Imagination*, 22–45. See also Spary, *Utopia's Garden*; Cooper, *Inventing the Indigenous*, 152–66; Margócsy, *Commercial Visions*, 200; Koerner, *Linnaeus*.

14. T. Porter, *Rise of Statistical Thinking*, 17, 44–45.

15. Schiebinger and Swan, "Introduction," 4–5; Liebersohn, *Travellers' World*, 77–138.

16. R. Porter, *Enlightenment*, 16; R. Porter, "Enlightenment in England"; Colley, *Britons*, 55–98.

17. On the British Museum, see Stearn, *Natural History Museum at South Kensington*, 10–14; Harris, *History of the British Museum Library*, 1–13. For naturalists as landowners, see Porter, "Gentlemen and Geology"; Miller, "'My Faivorite Studdys,'" 213–40; Rose, "Publishing Nature in the Age of Revolutions," 1134. On Kew, see Drayton, *Nature's Government*, 88–128; Gascoigne, *Science and the State*, 54–70. On funding private research programs, see Jonsson, *Enlightenment's Frontier*, 43–68.

18. On the Enlightenment, see Melton, *Rise of the Public in Enlightenment Europe*, 20. For Goldsmith, see Rousseau, *Oliver Goldsmith*, 21; Johns, *Piracy*.

19. Cain and Hopkins, *British Imperialism*, 50–51, 78.

20. For property and improvement, see Ritvo, "Possessing Mother Nature," 414–15. For social advancement in "polite society," see Guillory, "Literary Capital," 400.

21. Berg, *Luxury and Pleasure in Eighteenth-Century Britain*, 6; Shapin, *Social History of Truth*, 45. On ancien régime France, see Furet, *Interpreting the French Revolution*, 105–6.

22. Gascoigne, "Royal Society, Natural History and the Peoples of the 'New World(s),'" 545. Browne, "Botany for Gentlemen," 595.

23. Shapin, *Social History of Truth*; Cain and Hopkins, *British Imperialism*, 6, 82.

24. Schaffer et al., *Brokered World*, xi–xiv; Cooper, *Inventing the Indigenous*, 2–5.

25. Gascoigne, *Joseph Banks and the English Enlightenment*, 8; Drayton, *Nature's Government*, 95; Hoppit, "Sir Joseph Banks's Provincial Turn," 410–11.

26. White to Pennant, June 18, 1768, BL, Add MS 35138, item 10.

27. Benjamin White, "Copy Inventory of Rev. Gil. White's Effects" (1793), LH, Gilbert White Papers, HLH.

28. On White and conservatism, see Damrosch, "Gilbert White of Selborne," 29–46. On siblings who founded London businesses, see Earle, *Making of the English Middle Class*, 5–10.

29. On physicians, see McCormack, *William Hunter and His Eighteenth-Century Cultural Worlds*; Bynum and Porter, *William Hunter and the Eighteenth-Century Medical World*; McNeil, *Under the Banner of Science*; King-Hele, *Doctor of Revolution*. For Darwin's publishing, see Bewell, "Erasmus Darwin's Cosmopolitan Nature," 28.

30. Milam and Nye, "Introduction to Scientific Masculinities," 12.

31. Gascoigne, *Joseph Banks and the English Enlightenment*; Miller and Reill, *Visions of Empire*; Drayton, *Nature's Government*, 95; Hoppit, "Sir Joseph Banks."

32. Goodman, *Planting the World*.

33. For Pennant's "tours" in Wales and Scotland, see Constantine and Leask, *Enlightenment Travel and British Identities*. On Pennant's importance to natural history, see Lysaght, *Joseph Banks in Newfoundland and Labrador*, 93. The only biography of Pennant is an unpublished PhD dissertation: Evans, "Life and Work of Thomas Pennant." For Downing Hall, Flintshire, see Pennant, *History of the Parishes of Whiteford and Holywell*, 4–20; Evans, "Thomas Pennant and the Influences behind the Landscaping of the Downing Estate," 109–24.

34. See Dadswell, *Selborne Pioneer*; Carey, "Literary Gilbert White."

35. Daines Barrington to Thomas Pennant, July 8, 1775, Miscellaneous Manuscript Collection, box 2, folder 25, LWL.

36. Foster, *Gilbert White and His Records*; Menley, "Travelling in Place."

37. For a pioneering treatment of scale, see Latour, chaps. 5–6. The development of common standards has been attributed to the nineteenth-century emergence of climate science; see Coen, *Climate in Motion*, 16–20.

38. Daniel Solander, "Reports and Diary of Occurrences in the Nat. Hist.

Departments by Dr. Solander. Sept. 1764–Feb. 12th 1768," Add. MS.45874, 6, BL.

39. For Solander's manuscript slip catalogue, see Charmantier and Müller-Wille, "Carl Linnaeus's Botanical Paper Slips"; Rose, "Specimens, Slips and Systems," 219; Dietz, "Linnaeus's Restless System," 151.

40. On changing systems of classifications, see Müller-Wille, "Systems and How Linnaeus Looked at Them in Retrospect."

41. For the development of practices designed to manage factual information, see Poovey, *History of the Modern Fact*; Sherman, *Used Books*; Yeo, *Notebooks, English Virtuosi, and Early Modern Science*; Bittel, Leong, and van Oertzen, *Working with Paper*, 1–4.

42. On bringing information together, see Müller-Wille and Charmantier, "Natural History and Information Overload"; Delbourgo and Müller-Wille, "Introduction: Listmania"; Hodacs, "Local, Universal, and Embodied Knowledge," 94; Müller-Wille and Charmantier, "Lists as Research Technologies."

43. Mayhew and Withers, *Geographies of Knowledge*, 12–13; Coen, *Climate in Motion*, 16; Burnett, *Masters of All They Surveyed*, 71.

44. On the harmlessness of obvious anachronisms, see Jardine, "Uses and Abuses of Anachronism in the History of the Sciences."

45. te Heesen, "Notebook: A Paper Technology," 584–85. On the application of te Heesen's terminology, see Bittel, Leong, and von Oertzen, *Working with Paper*, 9.

46. Eddy, *Media and the Mind*, 7.

47. Charmantier and Müller-Wille, "Carl Linnaeus' Botanical Paper Slips."

48. For "paper empires," see Müller-Wille, "Names and Numbers," 114–19.

49. Schaffer et al., *Brokered World*, xiv. For reference to these knowledge exchanges as "ontologies," see Salmond, *Tears of Rangi*.

50. Brooks, "Imperial Structures, Indigenous Aim."

51. Shapin, *Social History of Truth*, 355–408.

52. Dietz, *Das System der Natur*. On correspondence, see Dietz, "Contribution and Co-production." On Osbeck, see Dietz, "What Is a Botanical Author?," 62. For Osbeck's and Linnaeus's reliance on Indigenous knowledge, see A. Cook, "Linnaeus and Chinese Plants."

53. See Hess and Mendelsohn, "*Paper Technology* und Wissensgeschichte."

54. Goldgar, *Impolite Learning*; Terrall, *Catching Nature in the Act*, 79–101. For the public sphere, see Melton, *Rise of the Public*, 11–15; Eley, "Nations, Publics, and Political Cultures," 299–305.

55. Stafleu, *Linnaeus and the Linnaeans*; Sloan, "Buffon-Linnaeus Controversy."

56. Linnaeus, *Systema Naturae*, 2.

57. For Linnaeus and the "natural system," see Müller-Wille, "Systems and How Linnaeus Looked at Them in Retrospect," 315.

58. James Edward Smith, *Selection of the Correspondence of Linnaeus, and Other Naturalists*, 389.

59. Watson, "Account on a Treatise in Latin," 558.

60. Rose, "Empire and the Theology of Nature in the Cambridge Botanic Garden."

61. C. E. Raven, *John Ray, Naturalist*, 322.

62. Lazenby, "Historia Plantarum generalis of John Ray," 344–45.

63. Quoted in Oswald and Preston, *John Ray's Cambridge Catalogue*, 133. For natural theology, see Mandelbrote, "Uses of Natural Theology in Seventeenth-Century England."

64. Lindroth, "Two Faces of Linnaeus"; Ray, *Wisdom of God*, 1.

65. Pennant, *History of Quadrupeds*, iii–iv. On gender, see Schiebinger, "Why Mammals are Called Mammals," 409.

66. Gilbert White to John White, May 26, 1770, in Holt-White, *Life and Letters of Gilbert White of Selborne*, 1:179.

67. Goldsmith, *History of the Earth and Animated Nature*, 1, x; Bales, "Literary Plagiarism and Scientific Originality"; Lynskey, "Scientific Sources of Goldsmith's 'Animated Nature.'"

68. C. E. Raven, *John Ray, Naturalist*, 324; Birkhead et al., "Willoughby's Ornithology," 269.

69. Ray, *Ornithology of Francis Willoughby of Middleton in the County of Warwick*, vi.

70. Ray, *Synopsis Methodica Animalium Quadrupedium*; C. E. Raven, *John Ray, Naturalist*, 379–80.

71. On print and objects, see Ogilvie, *Science of Describing*.

72. See Darnton, *Kiss of Lamourette*; J. Raven, *Business of Books*; Brewer, *Pleasures of the Imagination*; Sher, *Enlightenment and the Book*. For obscure authors, see J. Raven, *Judging New Wealth*, 58.

73. Kärin Nickelsen has confined analysis to the final published depictions, suggesting the scarcity of surviving original illustrations. All the main examples examined in this book survive in their entirety, from the original specimens and drawings to the printed images. For the interaction of scientific books with external sources, see Jardine and Frasca-Spada, *Books and the Sciences in History*, 7; Rudwick, *Bursting the Limits of Time*, 283–88.

74. Pennant, *British Zoology*, vol. 1, 10. For more on natural history and religion, see Paley, *Natural Theology*, xix–xx. For the relationship between Linnaean taxonomy and devotion in England, see Courtney Weiss Smith, *Empiricist Devotions*, 193.

75. Paley, *Natural Theology*, 277.

76. Spary, "Political, Natural and Bodily Economies," 179.

77. Turnovsky, *Literary Market*, 65. On Britain, see J. Raven, *Business of Books*, 212. The British government did not introduce more intrusive press censorship legislation, aimed at political pamphleteers and newspaper printers, until the French Revolution.

78. For Darnton's "communication circuit," see Darnton, "What Is the History of Books?"; Adams and Barker, "New Model for the Study of the Book," 14–15; Darnton, "What Is the History of Books? Revisited." Bordieu's diagram is in Bordieu, "Field of Cultural Production," 82.

79. Eddy, *Media and the Mind*, 7; Rudwick, "The Emergence of a Visual Language."

80. J. Raven, *Business of Books*, 60. Raven has noted that rumours of these individuals' wealth inspired many well-educated but impoverished authors. Kernan, *Samuel Johnson and the Impact of Print*, 102

81. Quoted in McCormach and Jungnickel, *Cavendish*, 251.

82. Miller, "Between Hostile Camps"; Miller, "'Into the Valley of Darkness.'"

83. For books as gifts, see Heal, *Power of Gifts*; Scott-Warren, *Sir John Harington and the Book as Gift*; Biagoli, *Galileo, Courtier*, 38–39; Ben-Amos, *Culture of Giving*. For early work on gifts and social obligations, see Mauss, *Gift*.

84. Cavagna, "Free Transmission of Knowledge."

85. Chartier, "Labourers and Voyagers," 51.

86. Pomian, *Collectors and Curiosities*, 9; Sher, *Enlightenment and the Book*: 8; Johns, "How to Acknowledge a Revolution," 116–18.

87. Similar points have been made in revisionist histories of the long eighteenth century. See Clark, *English Society*.

88. Mandler, *Aristocratic Government in the Age of Reform*, 116–17.

89. Gilbert White to Samuel Barker, March 26, 1781, in Holt-White, *Life and Letters of Gilbert White of Selborne*, 2:67.

Chapter 1: From Parish to Nation

1. White to Pennant, June 16, 1768, Add MS 35.138, 13, BL.

2. Milne and Gordon, *Indigenous Botany*; Cooper, *Inventing the Indigenous*.

3. Thomas Pennant, *British Zoology* (1776–1777), 2:750.

4. For national floras and faunas, see Browne, *Secular Ark*, 27–37. On the unification of Britain, see Colley, *Britons*, 84. On the increased differentiation of the British Isles, see Naylor, *Regionalizing Science*, 14–20; Morieux, *Channel*. On natural history and military rhetoric, see A. Secord, "Coming to Attention."

5. See Speck, *Concise History of Britain*, 4. For parishes, see Snell, *Parish and Belonging*, 13.

6. Janković, *Reading the Skies*, 96, 113.

7. Fox, "Printed Questionnaires, Research Networks, and the Discovery of the British Isles," 595; Yeo, *Encyclopaedic Visions*, 97; Hunter, "Robert Boyle and the Early Royal Society."

8. Mendyk, "Robert Plot"; te Heesen, "Boxes in Nature," 391; Plot, *Natural History of Stafford-Shire*; Lhwyd, *Archaeologia Britannica*.

9. Müller-Wille, "History Redoubled," 518.

10. Pennant, *Literary Life of the Late Thomas Pennant*, 17.

11. Merrett, *Pinax rerum naturalium Britannicarum*. Pennant's annotated copy is held at NHM, Special Collections, copy 3 (000209986).

12. Colclough, "Pocket Books and Portable Writing"; Jung, "Illustrated Pocket Diaries and the Commodification of Culture," 65. Jung suggests many of these bindings were made by wallet makers.

13. Quoted in Withers, "Geography, Natural History and the Eighteenth-Century Enlightenment," 152.

14. Boyle, "Other Enquiries Concerning Sea," 315; Pennant, *Tour in Scotland*, 287.

15. Pennant, *Literary Life of the Late Thomas Pennant*, 34; Pennant, *Of London*.

16. Briggs, "Thomas Pennant," 48; Low, *Fauna Orcadensis*.

17. See Daines Barrington to Thomas Pennant, July 8, 1775, MSS MISC, box 2, folder 26, LWL.

18. G. White, *Natural History and Antiquities of Selborne in the County of Southampton*, advertisement.

19. Mingay, *Parliamentary Enclosure in England*; Brewer, *Pleasures of the Imagination*, 625–29.

20. Ray, *Historia Plantarum*.

21. Browne, *Secular Ark*, 27.

22. Pennant, *British Zoology* (1776–1777), 2:750.

23. Willoughby and Ray, *Ornithology of Francis Willoughby*. Pennant's annotated copy can be found at shelf mark MS 22731iE, NLW. The notes formerly inserted between the pages make up "Thomas Pennant Ornithology notes," MS 22731iE, NLW.

24. David Pennant, "An Account Book" (1803–1807), MS 2584B, 20, NLW.

25. Pennant, *British Zoology* (1776–1777), 2:424–25.

26. Pennant, *British Zoology* (1776–1777), 1:xix.

27. Pennant, *British Zoology* (1776–1777), Anna and John Blackburne's copy, Zoology Special Collections, SB 72A o PEN, set 2. NHM.

28. Pennant, *British Zoology* (1768–1770), 2:198, White's copy, Manuscripts Department, Add. MS 46, 471, BL.

29. White's copy of Hudson, *Flora Anglica*, GEH *EC75.W5834.Zz762h, HLH.

30. Pennant's copy of Lightfoot, *Flora Scotica*, is held at Botany Special Collections, British Herbarium, SB 581.9(411) LIG (v1–2), NHM.

31. Pennant, *History of the Parishes of Whiteford and Holywell*, 152–53. See John Lightfoot, transcribed by Sigismund Bacstrom, "John Lightfoot's Journal of a Botanical Excursion in Wales," Library and Archives, MSS BANKS COLL LIG, NHM.

32. G. White, *Natural History and Antiquities of Selborne in the County of Southampton*, 1, 234–36; Browne, *Secular Ark*, 31.

33. Rose, "Gilbert White, John Ray and the Construction of the *Natural History of Selborne*"; Ray, *Synopsis Methodica Avium & Piscium*, Gilbert White's copy, STORE 54:6, Whipple Library, Cambridge.

34. Martyn, *Methodus Plantarum circa Cantabrigiam Nascentium*, Thomas Martyn's copy, Rare Books, CCE.47.49, CUL.

35. Gilbert White, "An Index in Latin," Papers of Gilbert White, bMS Eng731, HLH; Ray, *Synopsis Methodica Stripium Britannicarum*.

36. These three copies are held by LSL, MSS *914.20582 HUD.

37. Rösel von Rosenhof, *Der monathlich herausgegebnen Insecten-Belustigung*. For Pennant's copy, see Entomology Special Collections, Special Books, o R 12, NHM.

38. Rösel von Rosenhof, *Insecten-Belustigung*. Pennant's copy, vol. 1, opposite table iv, NHM.

39. David Pennant, "A Meteorological Record," 1784–1793, MS 2561B (ii), NLW.

40. Kington and Barker, *Weather Journals of a Rutland Squire*.

41. Golinski, *British Weather and the Climate of Enlightenment*, 109.

42. David Pennant, "A Meteorological Record," 1784–1793, MS 2561B (ii), NLW.

43. Foster, "Hon. Daines Barrington."

44. Barrington, *Naturalist's Journal*, preface.

45. Pennant, *British Zoology* (1776–1777), 2:750.

46. Gilbert White to John White, January 5, 1775, in Holt-White, *Life and Letters*, 1:275; Walford, *Scientific Tourist*, 15–20.

47. Pennant and de Beer, *Tour on the Continent*, 39.

48. R. Paul Evans, "Round Jump from Ornithology and Antiquity," 27. Portraits of Luss and Thompson by Moses Griffith are tipped into Pennant's extra illustrated copy of his *A Tour in Scotland and Voyage to the Hebrides* (1790), vol. 2, part 2, LLUN/PICTURES, (W6), NLW.

49. Thomas Pennant, "Expenses of my Different Works," CR2017/TP571, 8, WRO.

50. Chapman, *Johnson's Journey to the Western Islands of Scotland*, 192. I thank Nigel Leask for alerting me to this quote.

51. Pennant, *Literary Life of the Late Thomas Pennant*, 10, 32.

52. This reflects the treatment of laboratory staff. See Shapin, *Social History of Truth*, 355–408; Rose, "From the South Seas to Soho Square," 501.

53. Pennant, *Tour in Scotland and Voyage to the Hebrides* (1790), 421.

54. Sandwich to Daines Barrington, February 28, 1778, Thomas and David Pennant Collection, "Miscellaneous Pamphlets, 1770–1825," MS 2597 C, NLW.

55. Barrington to Pennant, October 15, 1776, MSS MISC, box 2, folder 28, LWL.

56. See Richard Sorrenson, "Ship as a Scientific Instrument," 226–28.

57. Pennant, *Tour in Scotland and Voyage to the Hebrides* (1790), 1:300; Ksiazkiewicz, "Geological Landscape as Antiquarian Ruin," 183-201.

58. Furniss, "'As if Created by Fusion of Matter after Some Intense Heat,'" 172–73.

59. Pennant, *Tour in Scotland and Voyage to the Hebrides* (1790), 300.

60. John Stuart to Thomas Pennant, received January 12, 1773, CR2017/TP369, 2a-2d, WRO.

61. Pennant to Banks, January 30, 1767, in Lysaght, *Joseph Banks in Newfoundland and Labrador*, 238.

62. Pennant, *Tour in Scotland and Voyage to the Hebrides* (1790), 1:iii.

63. Lightfoot, *Flora Scotica*, xiii.

64. A. Cook, "Linnaeus and Chinese Plants," 122; Witteveen, "Supressing Synonymy with a Homonym"; Cooper, *Inventing the Indigenous*: 168–69.

65. Ray, *Collection of English Proverbs*.

66. Cooper, *Inventing the Indigenous*, 78–79.

67. For Pennant's attribution to Morris, see Pennant, *British Zoology* (1776–1777), 2:731–49, 4:156–57.

68. Thomas Pennant, "Thomas Pennant Ornithology Notes, [1760s]," MS 22731iE, NLW.

69. Pennant, *British Zoology* (1768–1770), 3:345–351.

70. Thomas Pennant, "Notes by Thomas Pennant on Zoology, Ornithology, 1750–1798," MS 2549 B (i), folders 7–8, NLW.

71. Johann Reinhold Forster, *Catalogue of British Insects*; Müller-Wille, "Names and Numbers," 122; Müller-Wille, "Linnaean Paper Tools," 215–16. The octavo edition of *British Zoology* was originally published in 1768, 1769, and 1770. Parts of the volumes published in 1768 and 1769 were used to compile this notebook.

72. Johnson, *Journey to the Western Islands of Scotland*, 302.

73. Poovey, *History of the Modern Fact*, 256.

74. Pennant, *Tour in Scotland and Voyage to the Hebrides* (1790), 191.

75. Pennant, *Tour in Scotland and Voyage to the Hebrides* (1790), 193; Pennant, *British Zoology* (1776–1777), 3:103.

76. Pennant, *British Zoology* (1776–1777), 3:105; Pennant, *Tour in Scotland and Voyage to the Hebrides* (1790), 194.

77. Pennant, *Tour in Scotland and Voyage to the Hebrides* (1790), 2:194.

78. Pennant, *British Zoology* (1776–1777), 3:104–5.

79. Pennant, *British Zoology* (1776–1777), 3:42–43.

80. Pennant's extra illustrated *Tour in Scotland and Voyage to the Hebrides* (1790), vol. 2, part 2, LlGC De LLUN/PICTURES, (W6), 192, NLW.

81. Lightfoot to Pennant, October 21, 1777. Pennant's copy of *Flora Scotica*, Botany Special Collections, British Herbarium, SB 581.9(411) LIG (v1–2), tipped into vol. 1, NHM.

82. Thomas Pennant, "Expenses of my Different Works," CR2017/TP571, 19, WRO.

83. Lightfoot, *Flora Scotica*, Botany Special Collections, 938, NHM; Pennant, *History of the Parishes of Whiteford and Holywell*, 153.

84. Pennant, *Tour in Scotland and Voyage to the Hebrides* (1790), 2:215–16.

85. Pennant, *Tour in Scotland and Voyage to the Hebrides* (1790), 2:216.

86. Pennant, *Literary Life of the Late Thomas Pennant*, 15.

87. Pennant, *Tour in Scotland and Voyage to the Hebrides* (1790), 2:729.

88. Pennant, *British Zoology* (1776–1777), 2:750; Browne, *Secular Ark*: 33; te Heesen, "Accounting for the Natural World," 241.

89. Pennant, *Supplement to the Arctic Zoology*, 4.

90. Lightfoot, *Flora Scotica*, Pennant's copy, British Herbarium, SB 581.9(411) LIG (v1–2), NHM.

91. Joseph Banks, "Journal of an Excursion to Wales Etc.," MS Add. 6294, 135, CUL.

92. Grigson, *Menagerie*, 87–89.

93. See Kaeppler, *Holophusicon*.

94. Plumb, "Bird Sellers and Animal Merchants"; Rose and Mandelbrote, "Thomas Gray as a Reader and Writer on the Natural World."

95. Silliman, *Journal of Travels in England, Holland and Scotland*, 280.

96. Pennant, *Tour in Scotland and Voyage to the Hebrides* (1790), 1:13; Wystrach, "Ashton Blackburne's Place in American Ornithology," 607–10.

97. Pennant, *Synopsis of Quadrupeds*; Donald, *Picturing Animals in Britain*.

98. Rolfe, "William Hunter (1718–1783)," 277.

99. On extra illustration and collections of printed portraits, see Peltz, *Facing the Text*.

100. "Original watercolour, wash, and pencil drawings of birds, animals, fishes etc," Rare Books Room, QL255.P5 quarto, vol. I., 1.12.1, LHL. Paillou produced several copies of this image. Another from Banks's collection is now held by the British Museum: Peter Paillou (after George Stubbs), "Cheetah (? Feis jubata), after Stubbs, with black dotted tan coat, standing in profile to right, body colour, heightened with white (partly oxidised)," 1914,0520.230, BM.

101. Pennant, *Synopsis of Quadrupeds*, 125.

102. See McCormack. "Pennant, Hunter, Stubbs and the Pursuit of Nature"; Potts, "Natural Order and the Call of the Wild," 13.

103. Pennant, *British Zoology* (1768–1770), 1:vii.

104. A major product of these trips was Pennant's *Journey from Chester to London*.

105. Pennant's Outlines of the Globe is primarily in manuscript and held by the National Maritime Museum, Greenwich. The first four volumes were published as *The View of Hindoostan* (1790) and *The View of India, Extra Gangem, China and Japan* (1800).

106. Pennant and Banks, "Small notebook entitled by Pennant 'These queries I drew up for Mr. Banks' . . .," (1767), CR 2017/TP44, WRO.

107. Banks, "Journal," MS Add. 6294: 67–68, CUL.

108. Pennant, "Notes by Thomas Pennant on Zoology, Ornithology, etc." (1750–1798), MS 2549 B (ii), NLW.

109. Pennant and Banks, "Small Notebook," TP44, WRO. "Charlevoix" is a reference to Pierre François Xavier de Charlevoix (1682–1761), a Jesuit priest and traveler known as the first historian of New France.

110. Pennant, *Arctic Zoology* (1784–1785), 1:182.

111. Pennant, *British Zoology* (1776–1777), 1:xv–xvii.

112. Johann Reinhold Forster, *Catalogue of the Animals of North America*, title page.

113. Banks to Pennant, July 13, 1771, in Chambers, *Letters of Sir Joseph Banks*, 14.

114. Pennant, *Literary Life of the Late Thomas Pennant*, 13; Thomas Pennant, "Guide to the Animals Observed or Collected by Joseph Banks esqr. & Doctor Solander in the Voyage Round the World Begun August 25th 1768 Ended July 12th 1771" (c. 1771), Manuscripts Department, MS 9138; 70, NLA.

115. Thomas Pennant, "Notes by Thomas Pennant on Zoology, Ornithology, etc." 1750–1798, MS 2549 B (i), NLW.

116. On Solander's manuscript slips, see Rose, "Specimens, Slips and Systems"; Charmantier and Müller-Wille, "Carl Linnaeus' Botanical Paper Slips," 228–230. For Johnson's employment of these practices, see Reddick, *Making of Johnson's Dictionary*, 3–6. The copy of Johnson's *Dictionary* containing paper slips is held by the Beinecke Library, Yale University, shelf mark 2041002.

117. Joseph Banks, "Endeavour Journal" (1768–1771), SAFE/1/457, Banks Papers, series 3, vol. 1, 192, SLNSW.

118. Linnaeus, *Species Plantarum* . . . (1762–1763) [Bound in six volumes . . .], 582 LIN 109–114, Botany Special Collections, NHM; Linnaeus, *Systema Naturae* . . . (1766) [Bound in six volumes . . .], Zoology Special Collections, 4 o Li 62 (1–6), NHM.

119. Daniel Solander, "A Fair Copy of the Descriptions of Animals," Zoology Special Collections, NHM.

120. Thomas Pennant, "'Guide to the Animals . . .'" (c. 1771), Manuscripts Department, MS 9138, 70, NLA.

121. Daniel Solander, "Reports and Diary of Occurrences . . ." (1764–1768), Add. MS. 45874, 6–9, BL.

122. Dietz, "Contribution and Co-production," 551–52.

123. Pennant, *British Zoology* (1768–1770), 3:12.

124. Thomas Pennant, "Notes by Thomas Pennant on Zoology, Ornithology, etc." (1750–1798), MS 2549 B (i), NLW.

125. For Johnson's relationship with Boswell, see Leask, *Stepping Westward*, 145; Radner, "Constructing an Adventure and Negotiating for Narrative Control."

126. Durand-Guédy and Paul, *Personal Manuscripts*. On Winckelmann, see Décultot, "Reading *versus* Seeing?" For studies of published works, see Krajewski, *Paper Machines*; Décultot, "Between Reading and Writing"; Krämer, "Albrecht von Haller as an 'Enlightened' Reader-Observer."

Chapter 2: A New World for Natural History

Epigraph: James Cook, "Copies of Correspondence etc." (1768–1771), MS 2, NLA.

1. Pennant, *British Zoology* (1776–1777), 2:750.

2. Liebersohn, *Traveller's World*, 77–138; Koerner, "Purposes of Linnaean Travel"; Hodacs, "Linnaeans Outdoors."

3. Linnaeus to Osbeck, undated, in Osbeck, *Voyage to China and the East Indies*, 128.

4. Cook, "Copies of Correspondence etc.," MS 2, NLA; Forster, *Voyage Round the World*, xxi–xxii; V. Smith, *Intimate Strangers*, 61–62; Williams, "*Endeavour* Voyage," 3–18; Salmond, *Tears of Rangi*, 14.

5. Carter, *Sir Joseph Banks*, 61. See Gascoigne, *Science in the Service of Empire*, 189.

6. Bleichmar, *Visible Empire*, 84; Latour, *Science in Action*, 215–19.

7. Banks to Falconer 1768, quoted in Carter *Sir Joseph Banks (1743–1820)*, 61.

8. Daniel Solander, "Report to the Trustees of the British Museum," June 24, 1768, Central Archive, Original Papers, 225, BM.

9. For biogeography, see Browne, *Secular Ark*, 34–35. On paper technologies, see Charmantier and Müller-Wille, "Carl Linnaeus' Botanical Paper Slips," 216.

10. Forster, *Letter to the Right Honourable Earl of Sandwich*, 6.

11. Mariss, "Library in the Field"; Bleichmar, "Exploration in Print."

12. Banks, "Endeavour Journal" (1768–1771), SAFE 1/457, series 3, vol. 2, SLNSW.

13. Banks to Alströmer, November 16, 1784, in Chambers, *Letters of Sir Joseph Banks*, 78.

14. Similarities can be drawn between Joseph Banks and Johann Reinhold Forster. See Mariss, "Library in the Field," 53.

15. Ellis to Linnaeus, August 19, 1768, in J. E. Smith, *Selection of the Correspondence of Linnaeus*, 1:231.

16. Williams, *Naturalists at Sea*, 74; Drayton, *Nature's Government*, 105; Hoppit, "Sir Joseph Banks's Provincial Turn," 410–11; Carter, *Sir Joseph Banks*, 61.

17. Chambers, *Endeavouring Banks*, 45; Williams, *Naturalists at Sea*, 92.

18. Linnaeus, *Linnaeus' Philosophia Botanica*, 283.

19. Joseph Banks, "A Catalogue of Plants Collected at Madeira, Brazil, Tierra del Fuego and the Society Islands, Arranged for each Locality According to Linnaeus' Species Plantarum," Botany Special Collections, NHM.

20. Gascoigne, *Joseph Banks and the English Enlightenment*, 14.

21. Banks to Brown, April 8, 1803; Chambers, *Letters of Sir Joseph Banks*, 244.

22. Daniel Solander, [the original descriptions and systematic lists of the plants], "Plantæ Novæ Hollandiæ," vol. 4, 1768–1771, note inside the front cover, NHM.

23. Banks and Hooker, *Journal of the Right Hon. Sir Joseph Banks*, 267.

24. Solander, [the original descriptions and systematic lists of the plants], "Plantæ Novæ Hollandiæ," 65–66, NHM.

25. Banks to Alströmer, November 16, 1784, in Chambers, *Letters of Sir Joseph Banks*, 78–79.

26. Banks to Alströmer, November 16, 1784, in Chambers, *Letters of Sir Joseph Banks*, 78–79.

27. D. Robinson, "Sir Joseph Banks and the Lincolnshire Influence," 193–96.

28. "Sir Joseph Banks's interleaved copy of 'Thomas Tusser's Five Hundred Points of Good Husbandry (1610)," U951/Z38, Kent History and Library Centre, Maidstone, England.

29. Sorrenson, "Ship as a Scientific Instrument," 227.

30. Banks, "Endeavour Journal" (1768–1771), 158–59, SLNSW.

31. J. Cook, *Captain Cook's Journal during his First Voyage Round the World*, 38.

32. Solander to Linnaeus, December 1, 1768, in Duyker and Tingbrand, *Daniel Solander*, 283. Spöring's father was also Herman Diedrich Spöring (1701–1747), professor of medicine at the Turku (Åbo) Academy.

33. Rose, "From the South Seas," 501; B. Smith, *European Vision and the South Pacific*, 16.

34. Solander, "Reports and Diary of Occurrences," ADD MS 45874:8, BL; Rose, "Specimens, Slips and Systems," 227.

35. Schaffer, "In Transit," 85.

36. Joppien and Smith, *Art of Captain Cook's Voyages*, 42–43.

37. Williams, "'Devilish Fellows Who Test Patience to the Very Limit,'" 259–60.

38. Banks to Alströmer, November 16, 1784, in Chambers, *Letters of Sir Joseph Banks*, 79.

39. Rose, "Specimens, Slips and Systems," 225–28.

40. Joseph Banks refers to these as books in Banks, "Catalogue of Plants," Botany Special Collections, f.1, NHM.

41. Banks to Alströmer, November 16, 1784, in Chambers, *Letters of Sir Joseph Banks*, 79.

42. Woodward, *Brief Instructions for Making Observations in All Parts of the World*, 12.

43. Alexander Anderson to William Forsyth, August 1, 1786, f.99, RBGK.

44. [Madeira III], London, BM001121503, NHM; Werrett, *Thrifty Science*, 71–72; Fulford, Lee, and Kitson, *Literature, Science and Exploration*, 35–36.

45. A Paris foot measures 32.48 centimeters, a measure used in France before the French Revolution.

46. For attempts at preserving collections, see Carter, *Sir Joseph Banks*, 71. Evidence of incompletion is apparent from a "book" of plants and the numerous specimens from Cook's first voyage in the backlog of the NHM, London, including [Madeira III], bar code BM001121503, NHM.

47. Balston, *Whatmans and Wove Paper*, xiii; Feather, "John Clay of Daventry," 198.

48. Lysaght, *Sir Joseph Banks in Newfoundland*, 349.

49. Benton, *John Baskerville*, 16.

50. On Banks's use of Baskerville paper, see Lysaght, *Sir Joseph Banks in Newfoundland*, 349. Baskerville's working practices are described in Gaskell, *John Baskerville*, xxi, 34.

51. Solander, [the original descriptions and systematic lists of plants], 9 vols., Botany Special Collections, NHM; Solander, [the original manuscript descriptions of the animals], 4 vols., Zoology Special Collections, NHM. These notebooks were split up and rebound in the 1970s.

52. Linnaeus, *Species Plantarum* . . . (1762–1763) [bound in six volumes], 582 LIN 109–14, Botany Special Collections, NHM; Linnaeus, *Systema Naturae* . . . (1766) [bound in six volumes], Zoology Special Collections, 4 o Li 62 (1–6), NHM.

53. Winterbottom, *Hybrid Knowledge in the Early East India Company*, 143–46; te Heesen, *World in a Box*, 179; Müller-Wille and Charmantier, "Natural History and Information Overload," 10–13; Rose, "Empire and the Theology of Nature in the Cambridge Botanic Garden."

54. Müller-Wille, "Names and Numbers."

55. Linnaeus, *Linnaeus' Philosophia Botanica*, 13–15.

56. Kalm, *Travels into North America*, 2:46–47, 276; Jonsson, "Climate Change and the Retreat of the Atlantic."

57. Chambers, *Letters of Sir Joseph Banks*, 78.

58. Banks, "Endeavour Journal" (1768–1771), 1:158–59, SLNSW.

59. Bleichmar, "Exploration in Print," 139.

60. Sloane, *Voyage to the Islands Madera, Barbados, Nieves, S. Christophers and Jamaica*, copy held at NHM, Botany HCR; Plukenet, *Opera Omnia Botanica*, copy held at BL, shelf mark 441.g.11–14; Kaempfer, *Amoenitatum Exoticarum*, copy held at BL, shelf mark 440. k.1.

61. For Banks's copies, see de Buffon, *Histoire Naturelle*, copy held at BL, shelf mark 461.h.1-i11; Pennant, *British Zoology* (1768–1770), copy held at BL, shelf mark 990.k.11–14.

62. Bleichmar, "Exploration in Print"; Mariss, "Library in the Field."

63. Margócsy, "Refer to the Folio and Number."

64. Linnaeus, *Linnaeus' Philosophia Botanica*, 270; Müller-Wille, "History Redoubled," 525.

65. Müller-Wille, "Introduction," 33.

66. Muller Wille, "Names and Numbers," 117.

67. Banks, "Endeavour Journal" (1768–1771), 2:348, SLNSW.

68. Bleichmar, "Exploration in Print," 142.

69. William Roxburgh to Joseph Banks, December 30, 1790, Add MS 33979: 64, BL.

70. William Roxburgh to Joseph Banks, December 30, 1790, in Chambers, *Indian and Pacific Correspondence of Sir Joseph Banks*, 3:179–81.

71. Anderson to Forsyth, January 1776, f.10, RBGK.

72. For comparison with Forster, see Mariss, "Library in the Field," 45.

73. Whitehead, "Zoological Specimens from Captain Cook's Voyages."

74. Sloane, *Natural History of Jamaica*, copy held at NHM, Botany HCR. For examples of Banks's bookplates, see Carter, *Sir Joseph Banks*, plates 2–3, 160–61. Banks's copy of Sloane contains his earlier bookplate, showing that it was purchased before he moved to New Burlington Street in 1767.

75. Banks, "Endeavour Journal" (1768–1771), 1:34, SLNSW. Banks mentions Sloane's work on three occasions.

76. Sloane, *Natural History of Jamaica*, copy held at NHM, Botany HCR, table 1, figure 2.

77. Solander, [the original manuscript descriptions], "Pisces & Anim. Cætera Oceani Pacifici," Rare Books Room, NHM.

78. Linnaeus, *Systema Naturae . . .* (1766) [bound in six volumes], Zoology Special Collections, 4 o Li 62 (1–6), 3:194, NHM.

79. Parkinson, [Zoological illustrations by Sydney Parkinson and others], 183. (2.92), NHM; Wheeler, "Catalogue of the Natural History Drawings," 106.

80. It seems that the use of shading had a similar effect to the use of colour. See Baxandall, *Shadows and Enlightenment*.

81. Solander to Morton, December 1, 1768, in Duyker and Tingbrand, *Daniel Solander*, 277–79.

82. Francisco-Ortega et al., "Early British Collectors and Observers of the

Macronesian Flora"; Banks, "Catalogue of Plants," 8, Botany Special Collections, NHM; Banks, "Endeavour Journal" (1768–1771), vol. 1, SLNSW. An additional thirteen-folio manuscript titled "Plants of Madeira" has been inserted, between folios 33 and 34 in volume 1.

83. Banks, "Journal of an Excursion to Wales Etc.," MS Add. 6294, 6, CUL.

84. Banks, "Endeavour Journal" (1768–1771), 1:16–17, SLNSW.

85. McConnell, *Jesse Ramsden*, 160–63. For Banks's trip to Iceland, see Agnarsdóttir, *Sir Joseph Banks, Iceland and the North Atlantic*.

86. Berg, "Britain, Industry and Perceptions of China," 284–85; Lightman, McOuat, and Stewart, *Circulation of Knowledge between Britain, India and China*, 1–20.

87. Linnaeus, *Species Plantarum* . . . (1762–1763) [bound in six volumes], 582 LIN 109–14, Botany Special Collections, NHM.

88. Banks, "Endeavour Journal" (1768–1771), 2:255–56, SLNSW.

89. Banks, "Endeavour Journal" (1768–1771), 1:61, SLNSW.

90. Parsons and Murphy, "Ecosystems under Sail."

91. Banks, "Endeavour Journal" (1768–1771), 1:102–3, SLNSW.

92. Solander, (1768–1771), [the original descriptions and systematic lists of the plants], 9 vols., Botany Special Collections, NHM.

93. Bil, "Tangled Compositions."

94. J. White, *Ancient History of the Maori*, 5:120–22. I thank Professor Anne Salmond for alerting me to this quotation. The term used for Europeans is *tupa*, which Williams's Māori dictionary defines as "goblin, demon or object of terror." For Linnaeus's instructions, see Linnaeus in Hanson, Cormach and Hanson, eds., *Linnaeus Apostles*, 1: 206.

95. On Solander and the Tahitian language, see Rauschenberg, "Daniel Carl Solander," 14. Tupaia was a classic "go between." See V. Smith, "Banks, Tupaia, and Mai;" Salmond, *Tears of Rangi*, 27; Schaffer et al. *Brokered World*.

96. Linnaeus, *Critica Botanica*, 38.

97. Linnaeus, *Linnaeus' Philosophia Botanica*, 176; A. Cook, "Linnaeus and Chinese Plants," 128.

98. Solander, [the original descriptions and systematic lists of the plants] (1768–1771), "Plantæ Otaheitenses," inside the front cover, Botany Special Collections, NHM.

99. V. Smith, *Intimate Strangers*, 65; Bil, "Tangled Compositions," 3; Eckstein and Schearz, "Making of Tupaia's Map." For more of Solander, Parkinson's and Banks's lists of translations, see Solander, "Observationes de Otaheite, &c.," MS 12892, School of Oriental and African Studies (SOAS), University of London.

100. Salmond, *Tears of Rangi*, 18.

101. The active role of Indigenous peoples has recently been addressed in Fullagar and McDonnell, *Facing Empire*.

102. Müller-Wille, "Names and Numbers"; Müller-Wille and Charmantier, "Natural History and Information Overload," 10.

103. Solander, [the original descriptions and systematic lists of the plants] (1768–1771), "Plantæ Otaheitenses," f.63, Botany Special Collections, NHM.

104. V. Smith, "Banks, Tupaia and Mai," 139–141.

105. Solander, [manuscript descriptions of plants written on slips of paper] (1768–1771), 5:175, Botany Special Collections, NHM.

106. William Parry, "Omai (Mai), Sir Joseph Banks and Daniel Solander," c. 1775–1776, NPG 6652, National Portrait Gallery, London.

107. Solander, [manuscript descriptions of plants written on slips of paper] (1768–1771), 5:179, Botany Special Collections, NHM.

108. Parkinson, *Journal of a Voyage to the South Seas*, 37

109. Salmond, *Two Worlds*, 252–53.

110. Solander, [the original descriptions and systematic lists of the plants] (1768–1771), "Plantæ Australiæ Novæ Zelandia [New Zealand]," Botany Special Collections, NHM.

111. Salmond, *Tears of Rangi*, 44.

112. A. Cook, "Linnaeus and Chinese Plants," 128.

113. Fullagar, "Envoys of Interest," 248.

114. Solander, [manuscript descriptions of plants written on slips of paper], (1768–1771), 5:179, Botany Special Collections, NHM.

115. Parkinson, *Journal of a Voyage to the South Seas*, 37.

116. Linnaeus, *Species Plantarum* . . . (1762–1763) [bound in six volumes], 582 LIN 109–14, 1:221, Botany Special Collections, NHM.

117. See Staffan Müller-Wille, "Walnuts at Hudson's Bay," 46–47.

118. Banks, "Catalogue of Plants," 25, Botany Special Collections, NHM.

119. Linnaeus, *Species Plantarum* . . . (1762–1763) [bound in six volumes], 582 LIN 109–14, vol. 1, opposite 221, Botany Special Collections, NHM.

120. Nickelsen, *Draughtsmen, Botanists and Nature*; Daston and Galison, *Objectivity*, 59–63.

121. Banks, "Endeavour Journal" (1768–1771), 1:227, 229–30, 245–47, SLNSW.

122. Spary and White, "Food of Paradise," 77–78; Sivasundaram, "Natural History Spiritualised," 419.

123. Rumphius, *Herbarium Amboinense*, tables 33–34. Banks's annotated

copy can be found at BL, shelf mark 449.I.8–12; Dampier, *New Voyage Round the World*, 296. Banks's copy can be found at BL, shelf mark 979.f.7; Anson, *Voyage Round the World*, 409–29. Banks's copy can be found at BL, shelf mark 978.g.5.

124. Linnaeus, *Species Plantarum* . . . (1763–1763) [bound in six volumes], 582 LIN 109–14, vol. 1, opposite 1375, Botany Special Collections, NHM.

125. Banks, "Endeavour Journal" (1768–1771), 1:226–29, SLNSW; Salmond, *Aphrodite's Island*, 197.

126. Dampier, *New Voyage Round the World*, 296.

127. Rumphius, *Herbarium Amboinense* (1750), copy held at BL, 449.I.8–12, plates 33–34. The annotations from Banks's copy of Rumphius's work are similar to those in Banks's copy of Kaempfer's *Amoenitatum Exoticarum* (1712). Banks's copies of Dampier and Anson are not annotated, although they do contain his earlier bookplate.

128. Solander, [The original descriptions and systematic lists of the plants] (1768–1771), "Plantæ Otaheitenses," f.37, Botany Special Collections, NHM; Newell, "New Ecologies," 93–94. For the naming of the breadfruit, see Fielding, "'Correct Name for the Breadfruit.'"

129. Parkinson, (1768–1783) [original watercolor drawings and sketches made by S. Parkinson during Cook's first voyage (1768–1771), SI2/37, a-d, Botany Special Collections, NHM.

130. Parkinson, *Journal of a Voyage to the South Seas*, 36.

131. Parkinson, *Journal of a Voyage to the South Seas*, 284.

132. Linnaeus, *Linnaeus's Philosophia Botanica*, 283.

133. Nickelsen, *Draughtsmen, Botanists and Nature*, 73–83; On "typical" images, see Daston and Galison, *Objectivity*, 60–68.

134. Linnaeus, *Linnaeus' Philosophia Botanica*, 283.

135. Bleichmar, *Visible Empire*, 107–11; Nickelsen, *Draughtsmen, Botanists and Nature*, 165; Daston and Galison, *Objectivity*, 168–71.

136. Hawkesworth, *Account of the Voyages Undertaken by the Order of His Present Majesty*, 2, fig. 11.

137. Banks to Alströmer, November 16, 1784, in Chambers, *Letters of Sir Joseph Banks*, 79.

138. William Sheffield to Gilbert White, December 1772, in Lysaght, *Joseph Banks in Newfoundland and Labrador*, 253–55.

139. Thomas Martyn to Richard Pulteney, February 17, 1772, MS/2380/28/20, LSL.

Chapter 3: From Specimen to Print

1. On Fleet Street, see J. Raven, *Business of Books*, 163. Quoted in Nichols, *Literary Anecdotes of the Eighteenth Century*, 127; Noblett, "Pennant and His Publisher," 64–65.

2. Rogers, *Grub Street*; J. Secord, "Newton in the Nursery."

3. Brewer, *Pleasures of the Imagination*, 145.

4. Johns, *Nature of the Book*, 378. On the deregulation of the book trade, see J. Raven, *Judging New Wealth*; Darnton, "What Is the History of Books?," 67–69. For prints, see Brewer, *Pleasures of the Imagination*, 137.

5. Sher, *Enlightenment and the Book*, 4.

6. For on commission publishing, see St Clair, *Reading Nation*, 165.

7. Haynes, *Politics of Publishing in Nineteenth-Century France*, 25; Kilgour, *Evolution of the Book*, 111; Feather, *History of British Publishing*, 25.

8. J. Raven, *Business of Books*, 72. For Grub Street, see Griffin, *Literary Patronage in England*, 291.

9. Turnovsky, "Enlightenment Literary Market," 392.

10. For interests in publishing to publicize a collection, see Spary, "Scientific Symmetries," 9.

11. For Banks's comment, see Aiton, *Delineations of Exotick Plants*, 2; Nickelsen, *Draughtsmen, Botanists and Nature*, 71–102; Rudwick, "Picturing Nature in the Age of Enlightenment," 283–88; Roos, "Art of Science"; Kusukawa, "Drawings of Fossils by Robert Hooke and Richard Waller."

12. *Westminster Magazine*, 623–24.

13. Pennant, "Expenses of My Different Works," CR2017/TP571, 2, WRO. It must be noted that some presentation copies of the 1812 edition were printed in quarto. For differet editions, see Pennant, *British Zoology* (1766), Pennant, *British Zoology* (1768–1770), Pennant, *British Zoology* (1776–1770), Pennant, *British Zoology* (1812), and Pennant, *British Zoology* (1818). Very few of the original manuscripts of these survive apart from a significant portion of that written by Pennant for the 1768–1770 edition that includes several drafted title pages. See Thomas Pennant, "Zoology," MS. 15431 D, NLW.

14. Pennant, *British Zoology* (1776–1777), 4:2.

15. Newspaper cutting, "Lord and Lady Denbigh's Gift to the Nation," *Flint County Herald*, February 21, 1913, CR2017/TP551/6/11, WRO.

16. The British Museum (Natural History) removed these specimens from their original arrangements and cases when they were accessioned in 1913.

17. Newspaper cutting, "Pennant's Working Library," *London Morning Post*, May 9, 1913, CR2017/TP552/6/9, WRO.

18. Pennant, *British Zoology* (1766), 79–80; Pennant, *British Zoology* (1776–1777), vol. 1, opposite 245.

19. This method of arrangement was described shortly before the dispersal of Pennant's collection in 1911, see "Newspaper cutting," CR2017/TP55/6/11, WRO.

20. Pennant, *British Zoology* (1776–1777), 1:240.

21. Witteveen, "Supressing Synonymy." Similar specimens have been referred to as "classification types."

22. Pennant, "Expenses of My Different Works," CR2017/TP571, 2, WRO. Calculations of equivalent purchasing power of the prices quoted are sourced from the UK National Archives currency converter, https://www.nationalarchives.gov.uk/currency-converter/#currency-result.

23. Pennant, *British Zoology* (1776–1777), 1:241–45.

24. "Little woodpecker," 1912.12.30.92, Pennant Collection, NHM (Tring); "Red breasted goosander," 1912.12.30.92, Pennant Collection, NHM (Tring). Specimens were removed from their original cases by the British Museum (Natural History) between 1912 and 1913. Pennant, *British Zoology* (1776–1777), 2:421–23, 4:556–57.

25. Quoted in Tobin, *Duchess's Shells*, 186.

26. Pennant, *Tour on the Continent*.

27. Pennant to George Paton, May 24, 1776, ADV. MSS. 29.5.5 (2 vols.), 1:160–61, NLS. Pennant's surviving cabinets are held by the Mineral and Planetary Sciences Division, Department of Earth Sciences, NHM.

28. Pennant, *British Zoology* (1776–1777), 4:iv.

29. Pennant, "Expenses of My Different Works," CR2017/TP571, 20, WRO.

30. "Art VI. *British Zoology*," 275. On the separation of species between individual copperplates, see Spary, *Utopia's Garden*, 255–75.

31. See Pennant, *Histoire Naturelle des Oiseaux*, final unpaginated advertisement leaf; Gaskell, *New Introduction to Bibliography*, 178.

32. Pennant, "Expenses of My Different Works," CR2017/TP571, 20, WRO.

33. For Hixon, see "Obituary, with Anecdotes, of Remarkable Persons." For Pennant's articles, see Pennant, "Account of Some Fungitae," 513–16; Pennant, "Account of Different Species of Birds, Called Pinguins"; Pennant, "Account of Two New Tortoises." On authors' financing images for articles, see Csiszar, *Scientific Journal*, 55; Fyfe, McDougall-Walters, and Moxham, "Guest Editorial," 231.

34. Pennant, "Expenses of My Different Works," CR2017/TP571, 20, WRO.

35. Pennant, *British Zoology* (1776–1777), 4:112.

36. *Mytilus umbilicatus*, Pennant, *British Zoology* (1776–1777), Molluscs Department, 1912.12.30.1–2, NHM.

37. Kusukawa, "Drawings of Fossils by Robert Hooke and Richard Waller," 129–30; Baxandall, *Shadows and Enlightenment*, 2.

38. For "typical" or "average" images, see Daston and Galison, *Objectivity*, 69.

39. On synonyms, see Witteveen, "Supressing Synonymy with a Homonym," 144–47. Many specimens figured in *British Zoology* were listed as "type specimens" after their acquisition by the British Museum (Natural History) in 1912.

40. The term *large paper* refers to a folio publication that measures 52.8 × 37.2 cm.

41. Pennant, *Literary Life*, 8.

42. On Morrison, see Mandelbrote, "Publication and Illustration of Robert Morrison's *Plantarum historae universalis Oxoniensis*." For Pennant's charitable donations, see Pennant, *British Zoology* (1768–1770), vol. 2, advertisement leaf after index. For White's catalogue, see B. White, *Catalogue of a Large and Valuable Collection of Books in All Languages*, 23.

43. Pennant, *British Zoology* (1766), 51, 116.

44. Edwards to Pennant, November 21, 1763, CR2017/TP221/1–2, WRO.

45. In his annotated copy of *Arctic Zoology* Pennant stored a number of ephemeral items, including George Edwards's call card. On the recto it reads "Geo Edwards N°.1. Angel Street Near Newgate Street London," and on the verso Edwards inscribed "Mr Edwards at Plaistow in Essex. Dureing the summer season," under which Pennant has written "died Friday 23ᵈ July 1773 at Plaistow." This emphasizes that the images Pennant commissioned for *British Zoology* were among the last Edwards ever produced. See Thomas Pennant, *Arctic Zoology* (1792), "bound in 2 vols., 'ex libris Pennant,' with enclosures and extensive annotations by the author," MSS PEN B, NHM.

46. Thomas Pennant, "Notes by Thomas Pennant on British Zoology and Ornithology," MS 2545B, 29, NLW.

47. Keighren, Withers, and Bell, *Travels into Print*, 133.

48. Pennant, *British Zoology* (1766), 81.

49. Pennant, "Notes by Thomas Pennant on British Zoology and Ornithology," MS 2545B, 29, NLW.

50. Pennant, "Notes by Thomas Pennant on British Zoology and Ornithology," MS 2545B, 25–40, NLW.

51. Pennant, *Indian Zoology*, i.

52. These measurements are based on the plate mark. Paper often shrinks by up to 10 percent after intaglio printing. For domestic copper production, see Rose, "Publishing Nature in the Age of Revolutions," 11–12.

53. Gilbert White to John White, February 27, 1777, in Holt-White, *Life and Letters of Gilbert White of Selborne*, 2:7.

54. Benjamin White, "Copy Inventory of Rev. Gil. White's Effects" (1793), Gilbert White Papers, HLH.

55. Pennant, "Additions to Pennant's Outlines of the Globe" (c. 1798), MS 2544–2546B, loose material, 4–5, NLW.

56. Hills, *Papermaking in Britain*, 74–78.

57. John Curtis to David Pennant, July 6, 1799, CR2017/TP448, WRO.

58. Raithby, *Statutes of the United Kingdom of Great Britain and Ireland*, 101–2.

59. Peltz, *Facing the Text*, 220; Peltz, "Friendly Gathering," 45

60. Pennant, *History of the Parishes of Whiteford and Holywell*, 7.

61. Bridson, "Treatment of Plates in Bibliographical Description," 475.

62. Pennant, "Additions to Pennant's Outlines of the Globe," MS 23412E, loose material, 4, NLW.

63. Calculations of equivalent purchasing power of the prices quoted are sourced from the UK National Archives currency converter, https://www.na tionalarchives.gov.uk/currency-converter/#currency-result.

64. Pennant, "Additions to Pennant's Outlines of the Globe," MS 23412E, loose material, 4, NLW.

65. Henderson, *James Sowerby*, 19–20.

66. On the proof plates, see Griffiths, "Print Collecting in Rome, Paris and London," 44.

67. Fulford, Lee, and Kitson, *Literature, Science and Exploration*, 38; Bewell, "'On the Banks of the South Sea,'" 180–84; Veit, *Captain James Cook*, 84.

68. Gascoigne, *Joseph Banks and the English Enlightenment*, 62; Musgrave, *Multifarious Mr. Banks*, 173–75.

69. Browne, "Botany for Gentlemen," 614.

70. Joseph Banks to Thomas Pennant, May 4, 1783, in Chambers, *Scientific Correspondence of Sir Joseph Banks*, 2:81.

71. Gascoigne, *Joseph Banks and the English Enlightenment*, 97–98.

72. Pennant to Banks, May 11, 1798, in Chambers, *Scientific Correspondence of Sir Joseph Banks*, 2:83.

73. Anders Sparrman to Georg Forster, mid-December 1776, in Leuschner et al., *Georg Forsters Werke*, 27.

74. See Gooding, Mabberley, and Studholme, *Joseph Banks's Florilegium*,

297; Diment et al., "Catalogue of the Natural History Drawings," 9–10. On the lack of attention to the social dimensions of Banks's work, see Schaffer, "Visions of Empire," 344.

75. Quinby and Stephenson, *Catalogue of Botanical Books in the Collection of Rachel McMasters Miller Hunt.*

76. Nelson, "Some Publication Dates for Parts of William Curtis's *Flora Londinensis*," 638.

77. On Banks's rolling press, see Carter, *Guide to Biographical and Bibliographical Sources*, 334–35. For the intaglio engraves and printers, see Clayton, *English Print*, 213–14.

78. Clarke, *Strawberry Hill Press and Its Printing House*, 26–27; Brewer, *Pleasures of the Imagination*, 144–45; Harney, *Place-Making for the Imagination*, 17, 23, 27, 100.

79. Karen Wood, "Making and Circulating Knowledge," 90–92.

80. Lazarus and Pardoe, "Bute's *Botanical Tables*," 283–85; Miller, "'My Faivorite Studdys,'" 223–30.

81. For the distribution of Banks's and Bute's books, see chapter 4.

82. "III. Reliquiæ Houstouianæ." For Banks's purchase of Miller's books, see Meynell, "Books from Philip Miller's Library"; Baker and Leigh, *Catalogue of the Valuable Library of Philip Miller.*

83. See Houstoun, "Manuscript Catalogues and Drawings of Plants," Botany Manuscripts, MSS BANKS COLL HOU, NHM.

84. Banks to Bacstrom, June 16, 1791, in Chambers, *Scientific Correspondence of Sir Joseph Banks*, 4:61. For an inventory of Banks's plates, see Robert Brown, "Catalogue of Engraved Copper Plates in the Presses in the Engravers Room (under the Library or Herbarium)," Botany Special Collections, NHM.

85. H. H. Baber et al., "Inventory of Sir Joseph Banks's Library," (1820–1823) 460.g.1., 2:471, BL.

86. Banks to Bruce, December 24, 1789, in Chambers, *Scientific Correspondence of Sir Joseph Banks*, 3:250.

87. On Banks's relationship with Bulmer from the early 1770s, see MS. Montague. D. 6. 81, Bodleian Library, Oxford. For Banks's appointment of Bulmer as the Royal Society printer, see Bulmer to Banks, December 1792, ADD MS. 33982, 357, BL.

88. Csiszar, *Scientific Journal*, 55, 232; Moxham, "'Accoucheur of Literature,'" 22–23.

89. Quoted in Isaac, *William Bulmer*, 43.

90. Dryander, *Catalogus Bibliothecae Historico-Naturalis.*

91. Seba and Stein, *Cabinet of Natural Curiosities*, 16, 30; Margócsy, *Com-*

mercial Visions, 74–75. Examples of similar books include Edwards, *Natural History of Uncommon Birds*, and Pennant, *British Zoology* (1766).

92. J. Smith, *Selection of the Correspondence of Linnaeus*, 2:574.

93. Müller-Wille, "Collection and Collation," 20.

94. During the first large-scale attempt to publish these copperplates between 1980 and 1990 the French color-printing technique *à la poupée* was used. This involved working individual colored inks into the plate with twists of cloth, allowing for the successful color printing when the plate was passed through a rolling press. Alecto Historical Editions' desire to market this expensive product for public consumption made color printing more attractive than Banks's original intention to print in black. For the surviving monochrome copy of this book, see Printed Books, shelf mark, 10.Tab.42, BL; Botany Special Collections, NHM, interleaved between Sydney Parkinson's drawings. For Banks's production of Forster's *Icones Plantarum*, c. 1800, see Rose, "Publishing Nature in the Age of Revolutions."

95. Linnaeus, "*Critica Botanica*," 138.

96. Aiton, *Delineations of Exotick Plants*, iii.

97. Secord, "Botany on a Plate."

98. Quoted in Secord, "Botany on a Plate," 35.

99. On color and the loss of tonal effects, see Bridson, Wendel, and White, *Printmaking in the Service of Botany*, 53. On the use of shade, see Baxandall, *Shadows and Enlightenment*, 2.

100. Banks, "Endeavour Journal" (1768–1771), SAFE 1/457, series 3, vol. 1, 260, SLNSW.

101. Linnaeus, *Species Plantarum* . . . (1762–1763) [bound in six volumes], 582 LIN 109–14, Botany Special Collections, NHM.

102. Linnaeus, *Supplementum Plantarum*, 127.

103. Banks to Thunberg, June 1, 1781, Sir Jos. Banks, G300c, UUL. On copying, see Charmantier, "Notebooks, Files and Slips," 40–50.

104. Adams, *Flowering Pacific*, 142–43.

105. Bacstrom to Banks, March 1771, SAFE/Banks Papers/Series 06.141:450, SLNSW.

106. Solander, [the original descriptions] (1768–1771), "Plantæ Novæ Hollandiæ," 4:94, NHM.

107. Quoted in Adams, *Flowering Pacific*, 126.

108. Diment et al., "Catalogue of the Natural History Drawings," 11.

109. Bacstrom, "Catalogue of Drawings of Plants of Cook's 1st Voyage 1768–1771," Botany MSS BANKS COLL BAC, 19, NHM.

110. For Bauer, see Lack, *Bauers*, 209–11. On payments to the engrav-

ers, see Aiton, *Delineations of Exotick Plants*, shelf mark Bauer Unit shelf C4, notes on first endpaper, NHM; Britten, "Francis Bauer's 'Delineations of Exotick Plants,'" 181–83. For Banks's and Bauer's relationship, see Lack, *Bauers*, 188–225.

111. Quoted in Dickinson, Bruce, and Dowsett, "*Vivarium Naturae* by George Shaw," 330.

112. Bacstrom, "Catalogue of Drawings," NHM.

113. See "British Artists' Suppliers, 1650–1950–P," National Portrait Gallery, https://www.npg.org.uk/collections/research/programmes/directory -of-suppliers/suppliers-p.

114. Department of Prints and Drawings, Sarah Sophia Banks Collection, D,2.4116, BM.

115. Pontifex to Banks, 1796, Department of Prints and Drawings, Sarah Sophia Banks Collection, Banks, 85.128 (verso), BM; Stijnman, *Engraving and Etching*, 148; Griffiths, *Prints and Printmaking*, 326.

116. Banks to Marum, February 14, 1792, in Dawson, *Banks Letters*, 586.

117. Quoted in Thunberg, *Travels in Europe, Africa, and Asia*, 189–291.

118. Delbourgo, *Collecting the World*, 227; See Kaempfer, *Der 5.*

119. Kaempfer, "Delineationes et descriptiones plantarum Japonicarum et Persicarum: 1685, 1690. Japan: Delineationes et descriptiones plantarum Japonicarum ab E. Kæmpfero: 1," Sloane MS 2914, BL.

120. Rose, "Publishing Nature in the Age of Revolutions," 1143–44.

121. Solander, "'Reports and Diary of Occurrences in the Natural History Departments" (1764–1768), Add Ms. 45,874: 6-v7, BL; Rose, "Specimens, Slips and Systems," 225.

122. Baber, Walker, and Cary, "Inventory of Sir Joseph Banks's Library" (1820–1823), 460.g.1, 2:471, BL.

123. See Henrey, "Kaempfer's 'Icones,'" 104.

124. Chambers, *Joseph Banks and the British Museum*, 32.

125. St Clair, *Reading Nation*, 165.

126. For Banks's plates, see Brown, "Catalogue of Engraved Copper Plates . . .," Botany Special Collections, NHM. For Pennant's inventory, see Pennant, *Literary Life*, 38.

Chapter 4: From Print to Distribution

Epigraph: Banks to Lloyd, September 5, 1796, MS 12415C, item 45 (also labeled 31), NLW. I thank Mary-Ann Constantine for alerting me to the existence of this quote.

1. Brewer, *Pleasures of the Imagination*, 145.

2. See Colley, *Britons*, 375. For Anglophone interpretations of cameralism, see Drayton, *Nature's Government*, 99–100; Lindenfeld, *Practical Imagination*, 11–33.

3. Pennant, *British Zoology* (1776–1777), 1:xvii–xix.

4. See chapters 1 and 3.

5. For books as gifts, see Heal, *Power of Gifts*, 44–46; Scott-Warren, *Sir John Harrington*; Biagoli, *Galileo, Courtier*, 38–39; Ben-Amos, *Culture of Giving*, 195–241; Cavagna, "Free Transmission of Knowledge." On financial returns for authors, see J. Raven, *Judging New Wealth*, 58–60.

6. Turnovsky, *Literary Market*, 99–100.

7. Peltz, *Facing the Text*, 64.

8. Pennant, *Literary Life*, 8.

9. Pennant to Richard Bull, February 13, 1788, MS 5500C, 63, NLW.

10. Granger to Pennant, July 21, 1774, letter 1. Between the front cover and first flyleaf of vol. 4 in Pennant's copy of Granger's *Biographical History* (1775). Private collection.

11. Timaeus to Pennant, December 1, 1793, CR2017/TP374/1–2, WRO.

12. Pennant, *British Zoology* (1776–1777), 1:xi.

13. Pennant to Smith, June 14, 1798, GB-110/JES/COR/8/31, LSL. See Cantor, "Rise and Fall of Emmanuel Mendes da Costa."

14. Pennant, "Expenses of My Different Works," CR2017/TP571, f. 2, 9–10, WRO. For the conversion, see the UK National Archives currency converter, https://www.nationalarchives.gov.uk/currency-converter/#currency-result.

15. Kernan, *Samuel Johnson and the Impact of Print*, 10–11.

16. Goldsmith, *Miscellaneous Works of Oliver Goldsmith*, xliv; J. Raven, *Business of Books*, 347.

17. Pennant, *British Zoology* (1812).

18. Harley, "Bankruptcy of Thomas Jefferys" 47.

19. Pennant to Smith, June 14, 1798, GB-110/JES/COR/8/31, LSL.

20. Pennant, *Indian Zoology*, i.

21. Pennant, *Literary Life*, 4.

22. Pennant to Paton, June 30, 1771, ADV. MSS. 29.5.5, 1:6–7, NLS. This copy does not appear to have been recorded in Pennant's list of subscribers; see MS 2545B, fols. 25–27, NLW. On Pennant's working practices, see Briggs, "Thomas Pennant," 42.

23. On Bute's estates, see P. Brown, "Bute in Retirement," 242–53. Stillingfleet, *Miscellaneous Tracts Relating to Natural History*; Hill, *Vegetable System*.

24. Pennant, *Literary Life*, 6.

25. Pennant, *Literary Life*, 6; Pennant, *Tour on the Continent*, 76–77, 38–39. For Trew, see Nickelsen, *Draughtsmen, Botanists and Nature*, 20–23.

26. Pennant, *Literary Life*, 7–8.

27. See "Letters from Ferdinando Bassi at Bologna to Thomas Pennant, Downing," CR2017/TP171/1–8, WRO.

28. A. Blackburne to Pennant, January 11, 1778, CR2017/TP177/1–2, WRO. For Blackburne's collection, see Wystrach, "Anna Blackburne (1726–1793)."

29. Pennant, *Tour in Scotland*, 12.

30. Milam and Nye, "Introduction to Scientific Masculinities."

31. Pennant, *British Zoology* (1776–1777), Anna and John Blackburne's copy, Zoology Special Collections, SB 72A o PEN, set 2, NHM; Berkenhout, *Outlines of the Natural History of Great Britain and Ireland*. The Orford Hall copy is in a private collection. Anna Blackburne's annotations appear throughout volume 1, marking off shells in her collection or adding vernacular English names.

32. This resembles René Antoine Ferchault de Réaumur's household. See Terrall, "Masculine Knowledge," 1.

33. Ross, *Writing in Public*, 44.

34. Pennant, *Zoologia Britannica Tabulis Aeneis*; Pennant, *Literary Life*, 4–8.

35. Pennant, "Notes by Thomas Pennant on British Zoology and Ornithology," MS 2545B, fol. 85, NLW.

36. Pennant to Linnaeus, November 17, 1771, L4572, LSL.

37. Pennant, *British Zoology* (1768–1770), 2, advertisement leaf.

38. Pennant to Murray, September 15, 1769, 49 (8), 107, Blair Athol Castle Archives, Pitlochry, Scotland.

39. Pennant, *Literary Life*, 8. On Pennant's presentation of books, see Rees and Walters, "Library of Thomas Pennant," 141.

40. Pennant, *British Zoology* (1768–1770), 3:xii–xiii. Two out of seventeen presentation copies are accounted for: Printed Books, shelf mark 990.k.11–14, BL; Pennant and White [annotated by], "Thomas Pennant's British Zoology," Add. MS 46471–46472, 1768, BL.

41. Peter Pallas to Thomas Pennant, November 4, 1777, in Urness, *Naturalist in Russia*, 14–20.

42. Turnovsky, *Literary Market*, 100–101.

43. "British Zoology, *Vol. IV.*"

44. Pallas to Pennant, November 4, 1777, in Urness, *Naturalist in Russia*, 15.

45. Müller-Wille, "Names and Numbers," 115.

46. Pennant, *British Zoology* (1812), 3:362–63. David Pennant's copy with extensive additions. This copy was last sold in 2017 and is in a private collection. It had been sold, along with the rest of the Downing Hall library collection, in 1913. See W. M. Dew & Son, *Catalogue of Sale of the Remainder of the Downing Library*, 118, lot 1355.

47. Pennant, *British Zoology* (1768–1770), Printed Books, 990.k.11–14, BL.

48. Pennant, *British Zoology* (1768–1770), Add MS 46,471, 1, 244, BL.

49. White, *Natural History and Antiquities of Selborne*, 98.

50. Pennant, *British Zoology* (1776–1777), 1:402.

51. Pennant, *British Zoology* (1768–1770), Printed Books, 990.k.11–14, BL.

52. Joseph Banks, "A Journal of a Voyage up Great Britain's West Coast and to Iceland" (1772), shelf mark QH11 B36 1772, 69, MUL.

53. Pennant, *British Zoology* (1768–1770), 2:347; Pennant, *Arctic Zoology*, 2:462.

54. This copy is listed in the 1786 sale catalogue for the duchess's collection. See Lightfoot, *Catalogue of the Portland Museum*, vi.

55. Tobin, *Duchess's Shells*: 184–88. For Pennant and Portland's correspondence, see CR2017/TP149/1–8, WRO.

56. Tobin, *Duchess's Shells*, 218.

57. Pennant, *British Zoology* (1776–1777); "Typescript, with holograph annotations. Pennant's own working copy, heavily edited," Zoology Manuscripts, MSS PEN A, NHM. Another copy containing the original illustrations for volume 4 is held by Arader Galleries, New York, https://aradergalleries.com/products/jan-brandes-dutch-1743-1808-two-birds?_pos=1&_sid=10af b56e8&_ss=r. The copy held by the NHM was sold in 1913; see W. M. Dew & Son, *Downing Library*, 117.

58. Pennant, "Br. Zoology Given Away" (c. 1777), GB-110/LM/MA/PEN/2, LSL.

59. Pennant, "To Every Gentleman Desirous," 174.

60. It seems Pennant's aim was not realised by Ramsay, who died in 1778. See Withers, "Rev. John Walker."

61. Grenville's quarto copy of Pennant's *British Zoology* (1776–1777) can be found at Printed Books, shelf mark G.2803–6, BL. George III's quarto copy of Pennant's *British Zoology* (1776–1777) can be found at Printed Books, shelf mark 40.d.10–13, BL.

62. Banks's quarto copy of Pennant, *British Zoology* (1776–1777) can be found at shelf mark 458.a.2.(1.), 264–65, BL.

63. Porter, *Enlightenment*, 113; Colley, "Apotheosis of George III," 109.

64. George III's copy of Pennant's *British Zoology*, 40.d.10–13, BL.

65. Quoted in Jones, *Agricultural Enlightenment*, 16.

66. Pennant, "Presents of Hindoostan," MS. 23412E, loose item 1, NLW.

67. Pennant, "Additions to Pennant's Outlines of the Globe," MS 23412E, loose material 4, NLW. Pennant's spelling of this place differs between "Severndroog" and "Sawen Droog."

68. Pennant, "Thomas Pennant's Copy of his Outlines of the Globe" (1798), MS. 23412E, NLW; Thomas Maurice to David Pennant, c. 1800, CR 2017/TP482/1–2, WRO.

69. Hooker, *Journal of a Tour in Iceland* (1811), iii.

70. Hooker, *Journal of a Tour in Iceland* (1811), vii.

71. See Fawcett, "Some Aspects of the Norfolk Book-Trade," 388. Turner's large paper copy of Hooker, *Journal of a Tour in Iceland* (1811), is now held by Royal Botanic Gardens, Kew, Library and Archives, printed books, T1.28.

72. Smith's copy is held in a private collection. Dorothea Banks's copy can be found at shelf mark 791.e.2, BL.

73. Hooker, *Journal of a Tour in Iceland* (1813), 1, dedication page 2, private collection. For Turner and Banking, see Campbell, "Banker," 117. The copy shown has passed by direct decent through the Brightwen and Turner families. The Palgrave-Barker family sold the last privately owned parts of the Dawson Turner Collection on March 4, 2020. See *Printed Books, Maps, & Documents from the Library of Dawson Turner*, 110, lot 302.

74. Turner, *Fuci*, 1:i.

75. "Crier" is a reference to gossip in polite circles. Banks to Somerset, August 21, 1816, D/RA/A/13/4/22, Buckinghamshire Archives, Aylesbury, England. Banks refers to "Account of an Ancient Canoe Found in Lincolnshire," 244–45.

76. Moxham, "'*Accoucheur* of Literature,'" 25; Moxham and Fyfe, "Sociability and Gatekeeping," 199.

77. Biagoli, *Galileo, Courtier*, 389–99.

78. Rose, "Publishing Nature in the Age of Revolutions," 25–26.

79. Forster, *Ansichten vom Niederrhein*, 3:52.

80. Forster to Dohm, January 7, 1792, in Popp, *Georg Forsters Werke*, 22–23.

81. Silliman, *Journal of Travels in England, Holland and Scotland*, 292.

82. Sparrman to Forster, mid-December 1776, in Leuschner et al., *Georg Forsters Werke*, 27.

83. Rose, "Publishing Nature in the Age of Revolutions," 1142.

84. Banks to Dryander, September 30, 1786, in Chambers, *Scientific Correspondence of Sir Joseph Banks*, 4:207.

85. Thunberg, *Travels in Europe, Africa and Asia*, 4:292.

86. Thunberg, *Travels in Europe, Africa and Asia*, 5:289.

87. Two copies of the original black impressions of these images can be found at the BL and NHM. See Printed Books, 10.Tab.42, BL; Botany Special Collections, interleaved between Sydney Parkinson's drawings, NHM.

88. This is outlined in a note left by the surgeon Everard Home (1756–1832) in the front of his copy of Aiton, *Delineations of Exotick Plants*, Bauer Unit Shelf C 4, NHM.

89. Schaffer, "Visions of Empire," 343.

90. Banks to Roxburgh, August 9, 1798, in Chambers, *Indian and Pacific Correspondence of Sir Joseph Banks*, 4:527.

91. "III. Reliquiæ Houstouianæ," 254.

92. Dryander to Banks, October 1, 1782, in Chambers, *Scientific Correspondence of Sir Joseph Banks*, 2:22.

93. Pindar, *Peter's Prophecy*, frontispiece.

94. Silliman, *Journal of Travels in England, Holland and Scotland*, 234–35.

95. Saint-Fonde, *Travels in England, Scotland and the Hebrides*, 1:2.

96. For conversations at Soho Square, see J. Secord, *Visions of Science*, 57–58.

97. Forster to Jacobi, November 6, 1781, in Gervinus, *Georg Forsters Sämtliche Schriften*, 8:123.

98. Sher, *Enlightenment and the Book*, 242–43.

99. Banks to Somerset, August 31, 1816, quoted in Carter, *Guide to Biographical and Bibliographical Sources*, 153.

100. "Review of Agricultural Publications," 229. See Mackie to Banks, May 14, 1805, in Dawson, *Banks Letters*, 565.

101. Ryan to Banks, July 28, 1783, in Chambers, *Scientific Correspondence of Sir Joseph Banks*, 2:112. See also von Rohr to Banks, June 24, 1777, Add MS 8094, 185, BL.

102. For copies Bute distributed, see Lazarus and Pardoe, "Bute's Botanical Tables," 285–86. For Bute's premiership, see Miller, "My Faivorite Studdys," 220–25.

103. Gowin C. Knight, "Papers Relating to the British Museum" (1760), Add MS 4449, BL.

104. Draft letter from Banks to Cotta, at the foot of Cotta's original letter to Banks, dated July 8, 1791, Add MS 8097, 393–96, BL.

105. Draft letter from Banks to Cotta, at the foot of Cotta's original letter to Banks, dated July 8, 1791, Add MS 8097, 393–96, BL.

106. Mitchill, Miller, and Smith, *Medical Repository*, 413; Linnaeus and Re-

ichard, *Systema Plantarum* [interleaved and bound in ten volumes, containing copious annotations in the hands of Jonas Dryander, Samuel Törner, and others], Botany Special Collections, SPECIAL BOOKS 582 LIN 74, NHM.

107. Roxburgh to Banks, September 21, 1801, in Chambers, *Scientific Correspondence of Sir Joseph Banks*, vol. 4: 7.

108. Banks to Thunberg, September 13, 1787, G. 300, UUL.

109. On the French Revolution, see Topham, "Science, Print and Crossing Borders," 311–12; Watts, "Philosophical Intelligence."

110. Banks to Staunton, August 18, 1792, in Chambers, *Indian and Pacific Correspondence of Sir Joseph Banks*, 3:412.

111. Quoted in Lightman, McOuat, and Stewart, *Circulation of Knowledge*, 2–3.

112. Kitson, *Forging Romantic China*, 138; Goodman, *Planting the World*, 219–20.

113. Banks to Staunton, August 18, 1792, in Chambers, *Indian and Pacific Correspondence of Sir Joseph Banks*, 3:422.

114. Banks to Staunton, August 18, 1792, in Chambers, *Indian and Pacific Correspondence of Sir Joseph Banks*, 3:422–23.

115. Kaempfer, *History of Japan*.

116. Quoted in Goodman, *Planting the World*, 219–20.

117. Curtis and Banks, *Practical Observations on the British Grasses*.

118. For agricultural schools and Banks's involvement of the Board of Agriculture, see Nolan, "Agricultural Improvement in England and Wales," 143–144. For the foundation of the Horticultural Society, see Elliott, "Promotion of Horticulture," 122–29.

119. For Chinese tea, see Staunton, *Authentic Account of An Embassy*, 2:310. For Banks's production of Staunton's book, see Carter, *Sir Joseph Banks* (1988), 300–301.

120. William Bulmer, "Estimate of the Expense of Engraving . . .," SAFE/Banks Papers/Series 62.03, SLNSW. For Banks's receipt from Bulmer, see SAFE/Banks Papers/Series/62.04, SLNSW. Equivalent amount for £4,111/2/6 in 1797 could employ an artisanal worker for just over seventy-five years (27,407 days). See the UK National Archives currency convertor, https://www.nationalarchives.gov.uk/currency-converter/#currency-result. For contemporary purchasing power, see Hume, "Value of Money in Eighteenth-Century England."

121. Staunton, *Authentic Account of An Embassy*, preface.

122. "Sir George Staunton's Account of the Embassy to China," 251.

123. Berg, "Britain, Industry and Perceptions of China," 277.

124. J. Raven, *Judging New Wealth*, 29.

125. Klancher, *Making of English Reading Audiences*, 76–97; Johns, *Nature of the Book*, 355.

126. John Cochrane to David Pennant, December 2, 1813, CR2017/TP444, WRO; Pennant, *British Zoology* (1812).

127. Sutherland, "British Book Trade and the Crash of 1826."

128. Pennant, *British Zoology* (1818).

129. Bauer and Banks, *Strelitzia Depicta*.

130. Desmond, *Politics of Evolution*, 351.

131. See Fyfe, "Royal Society and the Free Circulation of Knowledge."

Chapter 5: The Use of Books

1. Richard D. Brown, *Knowledge Is Power*; Chartier, *Order of Books*; Brewer, *Pleasures of the Imagination*, 141–64; Topham, "Science, Print and Crossing Borders"; Topham, "Anthologising the Book of Nature."

2. For the eighteenth century as a transitional period, see Sher, *Enlightenment and the Book*. For changes in supply and demand, see St Clair, *Reading Nation in the Romantic Period*, 19–21. On the specialization of bookselling, see Jacobs, "Buying into Classes."

3. St Clair, *Reading Nation in the Romantic Period*, 103, 186–87.

4. Gorham, *Memoirs of John Martyn*, 173.

5. Plate titled "A Poisonous Fish Found on the Southern Coasts of Africa" is designed to identify this fish to avoid its consumption by ships' crews. Banks Copper Printing Plate Collection, NHM. An impression of this plate from Banks's collection can be found at Prints and Drawings, 1914,0520.752, BM.

6. For cinchona, see, Gänger, *Singular Remedy*; Crawford, *Andean Wonder Drug*; Jarco, *Quinine's Predecessor*.

7. "Account of a Book, Entitled, Gazophylacii Naturæ & Artis," 350; Kathleen Murphy, "James Petiver's 'Kind Friends.'"

8. For substitutes, see Chakrabati, "Empire and Alternatives," 81; Crawford, *Andean Wonder Drug*, 216; A. Thomas, "Establishment of Calcutta Botanic Garden"; Gänger, "World Trade in Medicinal Plants from Spanish America," 55.

9. Griffiths, *Print before Photography*, 326.

10. On Jones who supplied the copperplate, see "British Artist's Suppliers, 1650–1950–P," National Portrait Gallery, https://www.npg.org.uk/collections/research/programmes/directory-of-suppliers/suppliers-p. Two surviving impressions of this plate are known to exist. The first is held in the herbarium at the Royal Botanic Gardens, Kew, and is printed on steam-pressed laid paper. The

other is a later copy printed on Whatman's wove paper and held by the Hunt Institute for Botanical Research, Pittsburgh, HI1082. The latter has crude coloration, probably added in the nineteenth century.

11. On private plates, see Griffiths, *Print before Photography*, 326. On shadows, see Baxandall, *Shadows and Enlightenment*, 85; Daston and Galison, *Objectivity*, 63.

12. Banks's specimens of cinchona can be found in the Natural History Museum, London—for example, *Landenbergia macropiper*, sent by Casmiro Gómez Ortega in 1785, BM001008765; *Exostema angustifolium* (Sw.), sent by Olaf Swartz, BM000028154.

13. See Safier, *Measuring the New World*, 252–55; Condamine, "Sur L'Arbre du Quinquina."

14. Linnaeus, "*Critica Botanica*," 116–61; Daston and Galison, *Objectivity*, 59; Nickelsen, *Draughtsmen, Botanists and Nature*, 71–86.

15. Bleichmar, *Visible Empire*, 104; Nierto Olarte, "Remedies for the Empire," 90–91.

16. *Cinchona officinalis* (Herb Linn), Linnaean Herbarium, LINN 230.1, LSL. For the essential classificatory features, see Nickelsen, *Draughtsmen, Botanists and Nature*, 71–80.

17. Mutis to Linnaeus, September 24, 1764, in Smith, *Selection of Correspondence of Linnaeus*, 2:513; Linnaeus, *Species Plantarum* (1762), 244.

18. Joseph Banks to James Edward Smith, August 15, 1787, GB-119/JES/COR/1/51, LSL.

19. *Cinchona officinalis*, BM001008719, General Herbarium, NHM. According to a label written by William T. Stearn, the "binomial Cinchona officinalis L. was unaccompanied by any specific diagnosis when first published by Linnaeus (Sp. Pl. 1: 172; 1753) because the genus was monotype and hence the type of the specific name is that of the generic name Cinchona, for which the details given in his Genera Plantarum 5th ed. 79 n. 208 (1754) go back by way of the Genera 2nd ed. 527n. 1021 (1742) to the illustrations accompanying C. M. de la Condamine's paper 'Sur l'arbre de quinquina,' in Hist. Acad. Roy. Sci. Paris 1738: 226–243 pls 5–6 (1738). These illustrations thus represent the type-material from Loxa (Loja)."

20. Witteveen, "Supressing Synonymy with a Homonym," 140.

21. Miscellaneous botanical prints of cinchona, General Herbarium, RBGK; Banks Copper Printing Plate Collection, NHM.

22. On the global hunt for cinchona, see Crawford, "Empire's Extract"; Philip, "Imperial Science Rescues a Tree." On Wright, see Schiebinger, *Secret Cures of Slaves*, 10.

23. Wright, "Description of the Jesuits Bark Tree"; Schiebinger, *Secret Cures of Slaves*, 72. For Wright's experimental use of this species on the enslaved population, see: Wright, *Memoir of the Late William Wright*, 366, which contains direct reference to medical tests on enslaved people.

24. Wright to Banks, 1778, in Chambers, *Scientific Correspondence of Sir Joseph Banks*, 1:138.

25. For offprints, see Csiszar, *Scientific Journal*, 54; Chakrabati, "Empire and Alternatives," 75–94.

26. Everhard Home to Joseph Banks, October 15, 1783, in Chambers, *Scientific Correspondence of Sir Joseph Banks*, 2:177.

27. Anderson to Forsyth, March 1780, "Corresp with William Forsyth," Alexander Anderson, NRA 25004, RBGK.

28. "Officinal Plants," 615.10 HAW F CAT UNIT, Botany Special Collections, NHM.

29. Davidson, Monro, and Wilson, "Account of a New Species of the Bark-Tree."

30. Banks to Bacstrom, August 20, 1791, in Chambers, *Scientific Correspondence of Sir Joseph Banks*, 4:61.

31. For Bacstrom's annotations, see N. Thomas, "'Specimens of Bark Cloth'"; Bacstrom to Banks, August 18, 1791, in Chambers, *Scientific Correspondence of Sir Joseph Banks*, 3:269. For Bacstrom's participation in the Butterworth Squadron, see Pethick, *Nootka Connection*.

32. Bacstrom to Banks, August 18, 1791, in Chambers, *Scientific Correspondence of Sir Joseph Banks*, 3:270; Cole, "Sigismund Bacstrom's Northwest Coast Drawings," 64.

33. Banks to Bacstrom, August 20, 1791, in Chambers, *Scientific Correspondence of Sir Joseph Banks*, 4:61.

34. Banks to Bacstrom, August 20, 1791, in Chambers, *Scientific Correspondence of Sir Joseph Banks*, 4:61.

35. Bacstrom illustrations, WA MSS S-2405, Beinecke Rare Book and Manuscript Library, Yale University, New Haven, CT. For tea, see Hodacs, *Silk and Tea in the North*.

36. For tribes open to trading, see Berg, "Sea Otters and Iron," 53. For analysis of the clothes Cunnyha is wearing and a reproduction of Bacstrom's image, see Cole, "Sigismund Bacstrom's Northwest Coast Drawings," 70.

37. Lambert, *Description of the Genus Cinchona*, viii.

38. Bacstrom to Banks, November 18, 1796, in Chambers, *Indian and Pacific Correspondence of Sir Joseph Banks*, 4:422–24.

39. Rose, "Publishing Nature in the Age of Revolutions," 1156–57.

40. Müller-Wille, "Names and Numbers," 119–20.

41. Willdenow to Banks, September 27, 1799, Add. MS. 8099, 131, BL [original Latin]. I thank Robin Kreutel for assistance with the translation.

42. Willdenow to Banks, September 28, 1790, Add. MS. 8097, 357, BL [original Latin].

43. Willdenow to Banks, June 24, 1797, Add. MS. 8099, 130, BL [original Latin].

44. Rose, "From the South Seas to Soho Square," 510. *Systema Plantarum* (1779–1780) replaced Banks's interleaved copy of *Species Plantarum* (1762–1763) that had been annotated on the *Endeavour* voyage by Herman Spöring.

45. Linnaeus, *Systema Plantarum* (1779–1780) [interleaved and bound in ten volumes, containing annotations in the hands of Jonas Dryander, Samuel Törner, and others], Special Books 582 LIN 74, Botany Special Collections, NHM.

46. Thunberg, "Botanical Observations on the Flora Japonica." For one of Banks's copies of *Icones Selectæ Plantarum*, see Rare Books, 450.l.16, BL.

47. Stafleu, "Willdenow Herbarium: Herbarium Willdenow," 688; Müller Wille and Böhme, "'Jederzeit zu Diensten.'"

48. Willdenow to Banks, August 13, 1802, Add MS. 8099, 330, BL.

49. Smith donated volumes of Willdenow's *Species Plantarum* to the Linnean Society in 1799, 1800, 1801, 1802, 1805, 1806, 1811, and 1812. See L IV.753(797), LSL.

50. Linnaeus, *Species Plantarum*, Banks's interleaved copy, Special Books, 582 LIN, Botany Special Collections, NHM.

51. Linnaeus, *Species Plantarum*, interleaved and containing copious annotations by Robert Brown, L.IV.753(797), LSL.

52. Rose, "From the South Seas," 518.

53. Linnaeus, *Species Plantarum*, Banks's interleaved copy, annotation opposite 513, NHM.

54. Linnaeus, *Species Plantarum*, vol. 2, Banks's copy, opposite 42/1610, NHM. This activity is also exhibited by Robert Brown in his interleaved copy of Linné's *Species Plantarum*, L.IV.753(797), LSL. "Uicatier" translates to "vicariously."

55. Gage and Stearn, *Bicentenary History of the Linnean Society of London*, 18.

56. Banks to Fabbroni, February 4, 1785, in Chambers, *Scientific Correspondence of Sir Joseph Banks*, 2:20.

57. Linnaeus, *Species Plantarum* [Banks's copy], 2:779; Dryander also published on identifying synonyms. See Dryander, "In Genera and Species of Plants."

58. Witteveen, "Supressing Synonymy with a Homonym," 145; Witteveen and Müller-Wille, "Of Elephants and Errors."

59. Dietz, "Networked Names."

60. Linnaeus, *Species Plantarum* (1779–1780), Banks's interleaved copy, 1:513, NHM.

61. Banks, "Attempt to Ascertain the Time When the Potatoe Was First Introduced."

62. Banks, "Attempt to Ascertain the Time When the Potatoe Was First Introduced," 8–12; Linnaeus, *Species Plantarum*, vol. 1, Banks's copy, opposite 1033, NHM.

63. "Abolition of the Slave-Trade." I thank Oliver Ayers and Libby Collard for alerting me to this reference. See Ayres and Collard, *William Sancho.*

64. For Brown's copy, see Linnaeus, *Species Plantarum*, interleaved and containing copious annotations by Robert Brown, L.IV.753(797), LSL. Brown's discussion of his use of this book can be found in Vallance, Moore, and Groves, *Nature's Investigator*, 49; Mabberley, *Jupiter Botanicus*, 74; Moore, "Some Aspects of the Work Robert Brown."

65. Linné, *Species Plantarum* [Brown's copy], vol. 1, L.IV.753(797), opposite 1033, LSL.

66. See A. Secord, "Coming to Attention."

67. Margócsy, "'Refer to the Folio and Number,'" 65.

68. For Pennant's annotated copies of *British Zoology*, see Pennant, *British Zoology* (1768–1770), ZOOLOGY RBR 72A o PEN (set 1), Zoology Rare Books, NHM; Pennant, *British Zoology* (1776–1777) [typescript, with holograph annotations. Pennant's own working copy, heavily edited. All volumes contain loose notes, drawings, and correspondence addressed to Pennant's son David, who prepared the fifth posthumous edition of his father's work], MSS PEN A, Zoology Special Collections, NHM.

69. Pennant's annotated copy of *British Zoology* (1776–1777), vol. 1, MSS PEN A, first endpapers, Zoology Manuscripts, NHM.

70. For the role of Benjamin White, see chapters 3 and 4.

71. Pennant's annotated copy of *British Zoology* (1776–1777), vol. 1, MSS PEN A, flyleaf between 396 and 397, Zoology Manuscripts, NHM.

72. Pennant, *Arctic Zoology*, 1: iv–v.

73. Pennant's annotated copy of *British Zoology* (1776–1777), vol. 1, MSS PEN A, Zoology Manuscripts, NHM, notes on the first endpapers.

74. See Fyfe and Moxham, "Royal Society"; Dawson et al., *Science Periodicals*; Csiszar, *Scientific Journal.*

75. Pennant's annotated copy of *British Zoology* (1776–1777), MSS PEN A,

1:219, Zoology Manuscripts, NHM, shows annotations by Latham and David Pennant.

76. For Scopoli and White, see P. Foster, "Gibraltar Collections."

77. Latham to Pennant, November 15, 1812, CR2017/TP476/4, WRO.

78. Pennant's annotated copy of *British Zoology* (1776–1777), vol. 1, MSS PEN A, between 326 and 327, Zoology Manuscripts, NHM.

79. Pennant, *British Zoology* (1812), 444.

80. Kaempfer, *The History of Japan*, John Martyn's copy, CCA.47.31–32, 1:112–13, CUL.

81. Pennant, *British Zoology* (1768–1772), ZOOLOGY RBR 72A o PEN (set 1), 2:358, Zoology Rare Books, NHM.

82. Thomas Pennant, "Notes by Thomas Pennant on Zoology, Ornithology, etc., and an Account of Expenses Incurred in Connection with His Study of Natural History," MS 2550B, NLW.

83. Thomas Pennant [and others], "Plants of Africa," MS 2552B, n.p., NLW.

84. Pennant's annotated copy of *British Zoology* (1776–1777), MSS PEN A, 3:66–67, Zoology Manuscripts, NHM.

85. Pennant, *British Zoology* (1812), 3:87.

86. On teeth, see Di Gregorio, "In Search of the Natural System."

87. On Lacepède, see Emma C. Spary, "On the Ironic Specimen of the Unicorn Horn,"1044–45. La Cépède, *Histoire Naturelle des Cétacées*, 422.

88. Joseph Banks to David Pennant, October 20, 1811, in Pennant's annotated copy of *British Zoology* (1776–1777), MSS PEN A, 3:4–5, Zoology Manuscripts, NHM.

89. For handbills and ephemera, see Russell, *Ephemeral Eighteenth Century*, chap. 3; Fumerton, *Broadside Ballad in Early Modern England*; Murphy and O'Droscoll, *Studies in Ephemera*.

90. Pennant's annotated copy of *British Zoology* (1776–1777), MSS PEN A, 3:48–49, Zoology Manuscripts, NHM.

91. Pennant's version of this broadside is in a private collection. Another example can be found at Prints and Drawings, 1868,0808.5803, BM.

92. Examples include the women discussed by Thompson, "Women Travellers."

93. Pennant's annotated copy of *British Zoology* (1776–1777), MSS PEN A, 3:158–59, Zoology Manuscripts, NHM. The annotation is on a slip of marbled endpaper that has been torn out of another book or notebook. Matheson, "Thomas Pennant and the Morris Brothers."

94. Hamilton to Pennant, undated, c. 1811, in Pennant's annotated copy of

British Zoology (1776–1777), MSS PEN A, 3:242–43, Zoology Manuscripts, NHM.

95. Pennant, *British Zoology* (1776–1777), vol. 3, end of the volume, n.p.

96. William Eyres to Thomas Pennant, August 30, 1775, CR2017/TP154/4, WRO.

97. Thomas Martyn to John Strange, April 8, 1782, Egerton MS 1970, fols. 80r–80v, BL.

98. J. J. Rousseau, *Letters on the Elements of Botany Addressed to a Lady*, vii.

99. Pennant, *British Zoology* (1776–1777), 1:44–45, 1:359, private collection.

100. Hayes, "'Natural' Exhibitioner"; King, "New Evidence for the Contents of the Levarian Museum." Pennant's admission ticket is at CR2017/TP590/2, WRO. Pennant had an active interest in acquiring Lever's collection and purchased several lottery tickets in 1784. See CR2027/TP280/4, WRO.

101. John White to Gilbert White, April 6, 1774, in Holt-White, *Life and Letters of Gilbert White*, 1:251.

102. For Blackburne's copy of *British Zoology* (1776–1777), see chapter 1.

103. Way's copy of Pennant's *British Zoology* (1768–1771) can be found at Rare Books, shelf mark 7000.c.417–420, interleaved pages opposite 393, 420, 460, CUL.

104. For natural theology in universities, see Fyfe, "Reception of William Paley's Natural Theology."

105. Pennant, *British Zoology* (1812), x.

106. White to Churton, August 4, 1788, in Holt-White, *Life and Letters of Gilbert White*, 2:185; Gilbert White to Benjamin White, February 1788, in Holt-White, *Life and Letters of Gilbert White*, 2:178.

107. Quoted in Holt-White, *Life and Letters of Gilbert White*, 2:187.

108. Pennant's copy of White's *Natural History and Antiquities of Selborne* (1789) can be found at *EC75.W5834.798n (B), HLH.

109. White to Churton, August 4, 1788, in Holt-White, *Life and Letters of Gilbert White*, 2:185.

110. John Mulso to Gilbert White, December 15, 1788, in Holt-White, *Life and Letters of Gilbert White*, 2:190.

111. George Montagu to Gilbert White, May 21, 1789, in Holt-White, *Life and Letters of Gilbert White*, 2:198.

112. "Art V," 34.

113. White to Churton, May 20, 1789, in Holt-White, *Life and Letters of Gilbert White*, 2:196.

114. G. White, *Natural History and Antiquities of Selborne* (1822), Nor-

man's interleaved copy can be found at STORE 225:16–17, Whipple Library, Cambridge; Herbert George Henry Norman to Charles Darwin, November 30, 1866, in Burkhardt, *Correspondence of Charles Darwin*, 11: 401.

115. St Clair, *Reading Nation in the Romantic Period.*

116. John Latham to Thomas Pennant, May 21, 1822, MS 2591 E, NLW.

117. John Latham to David Pennant, November 30, 1823, MS 2591 E, NLW.

Conclusion

1. *London Morning Post*, February 17, 1816, 14065.

2. Maurice, *Observations on the Remains of Ancient Egyptian Grandeur and Superstition*, advertisement section on the final endpaper.

3. *London Morning Post*, April, 17, 1816, 14116.

4. On global collecting and natural systems, see Endersby, *Imperial Nature*, 213–15; Stevens, "Development of Biological Systematics"; A. Secord, "Coming to Attention."

5. For paper technologies, see Müller-Wille and Charmantier, "Natural History and Information Overload"; Charmantier and Müller-Wille, "Carl Linnaeus's Botanical Paper Slips"; Müller-Wille, "Names and Numbers"; te Heesen, *World in a Box*; Blair, *Too Much to Know*; Yeo, *Notebooks, English Virtuosi, and Early Modern Science*. For field science, see MacGregor, *Naturalists in the Field.*

6. See, for example, MacGregor, *Naturalists in the Field*; te Heesen, "Boxes in Nature"; Neve and Porter, "Alexander Catcott"; Bleichmar, *Visual Voyages*; Bleichmar, *Visible Empire*; Mariss, "Library in the Field"; Mariss, *Johann Reinhold Forster.*

7. Parkinson, *Journal of a Voyage to the South Seas*, 18.

8. Anne Salmond, *Trial of the Cannibal Dog*, 69.

9. R. Porter, "Gentlemen and Geology"; Shapin, *Social History of Truth.*

10. Pennant, *Literary Life*, 39. For a chronological account of those who worked on Banks's collection between 1771 and 1820, see Rose, "From the South Seas to Soho Square," 501.

11. Joseph Banks to Everhard Home, October 22, 1810, Dawson Turner Correspondence, vol. 18, fols. 88–89, NHM.

12. Darnton, *Literary Tour De France*; McKendrick, Brewer, and Plumb, *Birth of a Consumer Society.*

13. Banks to Somerset, August 21, 1816, D/RA/A/13/3/22, Buckinghamshire Archives, Aylesbury.

14. Pomian, *Collectors and Curiosities*, 9.

15. Darnton "What Is the History of Books?," 112; Topham, "Science, Print, and Crossing Borders."

16. Adams and Barker, "New Model for the Study of the Book," 14–15.

17. This view on the value of unpublished proof prints persisted into the nineteenth century. See Prescott, "Faraday."

18. Darnton, "What Is the History of Books?," 72; Turnovsky, "Enlightenment Literary Market."

19. Pennant, *Histoire Naturelle des Oiseaux*, final unpaginated page titled "A List of Mr. Pennant's Works."

20. Johns, "How to Acknowledge a Revolution"; J. Raven, *Business of Books*," 124.

21. Pennant, *British Zoology* (1812), 4:vii.

22. On the increased role of booksellers, see St Clair, *Reading Nation in the Romantic Period*, 166. On the crash, see Sutherland, "British Book Trade," 151.

23. On the foundation of societies, see Morrell and Thackray, *Gentlemen of Science*, 318; Platts, "In Celebration of the Ray Society," 6; Curle, *Ray Society*.

24. For the Paleontographical Society, see Allen, *Naturalist in Britain*, 88, 117–18.

25. On the donation of Pennant's collection to the British Museum, see "Correspondence between Sir Sidney Harmer and the Countess of Denbigh Concerning the Presentation of Thomas Pennant's Collections to the Museum in 1912: The Specimens Received Included Mammals, Birds, Shells and Other Invertebrates, Minerals and Fossils," DF 969, Archives, NHM.

26. Quoted in McOuat, "Cataloguing Power."

27. Pennant, *Literary Life*: 135; Banks to Hamilton, November 20, 1792, in Chambers, *Scientific Correspondence of Sir Joseph Banks*, 4:171; White to Marsham, January 2, 1793, in Holt-White, *Life and Letters of Gilbert White*, 2:255.

Bibliography

Manuscript Sources

British Library, London
British Museum, London
Cambridge University Library
Houghton Library, Harvard University, Cambridge, MA
Linda Hall Library, Kansas City, MO
Linnean Society of London, London
Lewis Walpole Library, Yale University, Farmington, CT
McGill University Library, Montreal
Natural History Museum, London
National Library of Australia, Canberra
National Library of Scotland, Edinburgh
National Library of Wales, Aberystwyth
Royal Botanic Gardens, Kew, London
State Library of New South Wales, Sydney
Uppsala University Library
Warwickshire Records Office, Warwick

Printed Sources

"III. Reliquiæ Houstouianæ: seu Plantarum in America meridionali a Gulielmo Houstoun, M. D. R. S. S. collectarum Icones manu propria ære ineisæ; cum descriptionibus e schedis ejusdem in bibliotheca Josephi Banks, Baroneti, R. S. P. asservatis." *London Medical Journal* 5 (1784): 253–56.

"Abolition of the Slave-Trade; with a Sketch of the Life of Ignatius Sancho, the ingenious African, and Father of Mr. W. Sancho, the Bookseller, Mews Gate." *New, Original and Complete Wonderful Museum and Magazine Extraordinary* 5 (1807): 2629–30.

Adams, Brian. *The Flowering Pacific: Being an Account of Joseph Banks' Travels in the South Seas and the Story of His Florilegium*. London: HarperCollins, 1986.

Adams, Thomas R., and Nicholas Barker. "A New Model for the Study of the Book." In *A Potencie of Life: Books in Society*, edited by Nicholas Barker, 5–43. London: British Library, 2001.

Agnarsdóttir, Anna, ed. *Sir Joseph Banks, Iceland and the North Atlantic, 1772–1820. Journals, Letters and Documents*. London: Taylor & Francis, 2017.

Aiton, Wiliam Townshend. *Delineations of Exotick Plants Cultivated in the Royal Gardens at Kew: Drawn and Coloured, and the Botanical Characters Displayed According to the Linnaean System*. London: William Bulmer for George Nichol, 1796.

Aiton, William Townshend. *An Epitome of the Second Edition of Hortus Kewensis, for the Use of Practical Gardeners; To Which Is Added of Esculent Vegetables and Fruits Cultivated in the Royal Gardens at Kew*. Vol. 6. London: Longman et al., 1814.

Allan, Alexander. *Views of Mysore Country*. London: Private Press, 1794.

Allen, David. *The Naturalist in Britain: A Social History*. London: Penguin Books, 1978.

Altick, Richard D. *The English Common Reader: A Social History of the Mass Reading Public, 1800–1900*. Chicago: University of Chicago Press, 1967.

"An Account of a Book, Entitled, Gazophylacii Naturæ & Artis." *Philosophical Transactions* 27 (1710–1712): 342–52.

Anson, George. *A Voyage Round the World, in the Years M DCC XL, I, II, III, IV*. London: Paul Knapton, 1748.

"Art V. The Natural History and Antiquities of Selborne, in the County of Southampton: With Engravings, and an Appendix." *Monthly Review or Literary Journal* 81 (1789): 33–40.

"Art VI. British Zoology. By Thomas Pennant, Esq; Vol. IV." *Monthly Review or Literary Journal* 57 (1777): 275–77.

Appel, Toby A. *The Cuvier-Geoffrey Debate: French Biology in the Decades before Darwin*. New York: Oxford University Press, 1987.

Ayres, Oliver, and Liberty Collard, *William Sancho: Bookseller at the Mews Gate*. Cambridge: Cambridge University Press, 2025.

Bales, Melissa. "Literary Plagiarism and Scientific Originality in the 'Trans-Atlantic Wilderness' of Goldsmith, Aikin and Barbauld." *Eighteenth-Century Studies* 49, no. 2 (2016): 265–79.

Bacon, Francis. *Sylvia Sylvarum, or a Natural History in Ten Centuries*. London: William Lee, 1670.

Baker, S., and G. Leigh. *A Catalogue of the Valuable Library of Philip Miller, F. R. S. and Gardener to the Botanic Garden at Chelsea Lately Deceased; Containing a good Collection of Miscellaneous Books, and a fine collection of Books in Natural History.* London: Baker and Leigh, 1774.

Balston, John. *The Whatmans and Wove Paper: Its Invention and Development in the West.* West Farleigh, UK: J. N. Balston: 1998.

[Banks, Joseph.] "Account of an Ancient Canoe Found in Lincolnshire. Transmitted to the Editor by the Right Hon. Sir Joseph Banks, Bart &c." *Journal of Science and the Arts, Edited at the Royal Institution of Great Britain* 1 (1816): 244–45.

Banks, Joseph. "An Attempt to Ascertain the Time When the Potatoe (*Solanum tuberosum*) was First Introduced into the United Kingdom." *Transactions of the Horticultural Society of London* 1 (1812): 147–56.

Banks, Joseph. *A Short Account of the Causes of the Diseases in Corn, called by Farmers the Blight, the Mildew, and the Rust.* London: H. D. Symonds, 1805.

Banks, Joseph. *Banks' Florilegium: A Publication in Thirty-Four Parts of Seven Hundred and Thirty-Eight Copperplate Engravings of Plants Collected on Captain James Cook's First Voyage Round the World in H.M.S. Endeavour 1768–1771.* London: Alecto Historical Editions in Association with the British Museum (Natural History), 1980–1990.

Banks, Joseph, and Joseph D. Hooker, eds. *Journal of the Right Hon. Sir Joseph Banks Bart. K. B., F. R. S.* London: Macmillan, 1896.

Barrington, Daines. *The Naturalist's Journal.* London: W. Sandby, 1767.

Batsaki, Yota, Sarah Burke Calahan, and Anatole Tchikne, eds. *The Botany of Empire in the Long Eighteenth Century.* Washington, DC: Dumbarton Oaks, 2016.

Batteux, C. et al., ed. *Mémoires concernant l'histoire, les sciences, les mœurs, les usages, &c. des Chinois: par les Missionaires de Pekin.* Paris: Chez Nyon, 1776–1791.

Bauer, Ferdinand, and Joseph Banks. *Strelitzia Depicta: Or Coloured Figures of the Known Species of the Genus Strelitzia from the Drawings in the Banksian Library.* London: Printed for Messrs. Arch. At the Lithographic Press of Moser and Harris, 1818.

Baxandall, Michael. *Shadows and Enlightenment.* New Haven, CT: Yale University Press, 1997.

Bayly, C. A. *The Birth of the Modern World, 1780–1914.* Oxford: Blackwell, 2004.

Bayly, C. A. *Imperial Meridian: The British Empire and the World, 1780–1830.* London: Longman, 1989.

Bayley, Harold. *The Tragedy of Sir Francis Bacon: An Appeal for Further Investigation and Research*. London: G. Richards, 1902.

Bellégo, Marine. "Delineating a Utopian Space: The Borders of the Calcutta Botanic Garden in the 19th Century." In *Reading(s)/Across/Borders: Studies in Anglophone Borders Criticism*, edited by Ciaran Ross, 55–71. Leiden: Brill, 2020.

Bellégo, Marine. *Enraciner l'empire: une autre histoire du jardin botanique de Calcutta (1860–1910)*. Paris: Muséum national d'histoire naturelle, 2021.

Ben-Amos, Ilana Krausman. *The Culture of Giving: Informal Support and Gift-Exchange in Early Modern England*. Cambridge: Cambridge University Press, 2008.

Bentley, G. E. "Blake's Heavy Metal: The History of Weight, Uses, Cost, and Makers of His Copperplates." *University of Toronto Quarterly* 76 (2007): 714–70.

Benton, Josiah H. *John Baskerville: Type-Founder and Printer, 1706–1775*. Boston: Private Press, 1914.

Beretta, Marco, and Alessandro Tosi, eds. *Linnaeus in Italy: the Spread of a Revolution in Science*. Sagamore Beach, MA: Science History Publications, 2007.

Berg, Maxine. *The Age of Manufactures, 1700–1920: Industry, Innovation, and Work in Britain*. 2nd ed. London: Routledge, 1994.

Berg, Maxine. "Britain, Industry and Perceptions of China: Matthew Boulton, 'Useful Knowledge' and the Macartney Embassy to China 1792–94." *Journal of Global History* 1 (2006): 269–88.

Berg, Maxine. *Luxury and Pleasure in Eighteenth-Century Britain*. Oxford: Oxford University Press, 2005.

Berg, Maxine. "Sea Otters and Iron: A Global Microhistory of Value and Exchange at Nootka Sound, 1774–1792." *Past & Present* 242, no. 14 (2019): 50–82.

Berkenhout, John. *Outlines of the Natural History of Great Britain and Ireland*. London: P. Elmsly, 1769–1772.

Bewell, Alan. "Erasmus Darwin's Cosmopolitan Nature." *ELH* 76, no. 1 (2009): 19–48.

Bewell, Alan. "'On the Banks of the South Sea': Botany and Sexual Controversy in the Late Eighteenth Century." In *Visions of Empire: Voyages, Botany and Representations of Nature*, edited by David Philip Miller and Peter Hans Reill, 173–95. Cambridge: Cambridge University Press, 1998.

Biagoli, Mario. *Galileo, Courtier: The Practice of Science in the Culture of Absolutism*. Chicago: University of Chicago Press, 1993.

Bil, Geoff. "Tangled Compositions: Botany, Agency, and Authorship Aboard HMS *Endeavour*." *History of Science* 60, no. 22 (June 2022): 1–28.

Bittel, Carla, Elaine Leong, and Christine von Oertzen. *Working with Paper: Gendered Practices in the History of Knowledge*. Pittsburgh: University of Pittsburgh Press, 2019.

Black, Jeremy. *The British and the Grand Tour*. London: Croom Helm, 1985.

Blair, Ann. *Too Much to Know: Managing Scholarly Information before the Modern Age*. New Haven, CT: Yale University Press, 2010.

Bleichmar, Daniella. "Exploration in Print: Books and Botanical Travel from Spain to the Americas in the Late Eighteenth-Century." *Huntington Library Quarterly* 70, no. 1 (2007): 129–51.

Bleichmar, Daniella. *Visible Empire: Botanical Expeditions and Visual Culture in the Hispanic Enlightenment*. Chicago: University of Chicago Press, 2012.

Bleichmar, Daniella. *Visual Voyages: Images of Latin American Nature from Columbus to Darwin*. Pasadena, CA: Huntingdon Library Press, 2017.

Boyle, Robert. "Other Enquiries Concerning Sea." *Philosophical Transactions* 1, no. 18 (1666): 315.

Blunt, Wilfred. *The Art of Botanical Illustration*. London: Collins, 1950.

Bolyanatz, Alexander H. *Pacific Romanticism: Tahiti and the European Imagination*. Westport, CT: Greenwood, 2004.

Bordieu, Pierre, "The Field of Cultural Production." In *The Book History Reader*, edited by David Finkelstein and Alistair McCleery, 77–99. London: Routledge, 2002.

Brener, Ellke, Ranven Kunstmann, Pryasha Mukhapadhyay, and Asha Rogers, eds. *The Global Histories of Books: Methods and Practices*. London: Palgrave Macmillan, 2017.

Brewer, John. *The Pleasures of the Imagination: English Culture in the Eighteenth Century*. London: HarperCollins, 1997.

Bridson, Gavin D. R. "The Treatment of Plates in Bibliographical Description." *Journal of the Society for the Bibliography of Natural History* 7 (1976): 469–88.

Bridson, Gavin D. R., Donald E. Wendel, and James J. White. *Printmaking in the Service of Botany: Catalogue of an Exhibition*. Pittsburgh: Hunt Institute for Botanical Documentation, 1986.

Briggs, C. Stephen. "Thomas Pennant: Some Working Practices of an Archaeological Travel Writer in Late Eighteenth-Century Britain." In *Enlightenment Travel and British Identities: Thomas Pennant's Tours in Scotland and Wales*, edited by Mary-Ann Constantine and Nigel Leask, 41–64. London: Anthem Press, 2017.

"British Zoology, *Vol. IV. By* Thomas Pennant, *Esq.* 8vo. 1l. 1s. White." *Critical Review or Annals of Literature* 44 (1777): 158–59.

Britten, James. "Francis Bauer's 'Delineations of Exotick Plants." *Journal of Botany, British and Foreign* 37, no. 9 (1899): 181–83.

Britten, James. *Illustrations of the Botany of Captain Cook's Voyage Round the World in HMS Endeavour in 1768–71.* London: Printed by order of the Trustees of the British Museum, 1900.

Broman, Thomas. "The Habermasian Public Sphere and "Science in the Enlightenment." *History of Science* 36 (1998): 123–50.

Brooks, Justin. "Imperial Structures, Indigenous Aims: Connecting Native Engagement in Scotland, North America and South Asia." In *Facing Empire: Indigenous Experiences in a Revolutionary Age*, edited by Kate Fullagar and Michael McDonnell, 281–302. Baltimore: Johns Hopkins University Press, 2018.

Brown, Peter D. "Bute in Retirement." In *Lord Bute: Essays in Reinterpretation*, edited by Karl W. Schweizer, 241–274. Leicester: Leicester University Press, 1988.

Brown, Richard D. *Knowledge Is Power: The Diffusion of Information in Early America, 1700–1865.* Oxford: Oxford University Press, 1989.

Browne, Janet. "Botany for Gentlemen: Erasmus Darwin and the *Loves of the Plants.*" *Isis* 80, no. 4 (1983): 593–621.

Browne, Janet. *The Secular Ark: Studies in the History of Biogeography.* New Haven, CT: Yale University Press, 1983.

Browne, Patrick. *The Civil and Natural History of Jamaica.* London: Benjamin White, 1789.

Bruce, James. *Travels to Discover the Source of the Nile: In the Years 1768, 1769, 1770, 1771, 1772 and 1773.* Edinburgh: G. G. J. and J. Robinson, 1790.

de Buffon, G. L. L., and Bernard Germain de Lacépède, *Histoire Naturelle énérale et particulière, avec la description du Cabinet du Roi.* Paris: De L'Imprimerie Royale, 1749–1804.

Birkhead, Tim, Paul J. Smith, Megan Doherty, and Isabelle Charmantier. "Willoughby's Ornithology." In *Virtuoso by Nature: The Scientific Worlds of Francis Willoughby FRS (1635–1672)*, edited by Tim Birkhead, 268–304. Leiden: Brill, 2016.

Burkhardt, Frederick, et al., eds. *The Correspondence of Charles Darwin.* Vol. 14. Cambridge: Cambridge University Press, 2004.

Burnett, Graham D. *Masters of All They Surveyed: Exploration, Geography, and a British El Dorado.* Chicago: University of Chicago Press, 2000.

Bynum, W. F., and Roy Porter, eds. *William Hunter and the Eighteenth-Century Medical World*. Cambridge: Cambridge University Press, 1985.

Cain, P. J., and A. G. Hopkins. *British Imperialism, 1688–2015*. 3rd ed. London: Routledge, 2016.

Calhoun, Craig, ed. *Habermas and the Public Sphere*. Cambridge: MIT Press, 1992.

Campbell, Jessie. "The Banker." In *Dawson Turner: A Norfolk Antiquary and His Remarkable Family*, edited by Nigel Goodman, 111–122. Chichester: Phillimore, 2007.

Cantor, Geoffrey. "The Rise and Fall of Emmanuel Mendes da Costa: A Severe Case of 'The Philosophical Dropsy?'" *English Historical Review* 116 (2001): 584–603.

Caputo, Sara. "Alien Seamen in the British Navy, British Law, and the British State, c. 1793–c. 1815." *Historical Journal* 62, no. 3 (2019): 685–707.

Carey, Brycchan. "The Literary Gilbert White." In *Birds in Eighteenth-Century Literature: Reason, Emotion, and Ornithology, 1700–1840*, edited by Brycchan Carey, Sayre Greenfield, and Anne Milne, 173–92. London: Palgrave Macmillan, 2020.

Carter, H. B., J. A. Diment, C. J. Humphries, and A. Wheeler. "The Banksian Natural History Collections of the *Endeavour* Voyage and their Relevance to Modern Taxonomy." *Archives of Natural History* 1 (1981): 61–70.

Carter, Harold B. *Sir Joseph Banks (1743–1820): A Guide to Biographical and Bibliographical Sources*. London: St Paul's Bibliographies in association with the British Museum (Natural History), 1987.

Carter, Harold B. *Sir Joseph Banks, 1743–1820*. London: British Museum (Natural History), 1988.

Cavagna, Anna Giulia. "A Free Transmission of Knowledge: The Literary Gifts and Reception of an Eighteenth-Century Scholar." In *Free Print and Non-Commercial Publishing since 1700*, edited by James Raven, 29–47. Farnham, UK: Ashgate, 2000.

Chakrabati, Pratik. "Empire and Alternatives: *Swietenia febrifuga* and the Cinchona Substitutes." *Medical History* 54 (2010): 75–94.

Chambers, Neil. *Endeavouring Banks: Exploring Collections from the Endeavour Voyage, 1768–1771*. London: Paul Holberton Publishing, 2016.

Chambers, Neil, ed. *The Indian and Pacific Correspondence of Sir Joseph Banks, 1768–1820*. 6 vols. London: Pickering & Chatto, 2009.

Chambers, Neil. *Joseph Banks and the British Museum: The World of Collecting, 1770–1830*. London: Pickering and Chatto, 2007.

Chambers, Neil, ed. *The Letters of Sir Joseph Banks: A Selection, 1768–1820.* London: Imperial College Press, 2000.

Chambers, Neil, ed. *The Scientific Correspondence of Sir Joseph Banks, 1765–1820.* 6 vols. London: Pickering & Chatto, 2006.

Chapman, R. W., ed. *Johnson's Journey to the Western Islands of Scotland: And, Boswell's Journal of a Tour to the Hebrides with Samuel Johnson.* Oxford: Oxford University Press, 1970.

Charmantier, Isabelle. "Notebooks, Files and Slips: Carl Linnaeus and His Disciples at Work." In *Linnaeus, Natural History and the Circulation of Knowledge,* edited by Hanna Hodacs, Kenneth Nyberg, and Stéphane Van Damme, 25–58. Oxford: Voltaire Foundation, 2018.

Charmantier, Isabelle, and Staffan Müller-Wille. "Carl Linnaeus's Botanical Paper Slips (1767–1773)." *Intellectual History Review* 24 (2014): 215–38.

Chartier, Roger. "Labourers and Voyagers: From the Text to the Reader." In *The Book History Reader,* edited by David Finkelstein and Alistair McCleery, 47–58. London: Routledge, 2002.

Chartier, Roger. *The Order of Books: Readers, Authors and Libraries in Europe between the Fourteenth and Eighteenth Centuries.* Stanford, CA: Stanford University Press, 1992.

Clark, J. C. D. *English Society, 1688–1832: Ideology, Social Structure, and Political Practice during the Ancien Régime.* Cambridge: Cambridge University Press, 1985.

Clark, J. C. D. *Revolution and Rebellion: State and Society in England in the Seventeenth and Eighteenth Centuries.* Cambridge: Cambridge University Press, 1986.

Clarke, Stephen. *The Strawberry Hill Press and Its Printing House: An Account and Its Iconography.* New Haven, CT: Yale University Press, 2011.

Clayton, Timothy. *The English Print.* New Haven, CT: Yale University Press, 1997.

Coen, Deborah R. *Climate in Motion: Science, Empire, and the Problem of Scale.* Chicago: University of Chicago Press, 2018.

Colclough, Stephen. "Pocket Books and Portable Writing: The Pocket Memorandum Book in Eighteenth-Century England and Wales." *Yearbook of English Studies* 45 (2015): 159–77.

Cole, Douglas. "Sigismund Bacstrom's Northwest Coast Drawings and an Account of His Curious Career." *British Columbia Studies* 46 (1980): 61–86.

Colley, Linda. "The Apotheosis of George III: Loyalty, Royalty and the British Nation, 1760–1820." *Past and Present* 102 (1984): 94–129.

Colley, Linda. *Britons: Forging the Nation, 1707–1837.* London: Pimlico, 2003.

Condamine, Charles Marie de la. "Sur L'Arbre du Quinquina." *Mémoires de l'Académie royale des sciences de Paris.* (1738): 226–43.

Constantine, Mary-Ann. "Thomas Pennant, Selected Works (1754–1804)." In *Handbook of British Travel Writing,* edited by Barbara Schaff, 199–212. Berlin: De Gruyter, 2020.

Constantine, Mary-Ann, and Nigel Leask, eds. *Enlightenment Travel and British Identities: Thomas Pennant's Tours of Scotland and Wales.* London: Anthem Press, 2017.

Cook, Alexandra. "Linnaeus and Chinese Plants: A Test of the Linguistic Imperialism Thesis." *Notes and Records of the Royal Society* 64 (2010): 121–38.

Cook, Harold J. *Matters of Exchange: Commerce, Medicine, and Science in the Dutch Golden Age.* New Haven, CT: Yale University Press, 2007.

Cook, James. *Captain Cook's Journal during his First Voyage Round the World, Made in H. M. Bark Endeavour, 1786–71, a Literal Transcription of the Original MSS.* Edited by W. J. L. Wharton. Cambridge: Macmillan and Co., 1893.

Cooper, Alix. *Inventing the Indigenous: Local Knowledge and Natural History in Early Modern Europe.* Cambridge: Cambridge University Press, 2009.

Cordiner, Charles. *Antiquities & Scenery of Northern Scotland, in a Series of Letters to Thomas Pennant Esq.* London: n.p., 1780.

Crawford, Matthew James. *The Andean Wonder Drug: Cinchona Bark and Imperial Science in the Spanish Atlantic, 1630–1800.* Pittsburgh: University of Pittsburgh Press, 2016.

Crawford, Matthew James. "Empire's Extract: Chemical Manipulations of Cinchona bark in the Eighteenth-Century Spanish Atlantic World." *Osiris* 29, no. 1 (2014): 215–29.

Csiszar, Alex. *The Scientific Journal: Authorship and the Politics of Knowledge in the Nineteenth Century.* Chicago: University of Chicago Press, 2018.

Curle, Richard. *The Ray Society: A Bibliographical History.* London: Ray Society, 1954.

Curtis, William, and Joseph Banks. *Practical Observations on the British Grasses, Especially Such as Are Best Adapted to the Laying Down or Improving of Meadows and Pastures: Likewise, an Enumeration of the British Grasses, to Which Is Now Added a Short Account of Diseases in Corn, Called by Farmers the Blight, the Mildrew, and the Rust.* London: H. D. Symonds and Curtis, 1805.

Dadswell, Ted. *The Selborne Pioneer: Gilbert White as Naturalist and Scientist.* Farnham, UK: Ashgate, 2002.

Dampier, William. *A New Voyage Round the World.* London: James Knapton, 1697.

Damrosch, Leopold. "Gilbert White of Selborne: Enlightenment Science and Conservative Ideal." *Studies in Burke and His Time* 19, no. 1 (1978): 29–46.

Daniell, Thomas, and William Daniell. *Oriental Scenery: One Hundred and Fifty Views of the Architecture, Antiquities, and Landscape Scenery of Hindoostan.* London: Private Press, 1796.

Darnton, Robert. *The Kiss of Lamourette: Reflections in Cultural History.* New York: W. W. Norton, 1990.

Darnton, Robert. *A Literary Tour de France: The World of Books on the Eve of the French Revolution.* Oxford: Oxford University Press, 2018.

Darnton, Robert. "What Is the History of Books?" *Daedalus* 111, no. 3 (1982): 65–83.

Darnton, Robert. "What Is the History of Books? Revisited." *Modern Intellectual History* 4, no. 3 (2007): 495–508.

Daston, Lorraine. "Taking Note(s)." *Isis* 95, no. 3 (2004): 443–48.

Daston, Lorraine, and Peter Galison. *Objectivity.* New York: Zone Books, 2007.

Davidson, George, Donald Monro, and George Wilson. "An Account of a New Species of the Bark-Tree, Found in the Island of St. Lucia." *Philosophical Transactions* 74 (1784): 452–56.

Dawson, Gowan, Bernard Lightman, Sally Shuttleworth, and Jonathan R. Topham, eds. *Science Periodicals in Nineteenth-Century Britain: Constructing Scientific Communities.* Chicago: University of Chicago Press, 2020.

Dawson, Warren R., ed. *The Banks Letters: A Calendar of the Manuscript Correspondence of Sir Joseph Banks Preserved in the British Museum (Natural History) and Other Collections in Great Britain.* London: British Museum, 1958.

de Buffon, G. L. L., and Bernard Germain de Lacépède. *Histoire Naturelle énérale et particulière, avec la description du Cabinet du Roi.* Paris: De L'Impimerie Royale, 1749–1804.

Décultot, Élisabeth. "Between Reading and Writing: Manuscript Collections of Excerpts in Eighteenth-Century Germany." In *Personal Manuscripts: Copying, Drafting, Taking Notes,* edited by David Durand-Guédy and Jürgen Paul, 85–115. Berlin: De Gruyter, 2023.

Décultot, Élisabeth. "Reading *versus* Seeing? Winckelmann's Excerpting Practice and the Genealogy of Art History." *Ber Wissenschaftsgesch* 43 (2020): 239–61.

Delbourgo, James. *Collecting the World: The Life and Curiosity of Hans Sloane.* London: Allen Lane, 2017.

Delbourgo, James, and Nicholas Dew. *Science and Empire in the Atlantic World.* London: Routledge, 2008.

Delbourgo, James, and Staffan Müller-Wille. "Introduction: Listmania." *Isis* 1, no. 3 (2012): 710–15.

Desmond, Adrian. *The Politics of Evolution: Morphology, Medicine, and Reform in Radical London*. Chicago: University of Chicago Press, 1989.

Dew, W. M. & Son. *Catalogue of Sale of the Remainder of the Downing Library (Formed by Thomas Pennant, the Well-Known Antiquary and Naturalist, 1726–98, and Augmented by His Son, David Pennant), Consisting of About 5,000 Volumes Comprising Rare and Valuable Works*. Bangor: North Wales Chronicle, 1913.

Di Gregorio, Mario A. "In Search of the Natural System: Problems of Zoological Classification in Victorian Britain." *History and Philosophy of the Life Sciences* 4, no. 2 (1982): 225–54.

Dickinson, Edward C., Murray D. Bruce, and Robert J. Dowsett. "*Vivarium Naturae or the Naturalist's Miscellany* (1789–1813) by George Shaw: An Assessment of the Dating of the Parts and Volumes." *Archives of Natural History* 33 (2006): 322–43.

Dietz, Bettina. "Aufklärung als Praxis: Naturgeschichte im 18. Jahrhundert." *Zeitschrift für Historische Forschung* 36, no. 2 (2009): 235–57.

Dietz, Bettina. "Contribution and Co-production: The Collaborative Culture of Linnaean Botany." *Annals of Science* 69, no. 4 (2012): 551–69.

Dietz, Bettina. *Das System der Natur: Die kollaborative Wissenkultur der Botanik im 18. Jahrhundert*. Cologne: Böhlau Verlag, 2017.

Dietz, Bettina. "Introduction: Special Issue: 'Translating and Translations in the History of Science." *Annals of Science* 73, no. 2 (2016): 117–21.

Dietz, Bettina. "Iterative Books: Posthumous Publishing in Eighteenth-Century Botany." *History of Science* 60, no. 2 (2022): 166–82.

Dietz, Bettina. "Linnaeus's Restless System: Translation as Textual Engineering in Eighteenth-Century Botany." *Annals of Science* 73, no. 2 (2016): 143–56.

Dietz, Bettina. "Making Natural History: Doing the Enlightenment." *Central European History* 43 (2010): 25–46.

Dietz, Bettina. "Natural History as Compilation: Travel Accounts in the Epistemic Process of an Empirical Discipline." In *Scholars in Action*, vol. 2, edited by André Holenstien, Hubert Steinke, and Martin Stuber, 703–19. Leiden: Brill, 2013.

Dietz, Bettina. "Networked Names: Synonyms in Eighteenth-Century Botany." *History and Philosophy of the Life Sciences* 41, no. 4 (2019): 1–20.

Dietz, Bettina. "Towards a History of Scientific Publishing." *History of Science* 60, no. 2 (2022): 155–65.

Dietz, Bettina. "What Is a Botanical Author? Pehr Osbeck's Travelogue and the Culture of Collaborative Publishing in Linnaean Botany." In *Linnaeus, Natural History and the Circulation of Knowledge*, edited by Hanna Hodacs, Kenneth Nyberg, and Stéphane Van Damme, 57–79. Oxford: Voltaire Foundation, 2018.

Diment, Judith A., Christopher J. Humphries, Linda Newington, and Elaine Shaughnessy. "Catalogue of the Natural History Drawings Commissioned by Joseph Banks on the *Endeavour* Voyage 1768–1771. *Bulletin of the British Museum (Natural History)* 11 (1984): 1–182.

Donald, Diana. *Picturing Animals in Britain, c. 1750–1850*. New Haven, CT: Yale University Press, 2007.

Drayton, Richard. *Nature's Government: Science, Imperial Britain and the "Improvement" of the World*. New Haven, CT: Yale University Press, 2000.

Dryander, Jonas. *Catalogus Bibliothecae Historico-Naturalis Josephi Banks*. London: William Bulmer, 1796–1800.

Dryander, Jonas. "On Genera and Species of Plants which Occur Twice or Three Times, under Different Names, in Professor Gmelin's Edition of Linnaues' *Systema Naturæ*." *Transactions of the Linnean Society of London* 2, no. 1 (1794): 212–35.

Dubald, Déborah, and Catarina Madruga. "Introduction: Situated Nature: Field Collecting and Local Knowledge in the Nineteenth Century." *Journal for the History of Knowledge* 3, no. 1 (2022): 1–11.

Durand-Guédy, David, and Jürgen Paul, eds. *Personal Manuscripts: Copying, Drafting, Taking Notes*. Berlin: De Gruyter, 2023.

Duyker, Edward, and Per Tingbrand, eds. and trans. *Daniel Solander: Collected Correspondence, 1753–1782*. Melbourne: Miegunyah Press, 1995.

Earle, Peter. *The Making of the English Middle Class: Business, Society, and Family Life in London, 1660–1730*. Berkeley: University of California Press, 1989.

Eckstein, Lars, and Anja Schearz. "The Making of Tupaia's Map: A Story of the Extent and Mastery of Polynesian Navigation, Competing Systems of Wayfinding on James Cook's *Endeavour*, and the Invention of an Indigenous Cartographic System." *Journal of Pacific History* 54, no. 1 (2019): 1–95.

Eddy, Matthew Daniel. *Media and the Mind: Art, Science, and Notebooks as Paper Machines, 1700–1830*. Chicago: University of Chicago Press, 2023.

Eddy, Matthew Daniel. "Tools for Recording: Commonplacing and the Space of Words in Linnaeus's *Philosophia Botanica*." *Intellectual History Review* 20, no. 2 (2010): 227–252.

Edwards, George. *A Natural History of Uncommon Birds, and of Some other Rare and Undescribed Animals*. London: Printed for the author, 1743–1751.

Eley, Geoffrey. "Nations, Publics, and Political Cultures: Placing Habermas in the Nineteenth Century." In *Habermas and the Public Sphere*, edited by Craig Calhoun, 289–339. Cambridge: MIT Press, 1992.

Elliott, Brent. "The Promotion of Horticulture." In *Sir Joseph Banks: A Global Perspective*, edited by R. E. R. Banks, B. Elliott, J. G. Hawkes, D. King-Hele, and G. L. Lucas, 104–17. London: Royal Botanic Gardens, Kew, 1994.

Emerson, Roger L. *Academic Patronage in the Scottish Enlightenment: Glasgow, Edinburgh and St Andrews Universities*. Edinburgh: Edinburgh University Press, 2008.

Endersby, Jim. *Imperial Nature: Joseph Hooker and the Practices of Victorian Science*. Chicago: University of Chicago Press, 2008.

Evans, R. J. W., and Alexander Marr, eds. *Curiosity and Wonder from the Renaissance to the Enlightenment*. London: Routledge, 2016.

Evans, R. Paul. "The Life and Work of Thomas Pennant (1726–1798)." PhD diss., University of Wales, 1993.

Evans, R. Paul. "A Round Jump from Ornithology and Antiquity: The Development of Thomas Pennant's *Tours*." In *Enlightenment Travel and British Identities: Thomas Pennant's Tours in Scotland and Wales*, edited by Mary-Ann Constantine and Nigel Leask, 15–38. London: Anthem Press, 2017.

Evans, R. Paul. "Thomas Pennant and the Influences behind the Landscaping of the Downing Estate." *Flintshire Historical Society Journal* 31 (1983–84): 109–24.

Fawcett, Trevor. "Some Aspects of the Norfolk Book-Trade, 1800–24." *Transactions of the Cambridge Bibliographical Society* 4, no. 5 (1968): 383–95.

Feather, John. *A History of British Publishing*. London: Routledge, 1988.

Feather, John. "John Clay of Daventry: The Business of an Eighteenth-Century Stationer." *Studies in Bibliography* 37 (1984): 198–209.

Fielding, Russell. "'The Correct Name for the Breadfruit': On Interdisciplinarity and the Artist Sydney Parkinson's Contested Contributions to the Botanical Sciences." *Notes and Records of the Royal Society* 78 (2024): 9–27.

Findlen, Paula, ed. *Empires of Knowledge: Scientific Networks in the Early Modern World*. London: Routledge, 2019.

Findlen, Paula. *Possessing Nature: Museums, Collecting, and Scientific Culture in Early Modern Italy*. Berkeley: University of California Press, 1994.

Forster, Georg. *Ansichten vom Niederrhein, von Brabant, Flandern, Holland, England und Frankreich im April, Mai und Juni 1790*. 4 vols. Berlin: Voss, 1791–1794.

Forster, Georg. *A Letter to the Right Honourable Earl of Sandwich: First Lord*

Commissioner of the Board of Admiralty, &c. from George Forster. London: G. Robinson, 1778.

Forster, Georg. *A Voyage Round the World.* 2 vols. Edited by Nicholas Thomas and Oliver Berghof. Honolulu: University of Hawai'i Press, 2000.

Forster, Johann Reinhold. *A Catalogue of British Insects.* London: Benjamin White, 1770.

Forster, Johann Reinhold. *A Catalogue of the Animals of North America.* London: Benjamin White, 1771.

Forster, Johann Reinhold. *Observations Made during a Voyage Round the World.* Edited by Nicholas Thomas, Harriet Guest, and Michael Dettelbach. Honolulu: University of Hawai'i Press: 1996.

Foster, Paul. "The Gibraltar Collections: Gilbert White (1720–1793) and John White (1727–1780), and the Naturalist and Author Giovanni Antonio Scopoli (1723–1788)." *Archives of Natural History* 34, no. 1 (2007): 30–46.

Foster, Paul G. M. *Gilbert White and His Records: A Scientific Biography.* London: Christopher Helm, 1988.

Foster, Paul G. M. "The Hon. Daines Barrington, F. R. S., Annotations on Two Journals Compiled by Gilbert White." *Notes and Records of the Royal Society* 41 (1986): 77–93.

Fox, Adam. "Printed Questionnaires, Research Networks, and the Discovery of the British Isles, 1650–1800." *Historical Journal* 53, no. 3 (2010): 593–621.

Francisco-Ortega, Javier, Arnoldo Santos-Guerra, Charlie E. Jarvis, Mark A. Carine, Miguel Menezes De Saqueira, and Mike Maunder. "Early British Collectors and Observers of the Macronesian Flora: From Sloane to Darwin." In *Beyond Cladistics: The Branching of a Paradigm,* edited by D. M. Williams and Sandra Knapp, 125–44. Berkeley: University of California Press, 2010.

Fulford, Tim, Debbie Lee, and Peter J. Kitson. *Literature, Science and Exploration in the Romantic Era: Bodies of Knowledge.* Cambridge: Cambridge University Press, 2004.

Fullagar, Kate. "Envoys of Interest: A Cherokee, a Ra'aiatean, and the Eighteenth-Century British Empire." In *Facing Empire: Indigenous Experiences in a Revolutionary Age,* edited by Kate Fullagar and Michael McDonnell, 239–55. Baltimore: Johns Hopkins University Press, 2018.

Fullagar, Kate, and Michael A. McDonnell, eds. *Facing Empire: Indigenous Experiences in a Revolutionary Age.* Baltimore: Johns Hopkins University Press, 2018.

Fumerton, Patricia. *The Broadside Ballad in Early Modern England. Moving Media, Tactical Publics.* Philadelphia: University of Pennsylvania Press, 2020.

Bibliography

Furet, François *Interpreting the French Revolution*. Cambridge: Cambridge University Press, 1981.

Furniss, Tom. "'As If Created by Fusion of Matter after Some Intense Heat': Pioneering Geological Observations in Thomas Pennant's Tours of Scotland." In *Enlightenment Travel and British Identities: Thomas Pennant's Tours in Scotland and Wales*, edited by Mary-Ann Constantine and Nigel Leask, 163–82. London: Anthem Press, 2017.

Fyfe, Aileen. "The Reception of William Paley's Natural Theology in the University of Cambridge." *British Journal of the History of Science* 30 (1997): 321–35.

Fyfe, Aileen. "The Royal Society and the Free Circulation of Knowledge." In *Old Traditions and New Technologies: The Pasts, Presents, and Futures of Open Scholarly Communications*, edited by M. P. Eve and J. Gray, 147–60. Cambridge: MIT Press, 2020.

Fyfe, Aileen, and Anna Gielas. "Introduction: Editorship and Editing of Scientific Journals, 1750–1950." *Centaurus* 62, no. 1 (2020): 1–16. https://onlinelibrary.wiley.com/doi/10.1111/1600-0498.12290.

Fyfe, Aileen, Julie McDougall-Walters, and Noah Moxham. "Guest Editorial: 350 Years of Scientific Periodicals." *Notes and Records of the Royal Society* 69 (2015): 227–39.

Fyfe, Aileen, and Noah Moxham. "The Royal Society and the Pre-History of Peer-Review, 1665–1965." *Historical Journal* 61, no. 4 (2018): 863–89.

Fyfe, Aileen, and Noah Moxham. "Sociability and Gatekeeping, 1770–1800." In *A History of Scientific Journals: Publishing at the Royal Society, 1665–2015*, edited by Aileen Fyfe, Noah Moxham, Julie McDougall-Walters, and Camilla Mørk Røstvik. London: UCL Press, 2022.

Fyfe, Aileen, Noah Moxham, Julie McDougall-Walters, and Camilla Mørk Røstvik. *A History of Scientific Journals: Publishing at the Royal Society, 1665–2015*. London, UCL Press, 2022.

Gage, Andrew Thomas, and William Thomas Stearn. *A Bicentenary History of the Linnean Society of London*. London: Academic Press, 1988.

Gänger, Stefanie. *A Singular Remedy: Cinchona across the Atlantic World, 1751–1820*. Cambridge: Cambridge University Press, 2020.

Gänger, Stefanie. "World Trade in Medicinal Plants from Spanish America, 1717–1815." *Medical History* 59, no. 1 (2015): 44–62.

Gascoigne, John. *Joseph Banks and the English Enlightenment*. Cambridge: Cambridge University Press, 1994.

Gascoigne, John. "The Royal Society, Natural History and the Peoples of the 'New World(s),' 1660–1800." *British Journal for the History of Science*. 42, no. 4 (2009), 539–62.

Gascoigne, John. *Science and the State: From the Scientific Revolution to World War II*. Cambridge: Cambridge University Press, 2019.

Gascoigne, John. *Science in the Service of Empire: Joseph Banks, the British State and the Uses of Science in an Age of Revolution*. Cambridge, Cambridge University Press, 1998.

Gaskell, Philip. *John Baskerville: A Bibliography*. Cambridge: Cambridge University Press: 1959.

Gaskell, Philip. *A New Introduction to Bibliography*. Oxford: Oxford University Press, 1972.

George, Sam. "The Cultivation of the Female Mind: Enlightened Growth, Luxuriant Decay and Botanical Analogy in Eighteenth-Century Texts." *History of European Ideas* 31 (2005): 209–23.

Georgi, Claudia. "Maria Graham, Travel Writing on India, Italy, Brazil, and Chile (1812–1824)." In *Handbook of British Travel Writing*, edited by Barbara Schaff and Claudia Georgi, 313–4. Berlin: De Gruyter, 2020.

Gervinus, Georg Gottfried, ed. *Georg Forsters Sämtliche Schriften*. 9 vols. Leipzig: Brockhaus, 1843.

Gibson, Susannah. *Animal, Vegetable, Mineral? How Eighteenth-Century Science Disrupted the Natural Order*. Oxford: Oxford University Press, 2015.

Gillispie, Charles C. *Science and Polity in France: The End of the Old Regime*. Princeton, NJ: Princeton University Press, 1981.

Goldgar, Anne. *Impolite Learning: Conduct and Community in the Republic of Letters, 1680–1750*. New Haven, CT: Yale University Press, 1995.

Goldsmith, Oliver. *An History of the Earth and Animated Nature*. 8 vols. London: J. Nourse, 1774.

Goldsmith, Oliver. *The Miscellaneous Works of Oliver Goldsmith with a Biographical Introduction*. Edited by David Masson. London: Macmillan and Co., 1869.

Golinski, Jan. *British Weather and the Climate of Enlightenment*. Chicago: University of Chicago Press, 2007.

Gooding, Mel, David Mabberley, and Joe Studholme. *Joseph Banks's Florilegium: Botanical Treasures from Cook's First Voyage*. New York: Thames and Hudson, 2017.

Goodman, Jordan. *Planting the World: Joseph Banks and His Collectors: An Adventurous History of Botany*. London: HarperCollins, 2020.

Gorham, George C., ed. *Memoirs of John Martyn, F. R. S., and of Thomas Martyn, B. D., F. R. S., F. L. S., Professors of Botany in the University of Cambridge*. London: Hatchard and Son, 1830.

Grass, Sean. *Autobiography, Sensation, and the Commodification of Identity in Vic-*

torian Narrative: Life upon the Exchange. Cambridge: Cambridge University Press, 2019.

Griffin, Dustin. *Literary Patronage in England, 1650–1800*. Cambridge: Cambridge University Press, 1996.

Griffiths, Anthony. *The Print before Photography: An Introduction to European Printmaking*. London: British Museum Press, 2016.

Griffiths, Anthony. "Print Collecting in Rome, Paris, and London in the Early Eighteenth Century." *Harvard University Art Museums Bulletin* 2, no. 3 (1994): 37–58.

Griffiths, Anthony. *Prints and Printmaking: An Introduction to the History and Techniques*. Berkley: University of California Press, 2016.

Griffiths, Anthony. "Proofs in Eighteenth-Century French Printmaking." *Print Quarterly* 21, no. 1 (2004): 3–17.

Guillory, John. "Literary Capital: Gray's 'Elegy,' Anna Laetitia Barbauld, and the Vernacular Canon." In *Early Modern Conceptions of Property*, edited by John Brewer and Susan Staves, 389–410. London: Routledge, 1995.

Grigson, Caroline. *Menagerie: The History of Exotic Animals in England*. Oxford: Oxford University Press, 2016.

Gunther, A. E. *The Founders of Science at the British Museum, 1753–1900*. Suffolk: Halesworth Press, 1981.

Hagglund, Betty. "The Botanical Writings of Maria Graham." *Journal of Literature and Science* 4, no. 1 (2011): 44–58.

Harley, J. B. "The Bankruptcy of Thomas Jefferys: An Episode in the Economic History of Eighteenth-Century Map-Making." *Imago Mundi* 20 (1966): 27–48.

Harney, Marion. *Place-Making for the Imagination: Horace Walpole and Strawberry Hill*. London: Routledge, 2016.

Harris, Philip Rowland. *A History of the British Museum Library*. London: British Library, 1998.

Hawkesworth, John. *An Account of the Voyages Undertaken by the Order of His Present Majesty for Making Discoveries in the Southern Hemisphere*. 3 vols. London: Strahan and Cadell, 1773.

Hawkins, John. "Account of A Species of Bark, the Original Quina-Quina of Peru, Sent over by Mons. De la Condamine to Cromwell Mortimer, Esq. Sec. R. Soc. about 1749" *Transactions of the Linnean Society of London* 1 (1797): 59–61.

Hayes, Clare. "A 'Natural' Exhibitioner: Sir Ashton lever and His Holosphusikon." *British Journal for Eighteenth-Century Studies* 24 (2001): 1–14.

Hayes, Erica Y., and Kacie L. Wills, "Sarah Sophia Banks's Coin Collection: Female Networks of Exchange." In *Women and the Art and Science of Collecting in Eighteenth-Century Europe*, edited by Arlene Leis and Kacie L. Wills, 79–92. New York: Routledge, 2021.

Haynes, Christine. *The Politics of Publishing in Nineteenth-Century France*. Cambridge, MA: Harvard University Press, 2010.

Heal, Felicity. *The Power of Gifts: Gift Exchange in Early Modern England*. Oxford: Oxford University Press, 2014.

Henderson, Paul. *James Sowerby: The Enlightenment's Natural Historian*. London: Kew Gardens, 2015.

Henrey, Blanche. "Kaempfer's 'Icones.'" *Journal of the Society for the Bibliography of Natural History* 3 (1955): 104.

Herder, F. von. "Verzeichniss von G. Forster's Icomes plantarum in itinere ad insulas maris australis collectarum." *Acta Horti Petropolitani* 9 (1886): 485–510.

Hess, Volker, and Andrew J. Mendelsohn. "*Paper Technology* und Wissensgeschichte." *N. T. M Zeitschrift für Geschichte der Wissenschaften, Technik und Medizin* 21 (2013): 1–10.

Hickman, Claire. *The Doctor's Garden: Medicine, Science, and Horticulture in Britain*. New Haven, CT: Yale University Press, 2021.

Hill, John. *The Vegetable System*. London: At the expense of the author, 1759–1775.

Hills, Richard L. *Papermaking in Britain, 1488–1988: A Short History*. London: Bloomsbury, 1988.

Hintz, Petra-Andrea. "The Japanese Plant Collection of Engelbert Kaempfer (1651–1716) in the Sir Hans Sloane Herbarium at the Natural History Museum, London." *Bulletin of the Natural History Museum, London* 31 (2001): 27–34.

Hodacs, Hanna. "Linnaeans Outdoors: The Transformative Role of Studying Nature 'on the Move' and 'Outside.'" *British Journal for the History of Science* 44, no. 2 (2011): 183–209.

Hodacs, Hanna. "Local, Universal, and Embodied Knowledge: Anglo-Swedish Contacts and Linnaean Natural History." In *Global Scientific Practice in an Age of Revolutions, 1750–1850*, edited by Patrick Manning and Daniel Rood, 90–104. Pittsburgh: University of Pittsburgh Press, 2016.

Hodacs, Hanna. "The Price of Linnaean Natural History: Materiality, Commerce and Exchange." In *Linnaeus, Natural History and the Circulation of Knowledge*, edited by Hanna Hodacs, Kenneth Nyberg, and Stéphane Van Damme, 81–112. Oxford: Voltaire Foundation, 2018.

Hodacs, Hanna. *Silk and Tea in the North: Scandinavian Trade and the Market* for *Asian Goods in Eighteenth-Century Europe*. London: Palgrave Macmillan, 2016.

Holt-White, Rashleigh, ed. *The Life and Letters of Gilbert White of Selborne*. 2 vols. London: John Murray, 1901.

Hooker, William Jackson. *Journal of a Tour in Iceland in the Summer of 1809*. Yarmouth: n.p., 1811.

Hooker, William Jackson. *Journal of a Tour in Iceland in the Summer of 1809*. 2 vols. London: Longman, Hurst, Rees, Orme and Brown, 1813.

Hoppit, Julian. "Sir Joseph Banks's Provincial Turn." *Historical Journal* 61, no. 2 (2018): 403–29.

Houstoun, William. *Reliquiæ Houstounianæ: seu America Meridionali a Guilielmo Houstoun M.D.R.S.S. collectarum icones manu propria ære incisæ; cum descriptionibus e schedis ejusdem In Bibliotheca Josephi Banks, Baroneti R.S.P. asservatis*. Edited by Joseph Banks. London: Privately printed, 1781.

Howard, Mary Thornton. *John Latham: Surgeon, Ornithologist and Antiquary: Romsey's First Historian*. Leicester: Matador, 2012.

Hume, Robert D. "The Value of Money in Eighteenth-Century England: Incomes, Prices, Buying Power and Some Problems in Cultural Economics." *Huntington Library Quarterly* 77, no. 74 (2015): 373–416.

Hudson, William. *Flora Anglica, Exhibens Plantas Per Regnum Angliae Sponte Crescentes, Distributas Secundum Systema Sexuale*. London: For the author, 1762.

Hunter, Michael. "Robert Boyle and the Early Royal Society: A Reciprocal Exchange in the Making of Baconian Science." *British Journal for the History of Science* 40, no. 1 (2007): 1–23.

Isaac, Peter. *William Bulmer: The Fine Printer in Context, 1757–1830*. London: Bain and Williams, 1993.

Jackson, H. J. *Marginalia: Readers Writing in Books*. New Haven, CT: Yale University Press: 2001.

Jacobs, Edward H. "Buying into Classes: The Practice of Book Selection in Eighteenth-Century Britain." *Eighteenth-Century Studies* 33, no. 1 (1999): 43–64.

Janković, Vladimir. *Reading the Skies: A Cultural History of English Weather, 1650–1820*. Manchester: Manchester University Press, 2000.

Jarco, Saul. *Quinine's Predecessor: Francesco Torti and the Early History of Cinchona*. Baltimore: Johns Hopkins University Press, 1993.

Jardine, Lisa, and Anthony Grafton. "'Studied for Action': How Gabriel Harvey Read his Livy." *Past & Present* 129 (1990): 30–78.

Jardine, Nicholas, and Emma C. Spary. "Introduction." In *Worlds of Natural*

History, edited by H. A. Curry, N. Jardine, J. A. Secord, and E. C. Spary, 3–16. Cambridge: Cambridge University Press, 2018.

Jardine, Nicholas, and Marina Frasca-Spada. *Books and the Sciences in History*. Cambridge: Cambridge University Press, 2000.

Jardine, Nicholas. "Uses and Abuses of Anachronism in the History of the Sciences." *History of Science* 38 (2000): 251–70.

Jarvis, Charlie, Mark Spencer, and Robert Huxley. "Sloane's Plant Specimens in the Natural History Museum." In *From Books to Bezoars: Sir Hans Sloane and His Collections*, edited by Alison Walker, Arthur MacGregor, and Michael Hunter, 137–57. London: British Library, 2012.

Johns, Adrien. "How to Acknowledge a Revolution." *American Historical Review* 107 (2002): 106–25.

Johns, Adrian. *The Nature of the Book: Print and Knowledge in the Making*. Chicago: University of Chicago Press, 1998.

Johns, Adrien. *Piracy: The Intellectual Property Wars from Gutenberg to Gates*. Chicago: University of Chicago Press, 2009.

Johnson, Samuel. *A Journey to the Western Islands of Scotland*. W. Strahan and T. Cadell, London, 1775.

Jones, Peter M. *Agricultural Enlightenment: Knowledge, Technology, and Nature, 1750–1840*. Oxford: Oxford University Press, 2016.

Jonsson, Fredrik Albritton. "Climate Change and the Retreat of the Atlantic: The Cameralist Context of Pehr Kalm's Voyage to North America, 1798–51." *William and Mary Quarterly* 72, no. 1 (2015): 99–126.

Jonsson, Fredrik Albritton. *Enlightenment's Frontier: The Scottish Highlands and the Origins of Environmentalism*. New Haven, CT: Yale University Press, 2013.

Joppien, Rüdiger, and Bernard Smith. *The Art of Captain Cook's Voyages*. Vol. 1. New Haven, CT: Yale University Press, 1985.

Jung, Sandro. "Illustrated Pocket Diaries and the Commodification of Culture." *Eighteenth-Century Life* 37, no. 2 (2012): 53–84.

Kaempfer, Engelbert. *Amoenitatum Exoticarum*. Lemgo, Germany: Aulæ Lippiacæ, 1712.

Kaempfer, Engelbert. *Der 5. Faszikel der „Amoenitates Exoticae" – die japanische Pflanzenkunde*. Edited by Brigitte Hoppe and Wolfgang Michel-Zaitsu. Hildesheim: Weidmannsche Verlagsbuchhandlung, 2019.

Kaempfer, Engelbert. *The History of Japan*. London: Printed for the translator, 1727.

Kaempfer, Engelbert. *Icones Selectæ Plantarum, quas in Japonia collegit et delineavit Engelbertus Kaempfer; ex archetypis in Museo Britannico asservatis*. Edited by Joseph Banks. London: Privately printed, 1791.

Kaeppler, Adrienne L. *Holophusicon—The Leverian Museum: An Eighteenth-Century English Institution of Science, Curiosity and Art.* Altenstadt: ZKF Publishers-Museum für Völkerkunde Wien, 2011.

Kalm, Pehr. *Travels into North America; Containing Its Natural History, and a Circumstantial Account of Its Plantations and Agriculture in General, with the Civil, Ecclesiastical and Commercial State of the Country.* 3 vols. Warrington, UK: William Eyres, 1771.

Keighren, Innes M., Charles W. J. Withers, and Bill Bell. *Travels into Print: Exploration, Writing, and Publishing with John Murray, 1773–1859.* Chicago: University of Chicago Press, 2015.

Kernan, Alvin B. *Samuel Johnson and the Impact of Print.* Princeton NJ: Princeton University Press, 1987.

Kilgour, Frederick G. *The Evolution of the Book.* New York: Oxford University Press, 1998.

King-Hele, Desmond. *Doctor of Revolution: The Life and Genius of Erasmus Darwin.* London: Faber and Faber, 1977.

King, J. C. H. "New Evidence for the Contents of the Leverian Museum." *Journal of the History of Collections* 8, no. 2 (1996): 167–86.

Kington, John, and Thomas Barker, eds. *The Weather Journals of a Rutland Squire: Thomas Barker of Lyndon Hall.* Oakham: Rutland Record Society, 1988.

Kitson, Peter J. *Forging Romantic China: Sino-British Cultural Exchange, 1760–1840.* Cambridge: Cambridge University Press, 2013.

Klancher, Jon P. *The Making of English Reading Audiences, 1780–1832.* Madison: University of Wisconsin Press, 1987.

Koerner, Lisbet. *Linnaeus: Nature and Nation.* Cambridge, MA: Harvard University Press, 1999.

Koerner, Lisbet. "Purposes of Linnaean Travel: A Preliminary Research Report." In *Visions of Empire: Voyages, Botany and Representations of Nature,* edited by David Philip Miller and Peter Hans Reill, 117–52. Cambridge: Cambridge University Press, 1998.

Krajewski, Markus. *Paper Machines: About Cards & Catalogues, 1548–1829.* Cambridge: MIT Press, 2011.

Krämer, Fabian. "Albrecht von Haller as an 'Enlightened' Reader-Observer." In *Forgetting Machines: Knowledge Management in Early Modern Europe,* edited by Alberto Cevdini, 224–42. Leiden: Brill.

Ksiazkiewicz, Allison. "Geological Landscape as Antiquarian Ruin: Banks, Pennant and the Isle of Staffa." In *Enlightenment Travel and British Identities: Thomas Pennant's Tours in Scotland and Wales,* edited by Mary-Ann Constantine and Nigel Leask, 183–201. London: Anthem Press, 2017.

Kühn, Sebastian, *Wissen, Arbeit, Freundschaft: Ökonomien und soziale Beziehungen an den Akademien in London, Paris und Berlin*. Göttingen: V&R Unipress, 2011.

Kusukawa, Sachiko. "Drawings of Fossils by Robert Hooke and Richard Waller." *Notes and Records of the Royal Society* 67 (2013): 123–38.

Kusukawa, Sachiko. *Picturing the Book of Nature: Image, Text, and Argument in Sixteenth-Century Human Anatomy and Medical Botany*. Chicago: University of Chicago Press, 2012.

Lacépède, Bernard-Germain de. *Histoire Naturelle des Cétacées*. Paris: Plassan, 1804.

Lacépède, Bernard-Germain de. *Histoire Naturelle des Poissons*. Paris, Plassan, 1798.

Lack, Hans Walter. *The Bauers: Joseph, Franz and Ferdinand: Masters of Botanical Illustration*. Munich: Prestel, 2015.

Lack, H. Walter. *Alexander von Humboldt and the Botanical Exploration of the Americas*. Munich: Prestel, 2018.

Lack, H. Walter, with David J. Mabberley. *The Flora Graeca Story: Sibthorp, Bauer, and Hawkins in the Levant*. Oxford: Oxford University Press, 1999.

Lambert, Aylmer Bourke. *A Description of the Genus Cinchona, Comprehending the Various Species of Vegetables from Which the Peruvian Bark and Other Barks of a Similar Quality Are Taken*. London: Benjamin and John White, 1797.

Larson, James L. *Interpreting Nature: The Science of Living from Linnaeus to Kant*. Baltimore: Johns Hopkins University Press, 1994.

Latour, Bruno. *Science in Action*. Milton Keynes, UK: Open University Press, 1987.

Lazarus, Maureen H., and Heather S. Pardoe. "Bute's *Botanical Tables*: Dictated by Nature." *Archives of Natural History* 36, no. 2 (2009): 277–98.

Lazenby, Elizabeth Mary. "The Historia Plantarum Generalis of John Ray: Book 1—A Translation and Commentary." PhD diss., Newcastle University, 1995.

Leask, Nigel. "Fingalian Topographies: Ossian and the Highland Tour, 1760–1805." *Journal for Eighteenth-Century Studies* 39, no. 2 (2016): 183–96.

Leask, Nigel. *Stepping Westward: Writing the Highland Tour, c. 1720–1830*. Oxford: Oxford University Press, 2020.

Leis, Arlene. "Sarah Sophia Banks: A 'Truly Interesting Collection of Visiting Cards and Co.'" In *Collecting the Past: British Collectors and their Collections from the 18th to the 20th Centuries*. Edited by Cynthia Johnson and Toby Burrows, 25–44. London: Routledge, 2018.

Leong, Elaine. *Recipes and Everyday Knowledge: Medicine and the Household in Early Modern England*. Chicago: University of Chicago Press, 2018.

Leuschner, Brigitte, Siegfried Scheibe, Horst Fiedler, Klaus-Georg Popp, and Annerose Schneider, eds. *BAND 18 Briefe an Forster: Georg Forsters Werke: Sämtliche Schriften, Tagebücher, Briefe*. Vol. 18. Berlin: Akademie-Verlag, 1958–2003.

Lhwyd, Edward. *Archaeologia Britannica: Giving Some Additional Account of What Has Hitherto Been Publish'd, of the Languages, Histories and Customs of the Original Inhabitants of Great Britain, from Collections and Observations in Travels through Wales, Cornwall, Bas-Bretagne, Ireland, and Scotland*. Oxford: Printed for the author, 1707.

Liebersohn, Harry. *The Travellers' World: Europe to the Pacific*. Cambridge, MA: Harvard University Press, 2009.

Lightfoot, John. *A Catalogue of the Portland Museum, Lately the Property of the Duchess Dowager of Portland, Deceased: Which Will be Sold by Auction by Mr. Skinner and Co*. London: Skinner & Co., 1786.

Lightfoot, John. *Flora Scotica: Or a Systematic Arrangement, in the Linnaean Method, of the Native Plants of Scotland and the Hebrides*. London: Benjamin White, 1777.

Lightman, Bernard, Gordon McOuat, and Larry Stewart. "Introduction: A Failure to Circulate." In *The Circulation of Knowledge between Britain, India and China: The Early Modern World to the Twentieth Century*, edited by Bernard Lightman, Gordon McOuat, and Larry Stewart, 1–20. Leiden: Brill, 2013.

Lightman, Bernard, Gordon McOuat, and Larry Stewart, eds. *The Circulation of Knowledge between Britain, India and China: The Early Modern World to the Twentieth Century*. Leiden: Brill, 2013.

Lindenfeld, David F. *The Practical Imagination: The German Sciences of State in the Nineteenth Century*. Chicago: University of Chicago Press, 1997.

Lindroth, Sten, "The Two Faces of Linnaeus." In *Linnaeus: The Man and His Work*, edited by Tore Frängsmyr, 1–62. Berkeley: University of California Press, 1983.

Linnaeus, Caroli. *Systema Naturae sive Regna Tria Naturæ Systematice Proposita per Classes, Ordines, Genera, & Species*. Lugduni Batavorum: Theodorum Haak, 1735.

Linnaeus, Carl. *The "Critica Botanica" of Linnaeus*. Translated by A. F. Hort. London: Ray Society, 1937.

Linnaeus, Caroli. *Flora Lapponica Exhibens Plantas per Lapponiam Crescentes, secundum, Systema Sexuale Collectas in Itiuere*. Amsterdam: Salomonem Schouten, 1737.

Linnaeus, Carl. *Linnaeus' Philosophia Botanica*. Translated by Stephen Freer. Oxford: Oxford University Press, 2005.

Linnaeus, Carolus. *Species Plantarum Exhibentes Plantas Rite Cognitas*. 2 vols. Stockholm: Laurentii Salvii, 1753.

Linnaeus, Carolus. *Supplementum Plantarum Systematis Vegetabilium*. Brunsvigæ: Impensis Orphanotrophei, 1781.

Linnaeus, Carlous. *Systema Plantarum: Secundum Classes, Ordines, Genera, Species cum Characteribus, Differentis, Nominibus, Synonymis Selectis et Locis Natalibus*. Edited by Johann Jacob Reichard. Frankfort: Varrentrapp, 1779–1780.

Linnaeus, Carolus. *Species Plantarum Exhibentes Plantas Rite Cognitas ad Genera Relatas cum Differentiis Specificis, Nominibus Trivialibus, Synonymis Selectis, Locus Natalibus: Secundum Systema Sexuale Digestas*. Edited by Carl Ludwig Willdenow. 6 vols. Berlin: Impensis G. C. Nauk, 1797–1824.

Linnaeus, Caroli, *Species Plantarum, Exhibentes Plantas Rite Cognitas, ad Genera Relatas, cum Differentiis Specificis*, 2nd Edn *Nominbus Trivalibus, Synonymis Selectis, Locis Natalibus, Secundum Systema Sexuale*. 2 vols. Stockholm: Laurentii Salvii, 1762–1763.

Linnaeus, Carlous. *Systema Naturae: per Regna Tria Naturæ Secundum Classes, Ordines, Genera, Species, cum Characteribus, Differentiis, Synonymis, Locis*. 11th ed. 2 vols. Stockholm: Laurentii Salvii, 1766.

London Gazette. December 20, 1817, 2575–2654.

Loveland, Jeff, and Stéphane Schmitt. "Poinsent's Edition of the *Naturalis Historia* (1771–1782) and the Revival of Pliny in the Sciences of Enlightenment." *Annals of Science* 72, no. 1 (2015): 2–27.

Low, George. *Fauna Orcadensis: Or the Natural History of the Quadrupeds, Birds, Reptiles, and Fishes of Orkney and Shetland*. Edinburgh: George Ramsay and Company, 1813.

Lynskey, Winifred. "The Scientific Sources of Goldsmith's 'Animated Nature.'" *Studies in Philology* 40, no 1. (1943): 33–57.

Lysaght, A. M. *Joseph Banks in Newfoundland and Labrador, 1766: His Diary, Manuscripts, and Collections*. Berkeley: University of California Press, 1971.

Mabberley, David. *Jupiter Botanicus: Robert Brown and the British Museum*. Braunschweig: Verlag Von J. Cramer and British Museum (Natural History), 1985.

MacGregor, Arthur, ed. *Naturalists in the Field: Collecting, Recording and Preserving the Natural World from the Fifteenth to the Twenty-First Century*. Leiden: Brill, 1971.

Mandelbrote, Scott. "The Publication and Illustration of Robert Morison's

Plantarum Historiae Universalis Oxoniensis." *Huntington Library Quarterly* 78 (2015): 349–79.

Mandelbrote, Scott. "The Uses of Natural Theology in Seventeenth-Century England." *Science in Context* 20, no. 3 (2007): 451–80.

Mandler, Peter. *Aristocratic Government in the Age of Reform: Whigs and Liberals, 1830–1852.* Oxford: Oxford University Press, 1990.

Manning, Patrick, and Daniel Rood, eds. *Global Scientific Practice in an Age of Revolutions, 1750–1850.* Pittsburgh: University of Pittsburgh Press, 2016.

Margócsy, Dániel. *Commercial Visions: Science, Trade, and Visual Culture in the Dutch Golden Age.* Chicago: University of Chicago Press, 2014.

Margócsy, Dániel. "'Refer to the Folio and Number': Encyclopaedias, the Exchange of Curiosities, and Practices of Identification before Linnaeus." *Journal of the History of Ideas* 71, no. 1 (2010): 63–89.

Mariss, Anne. *Johann Reinhold Forster and the Making of Natural History on Cook's Second Voyage, 1772–1775.* Lanham, MD: Lexington Books, 2019.

Mariss, Anne. "A Library in the Field: The Use of Books Aboard the Ship *Resolution* during Cook's Second Circumnavigation, 1772–1775." In *Understanding Field Science Institutions,* edited by Helena Ekerholm et al., 41–70. Sagamore Beach, MA: Science History Publications, 2018.

Marshall, John B. "The Handwriting of Joseph Banks, His Scientific Staff and Amanuensis." *Bulletin of the British Museum (Natural History)* 6 (1978): 1–85.

Martin, Edward A. *A Bibliography of Gilbert White: The Natural History and Antiquities of Selborne.* London: Roxburgh Club, 1897.

Martin, John. *A Bibliographical Catalogue of Books Privately Printed; Including Those of the Bannatyne, Maitland and Roxburghe Clubs, and of the Private Presses at Darlington, Auchinleck, Lee Priory, Newcastle, Middle Hill, and Strawberry Hill.* London: J. and A. Arch; Payne and Foss; J. Rodwell, 1834.

Martín, Nicolás Bas. *Spanish Books in the Europe of the Scientific Enlightenment (Paris and London).* Leiden: Brill, 2018.

Masson, Francis. *Stapeliæ Novæ or A Collection of Several New Species of that Genus Discovered in the Interior Parts of Africa.* London: William Bulmer for George Nicol, 1796.

Martyn, John. *Methodus Plantarum circa Cantabrigiam Nascentium.* London: Richard Reily, 1727.

Matheson, Colin. "Thomas Pennant and the Morris Brothers." *Annals of Science* 10, no. 3 (1954): 258–71.

Maurice, Thomas. *Observations on the Remains of Ancient Egyptian Grandeur, as Connected with Those in Assyria: Forming the Appendix to Observations on the Ruins of Babylon.* London: John Murray, 1818.

Maurice, Thomas. *Westminster Abbey; With Other Occasional Poems, and a Free Translation of the Oedipus Turannus of Sophocles.* London: Printed for the Author, 1813.

Mauss, Marcel. *The Gift: The Form and Reason for Exchange in Archaic Societies.* London: Cohen & West, 1966.

Mayhew, Robert J., and Charles W. Withers, eds. *Geographies of Knowledge: Science, Scale, and Spatiality in the Nineteenth Century.* Baltimore: Johns Hopkins University Press, 2020.

McConnell, Anita. *Jesse Ramsden (1735–1800) London's Leading Scientific Instrument Maker.* Farnham, UK: Ashgate, 2007.

McCormach, Russell, and Christa Jungnickel. *Cavendish: The Experimental Life.* Lewisburg, PA: Bucknell University Press, 1999.

McCormack, Helen. "Pennant, Hunter, Stubbs and the Pursuit of Nature." In *Enlightenment Travel and British Identities: Thomas Pennant's Tours in Scotland and Wales,* edited by Mary-Ann Constantine and Nigel Leask, 203–222. London and New York: Anthem Press, 2017.

McCormack, Helen. *William Hunter and His Eighteenth-Century Cultural Worlds: The Anatomist and the Fine Arts.* London: Routledge 2017.

McGann, Jerome. "The Socialization of Texts." In *The Book History Reader,* edited by David Finkelstein and Alistair McCleery, 39–46. London: Routledge, 2002.

McKendrick, Neil, John Brewer, and J. H. Plumb. *The Birth of a Consumer Society: The Commercialization of Eighteenth-Century England.* Bloomington: Indiana University Press, 1982.

McNeil, Maureen. *Under the Banner of Science: Erasmus Darwin and His Age.* Manchester: Manchester University Press, 1985.

McOuat, Gordon. "Cataloguing Power: Delineating 'Competent Naturalists' and the Meaning of Species in the British Museum." *British Journal for the History of Science* 34, no. 1 (2001): 1–28.

Melton, James Van Horn. *The Rise of the Public in Enlightenment Europe.* Cambridge: Cambridge University Press, 2001.

Mendyk, S. "Robert Plot: Britain's 'Genial Father of County Natural Histories.'" *Notes and Records of the Royal Society of London* 39, no. 2 (1985): 159–77.

Menley, Tobias. "Travelling in Place: Gilbert White's Cosmopolitan Parochialism." *Eighteenth-Century Life* 28, no. 3 (2004): 46–65.

Merrett, Christopher. *Pinax rerum naturalium Britannicarum, continens vegetabilia, animalia et fossilia, in hac insula reperta inchoatus.* London: T. Roycroft, 1667.

Meynell, Guy. "Books from Philip Miller's Library Later Owned by Sir Joseph Banks." *Archives of Natural History* 18 (1991): 379–89.

Milam, Erika Lorraine, and Robert A. Nye. "An Introduction to Scientific Masculinities." *OSIRIS* 30 (2019): 1–14.

Miller, David P. "'My Faivorite Studdys': Lord Bute as Naturalist." In *Lord Bute: Essays in Reinterpretation*, edited by Karl W. Schweizer, 213–40. Leicester: Leicester University Press, 1988.

Miller, David Phillip. "Between Hostile Camps: Sir Humphrey Davy's Presidency of the Royal Society of London, 1820–1827." *British Journal for the History of Science* 16 (1983): 1–47.

Miller, David Phillip. "'Into the Valley of Darkness': Reflections on the Royal Society in the Eighteenth Century." *History of Science* 27 (1989): 155–66.

Miller, David Phillip, and Peter Hanns Reill, eds. *Visions of Empire: Voyages, Botany, and Representations of Nature.* Cambridge: Cambridge University Press, 1998.

Milne, Colin, and Alexander Gordon. *Indigenous Botany; Or Habitations of English Plants.* London: W. Lowndes, 1793.

Mingay, G. E. *Parliamentary Enclosure in England: An Introduction to Its Causes, Incidence and Impact, 1750–1850.* London: Routledge, 1997.

Mitchill, Samuel L., Edward Miller, and Elihu H. Smith. *The Medical Repository.* 3rd ed. New York: T & J. Swords, 1805.

Mollendorf, M. A. "The World in a Book: Robert John Thornton's Temple of Flora (1797–1812)." PhD. diss., Harvard University, 2013.

Montagu, George. *Ornithological Dictionary: Or Alphabetical Synopsis of British Birds.* London: J. White, 1802.

Moore, D. T. "Some Aspects of the Work Robert Brown and the *Investigator* Naturalists in Madeira during August 1801." *Archives of Natural History* 28, no. 3 (2001): 383–94.

Morrell, Jack, and Arnold Thackray. *Gentlemen of Science: Early Years of the British Association for the Advancement of Science.* Oxford: Clarendon Press, 1981.

Morgan, Kenneth. "Sir Joseph Banks as Patron of the *Investigator* Expedition: Natural History, Geographical Knowledge and Australian Exploration." *International Journal of Maritime History* 26, no. 2 (2014): 235–64.

Morieux, Renaud. *The Channel: England, France and the Construction of a Maritime Border in the Eighteenth Century.* Cambridge: Cambridge University Press, 2016.

Morison, Robert. *Plantarum Historiae Universalis Oxoniensis Pars Secunda seu*

Bibliography

Herbarum Distributio Nova, per Tabulas Cognationis & Affinitatis Ex Libero Naturæ Observata & Detecta. Oxford: Theathro Sheldoniano, 1680.

Moxham, Noah. "'*Accoucheur* of Literature': Joseph Banks and the *Philosophical Transactions*, 1778–1820." *Centaurus* 62, no. 1 (2020): 21–37.

Müller-Wille, Staffan. "Collection and Collation: Theory and Practice of Linnaean Botany." *Studies in History and Philosophy of Science Part C: Studies in History and Philosophy of Biological and Biomedical Sciences* 38, no. 3 (2007): 541–62.

Müller-Wille, Staffan. "History Redoubled: The Synthesis of Facts in Linnaean Natural History." In *Philosophies of Technology: Francis Bacon and His Contemporaries,* edited by Claus Zittel, Gisela Engel, Romano Nanni, and Nicole C. Karafyllis, 515–38. Leiden: Brill, 2008.

Müller-Wille, Staffan. "Introduction." In *Musa Cliffortiana: Clifford's Banana Plant by Carl Linnaeus. Reprint and Translation of the Original Edition (Leiden, 1736),* edited and translated by Stephen Freer, 15–67. Liechtenstein, A. R. G. Gantner Verlag K. G, 2007.

Müller-Wille, Staffan. "Linnaeus' Herbarium Cabinet: A Piece of Furniture and Its Function." *Endeavour* 30, no. 2 (2006): 60–64.

Müller-Wille, Staffan. "Linnaean Paper Tools." In *Worlds of Natural History,* edited by H. A. Curry, N. Jardine, J. A. Secord, and E. C. Spary, 205–20. Cambridge: Cambridge University Press, 2018.

Müller-Wille, Staffan. "Names and Numbers: 'Data' in Classical Natural History, 1758–1859." *Osiris* 32 (2017): 109–28.

Müller-Wille, Staffan. "Systems and How Linnaeus Looked at Them in Retrospect." *Annals of Science* 70, no. 3 (2013): 305–17.

Müller-Wille, Staffan. "Walnuts at Hudson's Bay, Coral Reefs in Gotland: The Colonialism of Linnaean Botany." In *Colonial Botany: Science, Commerce, and Politics in the Early Modern World,* edited by Londer Schiebinger and Claudia Swan, 34–38. Philadelphia: University of Pennsylvania Press, 2005.

Müller-Wille, Staffan, and Isabelle Charmantier. "Lists as Research Technologies." *Isis* 103 (2012): 743–52.

Müller-Wille, Staffan, and Isabelle Charmantier. "Natural History and Information Overload: The Case of Linnaeus." *Studies in History and Philosophy of Biological and Biomedical Sciences* 43 (2012): 4–15.

Müller Wille, Staffan, and Katrin Böhme. "'Jederzeit zu Diensten': Karl Ludwig Willdenow's und Carl Sigismund Kunths Beiträge zur Pflanzengeographie Alexander von Humboldts." In *Alexander von Humboldt: Geographie der Pflanzen,* vol. 1, edited by Ottmar Ette and Ulrich Päßler, 75–108. Stuttgart: Metzler, 2020.

Murphy, Kathleen. "James Petiver's 'Kind Friends' and 'Curious Persons' in the Atlantic World: Commerce, Colonialism and Collecting." *Notes Records of the Royal Society* 74 (2020): 259–74.

Murphy, Kevin D., and Sally O'Droscoll. *Studies in Ephemera: Text and Image in Eighteenth-Century Print.* Lewisburg, PA: Bucknell University Press, 2013.

Musgrave, Toby. *The Multifarious Mr. Banks: From Botany Bay to Kew, the Natural Historian Who Shaped the World.* New Haven, CT: Yale University Press, 2020.

Naylor, Simon. *Regionalizing Science: Placing Knowledges in Victorian England.* London: Routledge, 2015.

Nelson, E. Charles. "Some Publication Dates for Parts of William Curtis's '*Flora Londinensis.*'" *Taxon* 29 (1980): 635–39.

Neve, Michael, and Roy Porter. "Alexander Catcott: Glory and Geology." *British Journal for the History of Science* 10, no. 1 (1977): 37–60.

Newell, Jennifer. "New Ecologies: Pathways in the Pacific, 1760s–1840s." In *Facing Empire: Indigenous Experiences in a Revolutionary Age,* edited by Kate Fullagar and Michael McDonnell, 91–114. Baltimore: Johns Hopkins University Press, 2018.

Nichols, John. *Literary Anecdotes of the Eighteenth Century; Comprising Biographical Memoirs of William Bowyer, Printer, F. S. A. and Many of His Learned Friends.* Vol. 3. London: Printed for the author, 1812.

Nickelsen, Kärin. *Draughtsmen, Botanists and Nature: The Construction of Eighteenth-Century Botanical Illustrations.* Dordrecht: Springer, 2006.

Noblett, William. "Pennant and His Publisher; Benjamin White, Thomas Pennant and *Of London.*" *Archives of Natural History* 11, no. 1 (1985): 61–68.

Nolan, Mack Thomas. "Agricultural Improvement in England and Wales and Its Impact on Government Policy, 1783–1801." PhD diss., Louisiana State University, 1977.

"Obituary, with Anecdotes, of Remarkable Persons." *Gentleman's Magazine and Historical Chronicle* 72 (1802): 1171.

O'Brian, Patrick. *Joseph Banks: A Life.* London: Harvill Press, 1997.

Ogilvie, Brian W. *The Science of Describing: Natural History in Renaissance Europe.* Chicago: University of Chicago Press, 2006.

Olarte, Nierto Mauricio. "Remedies for the Empire: The Eighteenth Century Spanish Botanical Expeditions to the New World." PhD. diss., Imperial College London, 1993.

Osbeck, Pehr, *A Voyage to China and the East Indies, by Peter Osbeck, Rector of Hasloef and Woxtorp, Member of the Academy of Stockholm, and of the Soci-*

ety of Uppsala. Translated by Johann Reinhold Forster. London: Benjamin White, 1771.

Oswald, P. H., and C. D. Preston, eds. and trans. *John Ray's Cambridge Catalogue (1660)*. London: Ray Society, 2011.

Paley, William. *Natural Theology: Or Evidence of the Existence and Attributes of the Deity, Collected from the Appearances of Nature*. Edited by Matthew Daniel Eddy and David Knight. Oxford: Oxford University Press, 2008.

Parkinson, Sydney. *A Journal of a Voyage to the South Seas, in His Majesty's Ship, the Endeavour: Faithfully Transcribed from the Papers of the Late Sydney Parkinson, Draughtsman to Joseph Banks Esq., on His Late Expedition with Dr. Solander, Round the World*. London: Stanfield Parkinson, 1773.

Parsons, Christopher M., and Kathleen S. Murphy. "Ecosystems under Sail: Specimen Transport in the Eighteenth-Century French and British Atlantics." *Early American Studies* 10, no. 3 (2012): 503–39.

Peltz, Lucy. *Facing the Text: Extra-Illustration, Print Culture, and Society in Britain, 1769–1840*. San Marino, CA: Huntingdon Library, 2017.

Peltz, Lucy. "A Friendly Gathering: The Social Politics of Presentation Books and Their Extra-Illustration in Horace Walpole's Circle." *Journal of the History of Collections* 19, no. 1 (2007): 33–49.

Pennant, Thomas. "Account of Different Species of Birds, Called Pinguins." *Philosophical Transactions* 58 (1768): 91–99.

Pennant, Thomas. "An Account of Some Fungitae and Other Curious Coralloid Fossil Bodies; by Thomas Pennant esq. Communicated by Mr. Henry Baker." *Philosophical Transactions* 49 (1756): 513–16.

Pennant, Thomas. "Account of Two New Tortoises; in a Letter to Matthew Maty, M. D. Sec. R. S." *Philosophical Transactions* 61 (1771): 266–73.

Pennant, Thomas. *Arctic Zoology*. 2 vols. London: Henry Hughs, 1784–1785.

Pennant, Thomas. *Arctic Zoology*. 2nd ed. 3 vols. London: Robert Faulder, 1792.

Pennant, Thomas. *British Zoology*. 4 vols. London: Cymmrodorion Society, 1766.

Pennant, Thomas. *British Zoology*. 4 vols. London: Benjamin White, 1768–1770.

Pennant, Thomas. *British Zoology*. 4 vols. London: Benjamin White, 1776–1777.

Pennant. Thomas. *British Zoology*. 4 vols. London: White and Cochrane et al., 1812.

Pennant, Thomas. *British Zoology*. 4 vols. Dublin: J. Christie, 1818.

Pennant, Thomas. *Catalogue of My Works*. London: Benjamin White, 1786.

Pennant, Thomas. *Histoire Naturelle des Oiseaux, par Le Comte De Buffon and*

Les Planches Enluminées, Systematically Disposed. London: Benjamin White, 1786.

Pennant, Thomas. *History of Quadrupeds.* 2 vols. London: Benjamin White, 1781.

Pennant, Thomas. *History of Quadrupeds.* 2 vols. London: B & J. White, 1793.

Pennant, Thomas. *The History of the Parishes of Whiteford and Holywell.* London: B & J. White, 1796.

Pennant, Thomas. *Indian Zoology.* 2nd ed. London: Robert Faulder, 1790.

Pennant, Thomas. *Journey from Chester to London.* London: Benjamin White, 1782.

Pennant, Thomas. *The Literary Life of the Late Thomas Pennant, esq. by Himself.* London: Benjamin and John White, 1793.

Pennant, Thomas. *Of London.* London: Robert Faulder, 1790.

Pennant, Thomas. *Supplement to the Arctic Zoology.* Henry Hughs, London, 1787.

Pennant, Thomas. *Synopsis of Quadrupeds.* London: Benjamin White, 1771.

Pennant, Thomas. *The View of Hindoostan.* London: Henry Hughs, 1798.

Pennant, Thomas. *The View of India Extra Gangem, China, and Japan.* London: John White, 1800.

Pennant, Thomas. "To Every Gentleman Desirous to Promote the Publication of an Accurate Account of the Antiquities, Present State, and Natural History of Scotland." *Scots Magazine* 34 (1772): 173–75.

Pennant, Thomas. *A Tour in Scotland and Voyage to the Hebrides, 1772.* Chester: John Monk, 1774.

Pennant, Thomas. *A Tour in Scotland and Voyage to the Hebrides, 1772.* 2 vols. London: Benjamin White, 1790.

Pennant, Thomas. *A Tour in Scotland, 1769.* 2nd ed. London: Benjamin White, 1772.

Pennant, Thomas. *Tour on the Continent, 1765.* Edited by Gavin de Beer. London: Ray Society, 1948.

Pennant, Thomas. *Zoologia Britannica Tabulis Aeneis CXXXII Illustrata.* Augsburg: Johan Jacob Haid und Sohn, 1771.

Pethick, Derek. *The Nootka Connection: Europe and the Northwest Coast, 1790–95.* Vancouver: Douglas & Mclyntyre, 1980.

Philip, Kavita, "Imperial Science Rescues a Tree: Global Botanic Networks, Local Knowledge and the Transcontinental Transplantation of Cinchona." *Environment and History* 1 (1995): 173–200.

Pindar, Peter. *Peter's Prophecy: Or, the President and Poet; An Important Epistle to Sir J. Banks.* London: Printed for G. Kearsley, 1788.

Platts, Elizabeth. "In Celebration of the Ray Society, Established 1844, and its Founder, George Johnston (1797–1855)." In *The Ray Society: 150th Anniversary*, 2–9. London: Ray Society, 1994.

Plot, Robert. *The Natural History of Stafford-Shire*. Oxford, 1687.

Plukenet, Leonard. *Opera Omnia Botanica*. London, 1720.

Plumb, Christopher. "Bird Sellers and Animal Merchants." In *Worlds of Natural History*, edited by H. A. Curry, N. Jardine, J. A. Secord and E. C. Spary, 255–70. Cambridge: Cambridge University Press, 2018.

Polehampton, Edward. *The Gallery of Nature and Art; Or, a Tour through Creation and Science*. London: R. Wilkes, 1815.

Pomian, Krzysztof. *Collectors and Curiosities: Paris and Venice, 1500–1800*. Cambridge: Polity, 1990.

Poovey. Mary. *A History of the Modern Fact: Problems of Knowledge in the Sciences of Wealth and Society*. Chicago: University of Chicago Press, 1998.

Popp, Klaus-Georg, ed. *Georg Forsters Werke: Sämtliche Schriften, Tagebücher, Briefe, Briefe 1792 bis 1794 und Nachträge*. Berlin: Akadrmie Verlag, 1989.

Porter, Roy. *English Society in the Eighteenth Century*. London: Penguin, 2001.

Porter, Roy. *Enlightenment: Britain and the Creation of the Modern World*. London: Penguin, 2001.

Porter, Roy. "The Enlightenment in England." In *The Enlightenment in National Context*, edited by Roy Porter and Mikuláš Teich, 1–18. Cambridge: Cambridge University Press, 1981.

Porter, Roy. "Gentlemen and Geology: The Emergence of a Scientific Career, 1660–1920." *Historical Journal* 21 (1978): 809–36.

Porter, Roy. *The Making of Geology: Earth Science in Britain, 1660–1815*. Cambridge: Cambridge University Press, 1977.

Porter, Theodore M. "Quantification and the Accounting Ideal in Science." *Social Studies of Science* 4 (1992): 633–51.

Porter, Theodore M. *The Rise of Statistical Thinking, 1820–1900*. Princeton, NJ: Princeton University Press, 1986.

Potts, Alex. "Natural Order and the Call of the Wild: The Politics of Animal Picturing." *Oxford Art Journal* 13 (1990): 12–33.

Pratt, Marie Louise. *Imperial Eyes: Travel Writing and Transculturation*. London: Routledge, 1992.

Prescott, Gertrude M. "Faraday: Image of the Man and the Collector." In *Faraday Rediscovered: Essays on the Life and Work of Michael Faraday, 1791–1867*, edited by David Gooding and Frank A. L. James, 15–32. London: Macmillan, 1985.

Printed Books, Maps, & Documents from the Library of Dawson Turner (1775–

1858) and the Reference Library of John Lawson, Bookseller. Cirencester: Dominic Winter Auctioneers, 2020.

Quinby, Jane, and Allan A. Stephenson, compilers. *Catalogue of Botanical Books in the Collection of Rachel McMasters Miller Hunt*. New York: Maurizio Martino, 1991.

Radner, John B., "Constructing an Adventure and Negotiating for Narrative Control: Johnson and Boswell in the Hebrides." In *Literary Couplings: Writing Couples, Collaborators, and the Construction of Authorship*, edited by Marjorie Stone and Judith Thompson, 59–78. Madison: University of Wisconsin Press, 2006.

Raithby, John ed. *The Statutes of the United Kingdom of Great Britain and Ireland*. Vol. 2. London: George Eyre, 1832.

Raman, Bhavani. *Document Raj: Writing and Scribes in Early Colonial South India*. Chicago: University of Chicago Press, 2012.

Rauschenberg, Roy Anthony. "Daniel Carl Solander: Naturalist on the 'Endeavour.'" *Transactions of the American Philosophical Society* 58, no. 8 (1968): 1–66.

Raven, C. E. *John Ray, Naturalist*. Cambridge: Cambridge University Press, 1940.

Raven, James. *The Business of Books: Booksellers and the English Book Trade, 1450–1850*. New Haven, CT: Yale University Press, 2007.

Raven, James. *Judging New Wealth: Popular Publishing and Responses to Commerce in England, 1750–1800*. Oxford: Clarendon Press, 1992.

Ray, John. *A Collection of English Proverbs: Digested into a Convenient Methods for the Speedy Finding Any One upon Occasion; With Short Annotations Whereunto Are Added Local Proverbs with Their Explications, Old Proverbial Rhythmes, Less Known or Exotick Proverbial Sentences and Scottish Proverbs*. Cambridge: W. Morden, 1670.

Ray, John. *Historia Plantarum Species hactenus editas aliasque insuper muitas noviter inventas & descriptas complectens*. London: Henry Faithorne, 1686–1704.

Ray, John. *The Ornithology of Francis Willoughby of Middleton in the County of Warwick*. London: John Martyn, 1678.

Ray, John. *The Ornithology of Francis Willoughby of Middleton in the County of Warwick*. London: Royal Society, 1678.

Ray, John. *Synopsis Methodica Animalium Quadrupedium et Serpentini Generis*. London: S. Smith & B. Walford, 1693.

Ray, John. *Synopsis Methodica Avium & Piscium*. London: William Innys, 1713.

Ray, John. *Synopsis Methodica Stripium Britannicarum*. 3rd ed. London: William and John Innys, 1724.

Ray, John. *The Wisdom of God Manifested in the Works of the Creation Being the Substance of Some Common Places Delivered in the Chappel of Trinity-College, in Cambridge*. London; Samuel Smith, 1691.

Reddick, Allen. *The Making of Johnson's Dictionary, 1746–1773*. Cambridge: Cambridge University Press, 1996.

Rees, Eiluned, and G. Walters. "The Library of Thomas Pennant." *Library* 25, no. 2 (1970): 136–49.

Rehbock, Phillip F. *The Philosophical Naturalists: Themes in Early Nineteenth-Century British Biology*. Madison: University of Wisconsin Press, 1983.

"Review of Agricultural Publications: A Short Account of the Cause of the Disease in Corn Called by Farmers the Blight, the Mildew and the Rust. By Sir Joseph Banks, Bart. London, 1805." *Farmer's Magazine* 6 (1805): 222–29.

Ritvo, Harriet. "Possessing Mother Nature: Genetic Capital in Eighteenth-Century Britain." In *Early Modern Conceptions of Property*, edited by John Brewer and Susan Staves, 413–26. London: Routledge, 1995.

Robinson, David N. "Sir Joseph Banks and the Lincolnshire Influence." In *Joseph Banks: A Global Perspective*, edited by R. E. R. Banks et al., 193–95. London: Royal Botanic Gardens, Kew, 1994.

Robinson, Tim. *William Roxburgh: The Founding of Indian Botany*. Chichester: Phillimore in Association with Royal Botanic Gardens Edinburgh, 2008.

Rogers, Pat. *Grub Street: Studies in a Subculture*. London: Methuen & Co Ltd., 1972.

Rolfe, W. D. Ian. "William Hunter (1718–1783) on Irish 'Elk' and Stubbs's 'Moose.'" *Archives of Natural History* 11, no. 2 (1983): 263–90.

Roos, Anna Marie. "The Art of Science: a 'Rediscovery' of the Lister Copperplates." *Notes and Records of the Royal Society* 66 (2012): 19–40.

Roos, Anna Marie. *Martin Lister and His Remarkable Daughters: The Art of Science in the Seventeenth Century*. Oxford: Bodleian Library, 2018.

Rose, Edwin D. "Empire and the Theology of Nature in the Cambridge Botanic Garden, 1760–1725." *Journal of British Studies* 62 (2023): 1011–42.

Rose, Edwin D. "From the South Seas to Soho Square: Joseph Banks's Library, Collection and Kingdom of Natural History." *Notes and Records: The Royal Society Journal of the History of Science* 73, no. 4 (2019): 499–526.

Rose, Edwin D. "Gilbert White, John Ray and the Construction of the Natural History of Selborne." *Archives of Natural History* 46, no. 1 (2019): 105–221.

Rose, Edwin D. "Natural History Collections and the Book: Hans Sloane's *A Voyage to Jamaica* (1707–1725) and his Jamaican Plants." *Journal of the History of Collections*, 30, no. 1 (2018): 15–33.

Rose, Edwin D. "Publishing Nature in the Age of Revolutions: Joseph Banks,

Georg Forster and the Plants of the Pacific." *Historical Journal* 63, no. 5 (2020): 1132–59.

Rose, Edwin D. "Specimens, Slips and Systems: Daniel Solander and the Classification of Nature at the World's First Public Museum." *British Journal for the History of Science* 51, no. 2 (2018): 205–37.

Rose, Edwin D., and Scott Mandelbrote. "Thomas Gray as a Reader and Writer on the Natural World." In *Thomas Gray among the Disciplines*, edited by Ruth Abbott and Ephraim Levinson, chap. 13. London: Routledge, 2025.

Rösel von Rosenhof, August Johann. *Der monathlich herausgegebnen Insecten-Belustigung*. Nuremburg: J. J. Fleischmann, 1746–1761.

Ross, Trevor. *Writing in Public: Literature and the Liberty of the Press in Eighteenth–Century Britain*. Baltimore: Johns Hopkins University Press, 2018.

Rousseau, G. S. *Oliver Goldsmith: The Critical Heritage*. London: Routledge, 1974.

Rousseau, G. S., and Roy Porter. *The Ferment of Knowledge: Studies in the Historiography of Eighteenth-Century Science*. Cambridge: Cambridge University Press, 1980.

Rousseau, J. J. *Letters on the Elements of Botany Addressed to a Lady. By the Celebrated J. J. Rousseau; Translated into English, with Notes, and Twenty-Four Additional Letter, Fully Explaining the System of Linnaeus*. Edited by Thomas Martyn. London: B. White, 1785.

Roxburgh, William. *Plants of the Coast of Coromandel; Selected From the Drawings and Descriptions Presented to the Hon. Court of Directors of the East India Company*. London: William Bulmer for George Nichol, 1795–1819.

Rudwick, Martin J. S. *Bursting the Limits of Time: The Reconstruction of Geohistory in the Age of Revolution*. Chicago: University of Chicago Press, 2005.

Rudwick, Martin J. S. "The Emergence of a Visual Language for Geological Science." *History of Science* 14 (1976): 149–95.

Rudwick, Martin J. S. "Georges Cuvier's Paper Museum of Fossil Bones." *Archives of Natural History* 27, no. 1 (2000): 51–68.

Rudwick, Martin J. S. *The Great Devonian Controversy: The Shaping of Scientific Knowledge among Gentlemanly Specialists*. Chicago: University of Chicago Press, 1985.

Rudwick, Martin J. S. "Picturing Nature in the Age of Enlightenment." *Proceedings of the American Philosophical Society* 149, no. 3 (2005): 279–303.

Rumphius, Georg Eberhard. *Herbarium Amboinense: Plurimas Conplectens Arbores, Frutices, Herbas, Plantas Terrestres, & Aquaticas, quae in Amboina et Adjacentibus Reperiuntur Insulis*. Amsterdam: Apud Franciscum Changuion, Joannem Catuffe, Hermannum Uytwerf, 1750.

Russell, Gillian. *The Ephemeral Eighteenth Century: Print, Sociability, and the Cultures of Collecting.* Cambridge: Cambridge University Press, 2020.

Safier, Neil. *Measuring the New World: Enlightenment Science and South America.* Chicago: University of Chicago Press, 2008.

Saint-Fonde, B. Faujas. *Travels in England, Scotland and the Hebrides; Undertaken for the Purpose of Examining the State of the Arts, the Sciences, Natural History and Manners, in Great Britain.* 2 vols. London: James Ridgeway, 1799.

Salmond, Anne. *Aphrodite's Island: The European Discovery of Tahiti.* Berkeley: University of California Press, 2010.

Salmond, Anne. *Tears of Rangi: Experiments across Worlds.* Auckland: Auckland University Press, 2017.

Salmond, Anne. *The Trial of the Cannibal Dog: The Remarkable Story of Captain Cook's Encounters in the South Seas.* New Haven, CT: Yale University Press, 2003.

Salmond, Anne. *Two Worlds: First Meetings between Maori and Europeans, 1642–1772.* Honolulu: University of Hawai'i Press, 1992.

Schaffer, Simon. "In Transit: European Cosmologies in the Pacific." In *The Atlantic World in the Antipodes: Effects and Transformations since the Eighteenth Century*, edited by Kate Fullagar, 70–93. Cambridge: Cambridge University Press, 2012.

Schaffer, Simon. "Visions of Empire: Afterward." In *Visions of Empire: Voyages, Botany and Representations of Nature*, edited by David Philip Miller and Peter Hans Reill, 335–52. Cambridge, Cambridge University Press, 1998.

Schaffer, Simon, Lissa Roberts, Kapil Raj, and James Delbourgo, eds. *The Brokered World: Go-Betweens and Global Intelligence, 1770–1820.* Sagamore Beach, MA: Science History Publications, 2009.

Schiebinger, Londa. *The Secret Cures of Slaves: People, Plants, and Medicine in the Eighteenth-Century Atlantic World.* Stanford, CA: Stanford University Press, 2017.

Schiebinger, Londa. "Why Mammals Are Called Mammals: Gender Politics in Eighteenth-Century Natural History." *American Historical Review* 98, no. 2 (1993): 382–411.

Schiebinger, Londa, and Claudia Swan. "Introduction. In *Colonial Botany: Science, Commerce, and Politics in the Early Modern World*, edited by Londa Schiebinger and Claudia Swan, 1–18. Philadelphia: University of Pennsylvania Press, 2005.

Scott-Warren, Jason. *Sir John Harington and the Book as Gift.* Oxford: Oxford University Press, 2001.

Scott-Waring, John. *An Epistle from Obera, Queen of Otahiti, to Joseph Banks, Esq.* London: John Almon, 1774.

Seba, Albertus. *Cabinet of Natural Curiosities: Locupletissimi Rerum Naturalium Thesauri 1734–1765.* Cologne: Taschen, 2003.

Secord, Anne. "Botany on a Plate: Pleasure and the Power of Pictures in Promoting Early Nineteenth-Century Scientific Knowledge." *Isis* 93, no. 1 (2002): 28–57.

Secord, Anne. "Coming to Attention: A Commonwealth of Observers during the Napoleonic Wars." In *Histories of Scientific Observation*, edited by Lorraine Daston and Elizabeth Lunbeck, 421–44. Chicago: University of Chicago Press, 2011.

Secord, Anne. "Natures Treasures: Dawson Turner's Botanical Collections." In *Dawson Turner: A Norfolk Antiquary and His Remarkable Family*, edited by Nigel Goodman, 43–66. Chichester, UK: Phillimore, 2007.

Secord, James A. "Knowledge in Transit." *Isis* 95, no. 4 (2004): 654–72.

Secord, James A. "Newton in the Nursery: Tom Telescope and the Philosophy of Tops and Balls, 1761–1838." *History of Science* 23 (1985): 127–51.

Secord, James A. *Victorian Sensation: The Extraordinary Publication, Reception, and Secret Authorship of Vestiges of the Natural History of Creation.* Chicago: University of Chicago Press, 2000.

Secord, James A. *Visions of Science: Books and Readers at the Dawn of the Victorian Age.* Oxford: Oxford University Press, 2015.

Seppel, Marten, and Keith Tribe, eds. *Cameralism in Practice: State Administration and Economy in Early Modern Europe.* Suffolk: Boydell and Brewer, 2017.

Shapin, Steven. "'Nibbling at the Teats of Science': Edinburgh and the Diffusion of Science in the 1830s." In *Metropolis and Province: Science in British Culture, 1780–1850*, edited by Ian Inkster and Jack Morell, 151–78. London: Routledge, 1983.

Shapin, Steven. *A Social History of Truth: Civility and Science in Seventeenth-Century England.* Chicago: University of Chicago Press, 1995.

Sher, Richard B. *The Enlightenment and the Book: Scottish Authors and Their Publishers in Eighteenth-Century Britain, Ireland, and America.* Chicago: University of Chicago Press, 2006.

Sherman, William H. *Used Books: Marking Readers in Renaissance England.* Philadelphia: University of Pennsylvania Press, 2008.

Sheteir, Ann B. "Botanical Dialogues: Maria Jacson and Women's Popular Science Writing in England." *Eighteenth-Century Studies* 23, no. 3 (1990): 301–17.

Silliman, Benjamin. *A Journal of Travels in England, Holland and Scotland, and of Two Passages over the Atlantic, in the Years 1805 and 1806: With Considerable Additions, Principally from the Original Manuscripts of the Author.* 3rd ed. New Haven, CT: S. Converse, 1810.

"Sir George Staunton's Account of the Embassy to China." *Monthly Review* 24 (1797): 241–51.

Sivasundaram, Sujit. "Natural History Spiritualised: Civilizing Islanders, Cultivating Breadfruit, and Collecting Souls." *History of Science* 39, no. 4 (2001): 417–43.

Sivasundaram, Sujit. *Waves across the South: A New History of Revolution and Empire.* London: William Collins, 2020.

Sloan. Phillip R. "The Buffon-Linnaeus Controversy." *Isis* 67 (1976): 356–75.

Sloane, Hans. *A Voyage to the Islands Madera, Barbados, Nieves, S. Christophers and Jamaica.* London: Printed for the author, 1707–1725.

Smith, Bernard. *European Vision and the South Pacific.* 2nd ed. New Haven, CT: Yale University Press, 1985.

Smith, Courtney Weiss. *Empiricist Devotions: Science, Religion and Poetry in Early Eighteenth-Century England.* Charlottesville: University of Virginia Press, 2016.

Smith, James Edward. *An Introduction to Physiological and Systematical Botany.* London: Longman et al., 1807.

Smith, James Edward, ed. *A Selection of the Correspondence of Linnaeus, and Other Naturalists, From the Original Manuscripts.* 2 vols. London: Longman, 1821.

Smith, Vanessa. "Banks, Tupaia, and Mai: Cross Cultural Exchanges and Friendship in the Pacific." *Parergon* 26, no. 2 (2009): 139–60.

Smith, Vanessa. *Intimate Strangers: Friendship, Exchange and Pacific Encounters.* Cambridge: Cambridge University Press, 2010.

Snell, K. D. M. *The Parish and Belonging: Community, Identity and Welfare in England and Wales, 1700–1950.* Cambridge: Cambridge University Press, 2006.

Sorrenson, Richard. "The Ship as a Scientific Instrument in the Eighteenth Century." *Osiris* 11, (1996): 221–36.

Sorrenson, Richard. "Towards a History of the Royal Society in the Eighteenth Century." *Notes and Records of the Royal Society of London* 50 (1996): 29–36.

Spary, Emma C. "Codes of Passion: Natural History Specimens as Polite Language in Late 18th-Century France." In *Sonderdruk aus Wissenschaft als kulturelle Praxis*, edited by Hans Erich Bödeker, Peter Hans Reill and Jürgen Schlumbohm, 105–35. Göttingen: Vandenhoeck & Ruprecht, 1999.

Spary, Emma C. "On the Ironic Specimen of the Unicorn Horn in Enlightened Cabinets." *Journal of Social History* 52, no. 4 (2019): 1033–60.

Spary, Emma C. "Political, Natural and Bodily Economies." In *Cultures of Natural History*, edited by N. Jardine, J. A. Secord, and E. C. Spary, 178–96. Cambridge: Cambridge University Press, 1996.

Spary, Emma C. "Rococo Readings of the Book of Nature." In *Books and the Sciences in History*, edited by Marina Frasca-Spada and Nick Jardine, 255–75. Cambridge: Cambridge University Press, 2000.

Spary, Emma C. "Scientific Symmetries." *History of Science* 42 (2004): 1–46.

Spary, Emma C. *Utopia's Garden: French Natural History from Old Regime to Revolution*. Chicago: University of Chicago Press, 2000.

Spary, Emma C., and Paul White. "Food of Paradise: Tahitian Breadfruit and the Autocritique of European Consumption." *Endeavour* 28 (2004): 75–80.

Speck, W. A. *A Concise History of Britain, 1707–1975*. Cambridge: Cambridge University Press, 1993.

St Clair, William. *The Reading Nation in the Romantic Period*. Cambridge: Cambridge University Press, 2004.

Stafleu, Frans A. *Linnaeus and the Linnaeans: The Spreading of Their Ideas in Systematic Botany, 1735–1789*. Utrecht: Oosthoek, 1971.

Stafleu, Frans A. "The Willdenow Herbarium: Herbarium Willdenow, Alphabetical Index by Paul Hiepko: Systematical Index to the Willdenow Herbarium." *Taxon* 21, no. 5/6 (1972): 605–88.

Staunton, George Leonard. *An Authentic Account of an Embassy from the King of Great Britain to the Emperor of China*. 3 vols. London: George Nichol, 1797.

Stearn, William T. *The Natural History Museum at South Kensington: A History of the British Museum (Natural History), 1753–1980*. London: Heinemann, 1981.

Stearn, William T. *Three Prefaces on Linnaeus and Robert Brown*. Weinheim, Germany: J. Cramer, 1962.

Stevens, Peter F. *The Development of Biological Systematics: Antoine-Laurent de Jussieu, Nature, and the Natural System*. New York: Columbia University Press, 1994.

Stijnman, Ad. *Engraving and Etching, 1400–2000: A History of the Development of Manual Intaglio Printmaking Processes*. London: Archetype Publications, 2012.

Stillingfleet. Benjamin. *Miscellaneous Tracts Relating to Natural History, Husbandry and Physic. To Which Is Added a Calendar of Flora*. 2nd ed. London: J and R. Dodsley, 1762.

Strasser, Bruno J. "Collecting Nature: Practices, Styles, and Narratives." *Osiris* 27 (2012): 303–40.

Sulloway, Frank J. "Darwin's Conversion: The *Beagle* Voyage and Its Aftermath." *Journal of the History of Biology* 15, no. 3 (1982): 325–96.

Sutherland, John. "The British Book Trade and the Crash of 1826." *Library* 6, no. 2 (1987): 148–61.

Taylor, C. ed., *Calmet's Dictionary of the Holy Bible: Historical, Critical, Geographical and Etymological*. London: Charles Taylor, 1800.

te Heesen, Anke. "Accounting for the Natural World: Double-Entry Bookkeeping in the Field." In *Colonial Botany: Science, Commerce, and Politics in the Early Modern World*, edited by Londer Schiebinger and Claudia Swan, 237–51. Philadelphia: University of Pennsylvania Press, 2005.

te Heesen, Anke. "Boxes in Nature." *Studies in History and Philosophy of Science* 31, no. 3 (2000): 381–403.

te Heesen, Anke. "The Notebook: A Paper Technology." In *Making Things Public: Atmospheres of Democracy*, edited by Bruno Latour and Peter Weibel, 582–89. Cambridge: MIT Press, 2003.

te Heesen, Anke. *The World in a Box: The Story of an Eighteenth-Century Picture Encyclopaedia*. Chicago: University of Chicago Press, 2002.

Terrall, Mary. *Catching Nature in the Act: Réaumur and the Practice of Natural History in the Eighteenth Century*. Chicago: University of Chicago Press, 2014.

Terrall, Mary. "Masculine Knowledge, the Public Good and the Scientific Household of Réaumur." *Osiris* 30 (2015): 182–201.

Thomas, Adrian P. "The Establishment of Calcutta Botanic Garden: Plant Transfer, Science and the East India Company, 1786–1806." *Journal of the Royal Asiatic Society* 16, no. 2 (2006): 165–77.

Thomas, Nicholas. "'Specimens of Bark Cloth, 1769': The Travels of Textiles Collected on Cook's First Voyage." *Journal of the History of Collections* 31, no. 2 (2019): 209–20.

Thompson, Carl. "Women Travellers, Romantic-Era Science and the Banksian Empire." *Notes and Records: The Royal Society Journal of the History of Science* 74, no. 4 (2019): 431–56.

Thunberg, Carl Peter. "Botanical Observations on the Flora Japonica." *Transactions of the Linnean Society of London* 2 (1794): 336–42.

Thunberg, Carl Peter. *Flora Japonica: sistens plantas insularum japonicarum secundum systema sexuale emendatum redactas ad 20 classes, ordines, genera et species*. Leipzig: J. G. Mülleriano, 1784.

Thunberg, Carl Peter. *Travels in Europe, Africa, and Asia, Made between the Years 1770 and 1779.* 4 vols. London: F. and C. Rivington, 1795.

Tobin, Beth Fowkes. *The Duchess's Shells: Natural History Collecting in the Age of Cook's Voyages.* New Haven, CT: Yale University Press, 2014.

Topham, Jonathan R. "Anthologising the Book of Nature: The Origins of the Scientific Journal and Circulation of Knowledge in Late Georgian Britain." In *The Circulation of Knowledge between Britain, India and China: The Early Modern World to the Twentieth Century,* edited by Bernard Lightman, Gordon McOuat, and Larry Stewart, 119–52. Leiden: Brill, 2013.

Topham, Jonathan R. "Biology in the Service of Natural Theology: Paley, Darwin, and the Bridgewater Treatises." In *Biology and Ideology from Descartes to Dawkins,* edited by Denis Alexander and Ronald L. Numbers, 88–113. Chicago: University of Chicago Press, 2010.

Topham, Jonathan R. *Reading the Book of Nature: How Eight Best Sellers Reconnected Christianity and the Sciences on the Eve of the Victorian Age.* Chicago: University of Chicago Press, 2022.

Topham, Jonathan R. "Science, Print, and Crossing Borders: Importing French Science Books into Britain, 1789–1815." In *Geographies of Nineteenth-Century Science,* edited by David Livingstone and Charles Withers, 311–44. Chicago: University of Chicago Press, 2011.

Topham, Jonathan R. "The Scientific, the Literary and the Popular: Commerce and the Reimagining of the Scientific Journal in Britain, 1813–1825." *Notes and Records of the Royal Society* 70 (2016): 305–24.

Turner, Dawson. *Fuci; Or Coloured Figures and Descriptions of the Plants Referred to by Botanists as the Genus Fucus.* 4 vols. London: John Arthur Arch, 1808–1819.

Turner, William. *Avium, Quarum Apud Plinium at Aristotelem Mention est, Historia.* Cologne: Gymnich, 1544.

Turnovsky, Geoffrey. "The Enlightenment Literary Market: Rousseau, Authorship and the Book Trade." *Eighteenth-Century Studies* 36, no. 3 (2003): 387–410.

Turnovsky, Geoffrey. *The Literary Market: Authorship and Modernity in the Old Regime.* Philadelphia: University of Pennsylvania Press, 2010.

Tusser, Thomas. *Five Hundred Points of Good Husbandry.* London: Company of Stationers, 1610.

Urness, Carol Louise, ed. *A Naturalist in Russia: Letters from Peter Simon Pallas to Thomas Pennant.* Minneapolis: University of Minnesota Press, 1967.

Vallance, T. G., D. T. Moore, and E. W. Groves, compilers. *Nature's Investigator:*

The Diary of Robert Brown in Australia, 1801–1805. Canberra: Australian Biological Resources Study, 2001.

Veit, Walter. *Captain James Cook: Image and Impact, South Seas Discoveries and the World of Letters.* Melbourne: Hawthorn Press, 1972.

von Arburg, Hans Georg. "Lichtenberg, das Exzerpieren und das Problem der Originalitt." In *Lesen, Kopieren, Schreiben: Lese- und Exzerpierkunst in der europäischen Literatur des 18. Jahrhundert,* edited by Elisabeth Décultot, 161–86. Berlin: Ripperger and Kremers, 2014.

Wakefield, Andre. *The Disordered Police State: German Cameralism as Science and Practice.* Chicago: University of Chicago Press, 2009.

Walchester, Kathryn. *Travelling Servants: Mobility and Employment in British Travel Writing, 1750–1850.* London: Routledge, 2019.

Walford, Thomas. *The Scientific Tourist through England, Wales, Scotland.* London: J. Booth, 1818.

Watson, William, "An Account on a Treatise in Latin, Entitled Caroli Linnæi Serenissimae regiæ majestatis Sueciæ Archiatri regae . . . with Remarks." *Gentleman's Magazine and Historical Chronicle* 24 (1754): 555–58.

Watts, Iain P. "Philosophical Intelligence: Letters, Print, and Experiment during Napoleon's Continental Blockade." *Isis* 106 (2015): 749–70.

Werrett, Simon. *Thrifty Science: Making the Most of Materials in the History of Experiment.* Chicago: University of Chicago Press, 2019.

The Westminster Magazine; Or the Pantheon of Taste. London: T. Wright, 1777.

Wheeler, Alwyne. "Catalogue of the Natural History Drawings Commissioned by Joseph Banks in the *Endeavour* Voyage, 1768–1771, held in the British Museum (Natural History)." Part 3. *Bulletin of the British Museum (Natural History)* 13 (July 1986): 1–172.

White, Benjamin. *A Catalogue of a Large and Valuable Collection of Books in All Languages, and in Every Class of Literature.* London: Benjamin and John White, 1793.

White, Benjamin. *A Catalogue of a Valuable and Extensive Collection of Books, Containing Many Rare and Curious Articles in the Finest Preservation and Most Splendid Bindings; Also a Large Assortment in Every Branch of Scientific and Polite Literature.* London: Benjamin White, 1784.

White, Benjamin. *A Catalogue of a Valuable and Extensive Collection of Books Containing Many Rare and Curious Articles in the Finest Preservation and the Most Splendid Bindings; Also a Large Assortment in Every Branch of Scientific and Polite Literature.* London: Benjamin White and Son, 1785.

White, Benjamin. *A Catalogue of a Valuable Collection of Books Consisting of Several Libraries and Particularly Those of the Rev. Thomas Negus, D. D. Rector of St Mary's Rotherhithe, and Mr. William Price, a Very Ingenious Painter of Glass.* London: Benjamin White, 1766.

White, Benjamin, and John White. *A Catalogue of a Large and Valuable Collection of Books in All Languages, and in Every Class of Literature; Including the Library of Dr. William Pitcairn, Late Treasurer of St Bartholomew's Hospital.* London: Benjamin and John White, 1793.

White, Gilbert. *The Natural History and Antiquities of Selborne in the County of Southampton.* London: Benjamin White, 1789.

White, Gilbert. *The Natural History and Antiquities of Selborne: To Which Are Added, the Naturalists' Calendar, Miscellaneous Observations, and Poems.* London: J. and J. Arch; Longman, Hurst, Rees, Orme and Brown et al., 1822.

White, Gilbert. *The Natural History of Selborne.* Edited by Anne Secord. Oxford: Oxford University Press, 2013.

White, John. *The Ancient History of the Maori, His Mythology and Traditions: Tai-Nui.* Vol. 5. Wellington: By Authority, 1888.

Whitehead, P. J. P. "Zoological Specimens from Captain Cook's Voyages." *Journal of the Society of the Bibliography of Natural History* 5, no. 3 (1969): 161–201.

William Borlase. *The Natural History of Cornwall.* London: Sandby, 1758.

Williams, Glyn. "'Devilish Fellows Who Test Patience to the Very Limit': Naturalists in the Pacific in the Age of Cook." In *Naturalists in the Field: Collecting, Recording and Preserving the Natural World from the Fifteenth to the Twenty First Century,* edited by Arthur MacGregor, 258–78. Leiden: Brill, 2018.

Williams, Glyn. *Naturalists at Sea: Scientific Travellers from Dampier to Darwin.* New Haven, CT: Yale University Press, 2013.

Williams, Glyndwr. "The *Endeavour* Voyage: A Coincidence of Motives." In *Science and Exploration in the Pacific: European Voyages to the Southern Ocean in the Eighteenth Century,* edited by Margarette Lincoln. 1–18. London: Boydell Press, 2001.

Wills, Hannah. "Joseph Banks and Charles Blagden: Cultures of Advancement in the Scientific Worlds of Late Eighteenth-Century London and Paris." *Notes and Records: The Royal Society Journal of the History of Science* 73, no. 4 (2019): 477–98.

Winterbottom, Anna. *Hybrid Knowledge in the Early East India Company.* London: Palgrave Macmillan, 2016.

Withers, Charles W. J. "Geography, Natural History and the Eighteenth-Century Enlightenment: Putting the World in Place." *History Workshop Journal* 39 (1995): 136–63.

Withers, Charles W. J. "The Rev. John Walker and the Practice of Natural History in Late Eighteenth Century Scotland." *Archives of Natural History* 18, no. 2 (1991): 201–20.

Witteveen, Joeri. "Supressing Synonymy with a Homonym: The Emergence of the Nomenclatural Type Concept in Nineteenth Century Natural History." *Journal of the History of Biology* 49 (2016): 135–89.

Witteveen, Joeri, and Staffan Müller-Wille. "Of Elephants and Errors: Naming and Identity in Linnaean Botany." *History and Philosophy of the Life Sciences* 42 (2020): 1–34.

Wood, Frances. "Britain's First View of China: The Macartney Embassy, 1792–1794." *RSA Journal* 142 (1994): 59–68.

Wood, Karen. "Making and Circulating Knowledge through Sir William Hamilton's 'Campi phlegri.'" *British Journal for the History of Science* 39, no. 1 (2006): 67–96.

Woodward. John. *Brief Instructions for Making Observations in All Parts of the World: As Also for Collecting, Preserving, and Sending Over Natural Things.* London: Richard Wilkin, 1696.

Wragge-Morley, Alex. *Aesthetic Science: Representing Nature at the Royal Society of London, 1650–1720.* Chicago: University of Chicago Press, 2020.

Wright, William, *Memoir of the Late William Wright M. D., With Extracts from His Correspondence, and a Selection of his Papers on Medical and Botanical Subjects.* Edinburgh: Blackwood, 1828.

Wright, William. "XXVII. Description of the Jesuits Bark Tree of Jamaica and the Caribbees." *Philosophical Transactions* 67 (1777): 504–6.

Wystrach, V. P. "Anna Blackburne (1726–1793)—A Neglected Patroness of Natural History." *Journal of the Society of the Bibliography of Natural History* 8, no. 2 (1977): 148–68.

Wystrach, V. P. "Ashton Blackburne's Place in American Ornithology." *Auk* 92, no. 3 (1975): 607–10.

Yale, Elizabeth. *Sociable Knowledge: Natural History and the Nation in Early Modern Britain.* Philadelphia: University of Pennsylvania Press, 2016.

Yeo, Richard. *Encyclopaedic Visions: Scientific Dictionaries and Enlightenment Culture.* Cambridge: Cambridge University Press, 2010.

Yeo, Richard. *Notebooks, English Virtuosi, and Early Modern Science.* Chicago: University of Chicago Press, 2014.

Index

Index

Index

butterfly, 48, 264
Butterworth Squadron, 245–46
Bychton Hall, 138

cabinets, 92, 131, 133, 137, 176, 315n27
Cadell, Thomas. *See* Strahan and Cadell
Caernarfon and Caernarfonshire, 63, 265
Calcutta, 222–23
Cambridge, 26, 47, 269; Botanic Garden, 269; University of, 22, 122, 151, 236, 268, 271
cameralism, 6, 7, 184, 321n2
Campbell, Lady Frederick, 52, 54, 58, 59, 60
Canary Islands, 100, 264
Canna indica, 102
Canton, 79, 230, 247
Cape Horn, 80, 81, 98, 245, 246
Cape of Good Hope, 104, 121, 158, 251
cases, for specimens. *See* cabinets
catalogues, compilation of, 19, 37, 74, 85, 89, 286–87. *See also* Solander
Catherine the Great, Empress of Russia, 194, 218, 220
Cavendish, Henry, 30, 218
censorship. *See* publishing
cetaceans: classification of, 266; oil extraction, 69
Ceylon, 141, 190
Charlevoix, Pierre François Xavier de, 68, 305n109
Charlotte of Mecklenburg-Strelitz, Queen of Great Britain, 65, 66, 122–23, 220
Charmantier, Isabelle, 16, 112
cheetah, 66–67
chemistry, 30, 66, 205, 218
Chesham, Francis, 143, 145, 207
Chester, 67, 128
China, 19, 79, 101–2, 165, 226–30, 245, 247

Christiansted, 220. *See also* Royal Botanic Gardens, Kew
Churton, Richard, 273
Cinchona, 238, 244, 245, 328n12; *Cinchona officinalis*, 181, 237–40, 241, 242–47, 274, 328n19; *Cinchona jamaicensis*, 243–44
classification, systems of, 4, 11–12, 14–16, 20–25, 33, 36, 184; and paper technologies, 72–73; changes to, 232, 276
clergy, 10, 37–38, 60, 186, 187, 271; records kept by, 38
Cleverley, John, 54–55
Clifford, George, 156
Clive, Lord Robert, 191, 195
Cochrane, John. *See* White and Cochrane
coffeehouses, 20, 284
collaboration, 18, 29, 38–39, 184, 280; collecting on expeditions, 56–58, 62, 86–109, 279; organizing information, 72, 250–56; producing books, 5, 25, 55, 126; collections, 19, 20
collections: compilation of, 35–36; organization of, 12, 16, 26, 32, 35, 124, 275; relationship with books, 25, 129, 146, 200, 236, 282, 283, 250–22; visits to, 35, 36, 39, 50, 52–53, 194, 212, 281, 287
Collinson, Peter, 193, 198
commerce, 25, 29, 125, 207, 218–20, 281; rejection of, 183, 198, 209–14, 232, 273–74
commonplace books, 38, 39, 104
Condamine, Charles Marie de la, 240–41, 328n19
congers. *See* publishers
Convolvulus alatus (Decalobanthus peltatus), 104–14, 115–16
Cook, James, 3, 82, 83, 270; first voyage

Index

Index

Index

174; *Journal of a Voyage*, 113–15,
202; *Scober serpens*, 100; *Sitodium
altile* / breadfruit (*Artocarpus
altilis*), 120–21, 175, 237

Parliament. *See* British Government

Parry, Henry, 200–201, visit to Banks's
library, 267

Parry, William, 113

Paton, George, 190

patronage, of natural history, 7, 51,
67–68, 152, 185, 190, 215

Pennant, David, 43, 48–50, 140, 183,
275; *British Zoology*, 189, 200–201,
231–32, 259–68, 285, 323n46;
Outlines of the Globe, 207; *View of
India Extra Gangem, China and
Japan*, 142–43; visit to Banks's
library, 267

Pennant, Thomas, 3, 6, 10–13, 37,
40–77, 115, 124, 166, 171, 183–86,
215–87

— *Arctic Zoology*, 13, 63, 66, 68, 69,
147, 261, 316n45; as an author, 186,
189, 199; Banks's voyage to Newfoundland and Labrador, 68–70

— *British Zoology*, 43–44, 56–96,
127–28, 132–37, 170–77, 186,
198–206, 231, 268, 270, 285,
304n71; annotated copies of, 182,
231–232, 259–274; editions, 206,
231–32, 264, 276, 314n13; gifts
of, 196–97, 199, 204–5, 259; price
of, 204; print run, 134–35, 203;
profitability of, 140, 263; relationship with Pennant's collection, 128,
131; solidifying Pennant's network,
191–96, 200, 259; subscribers,
190–94; translation of, 197

— Cook's Pacific voyages, 35, 54,
70–75, 122, 267

— Downing Hall, 10, 13, 39, 43–46,
53, 58, 140, 205, 231, 323n46;

library and collection at, 55, 68,
129–31, 137; plan of, 130; visitors
to, 50, 68, 70

— family members' contributions, 195,
259; *History of Quadrupeds*, 13,
66, 76, 68; *History of the Parishes
of Whiteford and Holywell*, 46, 171;
Indian Zoology, 13, 68, 141, 190;
Lightfoot's *Flora Scotica*, 55, 60–65
145; Linnaean system, 60, 129, 133,
136, 145; *Literary Life*, 187; natural
history collection, 129–31; notebooks, 69–71; 264–65; *Of London*,
41, 232; *Outlines of the Globe*, 68,
140, 142–45, 207, 265, 305n105;
questionnaires, 41, 51, 205; Ray's
system, 43, 129, 131, 132, 139, 184;
relationship with Banks, 70–75,
147–48, 280; religious views,
26, 272; Sheriff of Flintshire, 42;
Synopsis of Quadrupeds, 66; *Tour in
Scotland and Voyage to the Hebrides
(1774)*, 55–56, 59–60, 128, 187–
89; tour of Scotland (1769), 35–36,
51, 58, 68; *Tour of Scotland (1769)*,
53, 195, 232; *Tour on the Continent*, 52, 66, 133, 193–94; *View of
Hindoostan*, 140–42, 145–46, 171,
206–7, 230, 277

— voyage to the Hebrides (1772), 36,
39, 51, 52, 54–55, 279, expenses of,
53, 187

periodicals, 96–97, 150, 211; *Edinburgh
Review*, 219; *Farmer's Magazine*,
219; *Gentleman's Magazine*, 97;
London Medical Journal, 215;
London Morning Post, 131, 277;
Monthly Review, 133, 199, 274;
*Transactions of the Horticultural So-
ciety of London*, 211, 255; *Westmin-
ster Magazine*, 127; *Philosophical
Transactions, see Royal Society*

Index

Petiver, James, 238

physicians, 9, 11, 216, 238, 242, 268, 297n29

Pigot, Sir George, 66–67

Pindar, Peter, 216, 218

Piper myristicum, 113–14

Plot, Robert, 38, 40, 42

poetry, 25, 126, 183, 282

politicians, 185, 212, 216, 218

Polynesia, 18, 279, 280; languages, 114, 120

Pontifex, William, 164, 166

Pope, Alexander, 30, 185

Porter, Roy, 7, 280

Portland, Margaret Bentinck, Duchess of, 52, 65, 203, 205, 220

Possession Island, 86

potatoes, 255

Poverty Bay / Tūranganui-a-Kiwa, 82, 113–14

preservation of specimens, 102–3, 137

print revolution, 31, 282, 284

print runs, 28, 31, 125, 147–48, 150, 185, 203, 283

printers: intaglio, 29, 222, 234–37, 150, 155, 157, 168, 215, 222; letterpress, 143, 261, 267, 152–55

printing press. *See* printers

printing, 20, 26, 154, 281

private libraries, 19, 268, 275

private printing, 28, 124, 148, 150, 163, 206–10, 212, 239

production, 163–64, 166–68; proofs, 146, 178; quantity of, 171–72; size of, 149

profits, 53, 140, 188, 189, 263, 281

property. *See* landownership

public sphere, 20, 148, 152

public, 7, 41; reading public, 186, 235, 270; museums, 5, 15, 220; libraries, 268

publication, 5

publishers, 25, 28, 31, 221; commercial publishing, 28, 31, 33, 125, 137, 210–18, 234, 274, 280; on commission, 125, 170, 199, 231; congers, 189, 209, 231; distribution process, 126, 134, 135, 149, 235; markets, 8, 219, 277–78; press censorship, 8, 300n77; publishing, 4, 19, 285; subscribers, 149, 190

publishing models, 25, 33, 126, 234; diagrams showing, 27–29, 75, 136, 234, 276, 277, 278, 280, 283,

Pulteney, Richard, 47, 236

Qianlong, Emperor of China, 230

questionnaires, 40–41, 68–69, 51, 70, 205

Ramsay, Robert, 204–5, 323n60

Ranza, Loch, 59–60

Raven, James, 25, 231, 283

Ray Society, 285

Ray, John, 13,

reading public. *See* public

Reichard, Johann Jacob: *Systema Plantarum*, 168, 250–52

religion, 26, 32, 287. *See also* natural theology

Republic of Letters, 20, 220, 230

reviewing, 260, 261–62; reviews, 185, 199, 213, 219, 230

Revolutionary Wars, 226, 235, 248

Richmond, Thomas, 83, 88

Rio de Janeiro, 89, 97, 103

Ritter, Joseph, 53, 76

Roberts, James, 83, 123

rolling press. *See* copperplate images

Rösel, August Johann, 48–49, 264

Roxburgh, William, 98, 154, 214, 215, 222–23, 232, 244, 285

Royal Academy of Arts, 66, 67

Royal Institution, 211

Index

Royal Society of Edinburgh, 154, 204,
Royal Society of London, 10, 79, 100,
114, 134, 155, 170, 205–6, 212,
216–20, 233, 255; offprints, 244;
Pennant's articles in, 135; *Philo-
sophical Transactions*, 13, 40–41,
49, 97, 154, 211, 242–43, 244
Rudwick, Martin, 29, 127
Rumphius, Georg Eberhard, 117; *Her-
barium Amboinense*, 119
Russia, 194, 195, 200, 218, 220, 261;
Kamchatka, 41
Ryan, John, 219–20

Saint Helena, 104, 122, 254
Salmond, Anne, 109, 279, 298n49,
311n94
Samalkota, 98, 223
Sancho, Ignatius, 256
Sancho, William, 256
Sandwich, Earl of, 54
Scandinavia, 6, 242
Schulzen, Frederick, 252–54
scientific instruments, 49–50, 89;
barometer, 49; electrical machine,
101; thermometer, 49; weather-
vane, 49
Scotland, 3, 10, 13, 33–65, 51, 68,
190–93, 270, 280; universities, 271
Seba, Albertus, 155, 156
secretaries, 53, 76, 83, 89, 90, 148, 159,
250–258; social standing of, 281
Senegal, 264, 265
servants, 53–54, 81
Shapin, Steven, 9; "invisible techni-
cians," 19
Sheffield, William, 122, 123
shells, 65, 68, 130–33, 145, 261, 268,
271
Shetland, 41, 63
Siberia, 195, 261
Silliman, Benjamin, 66, 212, 218

slave trade, 265
slavery. *See* enslaved peoples
Sloane, Hans, 65, 90, 151, 166; *Natural
History*, 96, 100, 156
Smith, James Edward, 21, 47, 143, 203,
206, 209, 251, 254–55
Society Islands, 85, 87, 104, 112, 113,
117
Soho Square. *See* Banks, Joseph
Solander boxes, 73, 111
Solander, Daniel: annotations, 96,
100, 119–20; British Museum,
14–15, 22, 84, 90, 92, 98, 111, 148,
165–66; Cook's first Pacific voyage,
33, 70–72, 80–123, 158–62, 185,
249, 254, 256; correspondence with
Linnaeus, 89; death, 165; familiarity with the Linnaean system, 84,
88, 94, 98; field notebooks, 74–75,
86–87, 104–17, 159, 162–63, 214,
280; Linnaean apostle, 79; managing scales of geography and time,
80, 90; manuscript slips, 15, 70,
71–76, 84, 111–20, 159; moving
between collections, 15–16, 72;
obituary of, 87, 122; printing the
South Seas plants, 155–60; vernacular names, 105–15, 112; voyage
to Iceland, 160, 202; zoological
notebook, 100
Solanum tuberosum. *See* potato
Somerset, Duchess of, 211, 219
South Seas, 74, 81, 212, 258
Sowerby, James, 143, 145–46
Spain, 215, 238, 242; fort at Nootka
Sound, 247; language, 242; Spanish trading monopolies, 238,
Sparrman, Anders, 149, 200, 213
species, 3, 5, 95, 132, 260. *See* new
species
specimen cabinets. *See* cabinets
specimens in books, 260, 263–64

Index

Spöring, Herman, 71, 82–90, 95, 100, 109, 121, 159; interleaved copy of *Species Plantarum*, 115–16, 149, 158, 250, "Primitæ Floræ" manuscripts, 117, 149; use of manuscript slips, 113–15

Sri Lanka. *See* Ceylon

St. Claire, William, 125, 235

St. Croix, 219–20

St. Lawrence, 68–69

St. Lucia, 244

St. Petersburg, 194, 261

Staffa, Isle of, 54; Fingal's Cave, 55, 187

Staffordshire, 38, 40, 63, 67, 195

stand-alone plates, 237; use in a collection, 48, 100; use of color, 156–57, 178, 180, 197–98, 240, 319n94

stand-alone plates. *See* copperplate images

states: centralization, 58; definition of, 64, 278; formation of, 6–7, 37; funding, 7, 30, 79; regulation, 8; state appointments for naturalists, 7, 80

Staunton, George Leonard, 226–27; *Authentic Account*, 228–30, 285

Stillingfleet, Benjamin, 35, 192, 198

Strahan and Cadell, 146, 188, 219

Strawberry Hill. *See* Walpole, Horace

Stuart, John, 52, 55–56

Stubbs, George, 66, 67, 305n100

subscribers, *see publishing*.

subscription payments, 137, 190–91

substitutes, 223, 237–38, 242–44

sugar, 97, 114

Swan, John, 252–254, 256

Sweden, 64, 89, 213, 261

synonymy, 15, 19, 56, 96–98, 119, 254–55, 263

system of classification, 12–15, 21, 22–23, 72–73, 118–19, 129–31, 184, 279; *Historia Plantarum*, 22, 42, 56; *Ornithology*, 24, 39, 42–43, 56; *Synopsis Methodica Animalium Quadrupedium*, 25; *Synopsis Methodica Avum et Piscium*, 24, 47, 267

systems of classification, 20–25

Tahiti, 79, 101, 104, 105,107–17, 146, 174; illustration of flora and fauna, 117–21, 279–80; Tahitian language, 95, 107, 109

tax. *See* duties

taxidermy, 130–32, 137, 139–40, 159; relationship with Ray's system, 132

taxonomy, 26, 231

Te Horetā, also known as Te Taniwha, 106–7

tea: American, 246; Chinese, 228–229; Japanese, 227

Testacea. *See* shells

Thompson, Archibald, 52, 54, 58, 60

Thunberg, Carl Peter, 159, 165, 169, 213–14, 223; *Flora Japonica*, 168, 226, 249, 250

Tierra Australis Incognita. *See* New Holland

Tierra del Fuego, 85, 88, 95, 104, 122, 163, 246

Törner, Samuel, 250–254

trade, 5–7, 33, 247; laissez faire, 8

transit of Venus, 79, 89

translation, between cultures, 18; languages, 36, 55–58, 73, 104, 108–9, 186, 237

Tupaia, 107–8, 112–14, 121; Polynesian and Māori languages, 113–14, 280, 311n95

Turner, Dawson, 207–9, 324n73

Turner, Mary, 208–9

Turnovsky, Geoffrey, 126, 199

Index